PROJECTIVE GEOMETRY
and
PROJECTIVE METRICS

PURE AND APPLIED MATHEMATICS

A Series of Monographs and Textbooks

Edited by

PAUL A. SMITH and SAMUEL EILENBERG

Columbia University, New York

In preparation

Projective Geometry
and Projective Metrics

BY HERBERT BUSEMANN

University of Southern California

and

PAUL J. KELLY

University of California
Santa Barbara College

1953

ACADEMIC PRESS INC., PUBLISHERS

NEW YORK, N.Y.

ACADEMIC PRESS INC.
111 FIFTH AVENUE
NEW YORK 3, N. Y.

United Kingdom Edition
Published by
ACADEMIC PRESS INC. (LONDON) LTD.
BERKELEY SQUARE HOUSE, LONDON W. 1

Library of Congress Catalog Card Number: 52-13363

First Printing, 1953
Second Printing, 1962

PRINTED IN THE UNITED STATES OF AMERICA

PREFACE

The present book differs widely in content, methods, and point of view from traditional presentations of the subject. To a great extent, this departure is due to the changed attitude of contemporary, in particular American, mathematicians toward geometry. Although reluctantly, geometers must admit that the beauty of synthetic geometry has lost its appeal for the new generation. The reasons are clear: not so long ago synthetic geometry was the only field in which the reasoning proceeded strictly from axioms, whereas this appeal—so fundamental to many mathematically interested people—is now made by many other fields. Moreover, much research is taking place in the new fields, but very little in synthetic geometry. There is an additional reason, more peculiar to the United States: individually attractive results, in which projective geometry abounds, are not appreciated, with very few exceptions as in number theory, because there is a tendency to use generality as the only criterion.

Nevertheless, the basic results, and even more the methods, of projective and non-Euclidean geometry are as indispensable for the geometer as calculus is for the analyst. The present book represents an effort to emphasize this fact. Many special terms, like "complete quadrilateral," and "trilinear polarity," will not be found; often a whole chain of beautiful theorems is represented by a single example. Special results are discussed in greater number only when they are needed to develop a feeling for the subject, as in the section on synthetic hyperbolic geometry, or when they illustrate a general method, as in the section on linear line loci.

On the other hand, more space than usual is devoted to the discussion of the basic concepts of distance, motion, area and perpendicularity. In fact, the non-Euclidean geometries are reached via general metric spaces and the Hilbert problem of finding those geometries in which straight lines are the shortest connections. Of course, the general problem is only formulated here; but this leads naturally to the consideration of geometries other than the Euclidean and two non-Euclidean ones, and thus to the modern view in which the three classical geometries are seen as very special, and closely related, cases of general geometric structures. Experience with students has shown that, at that point, some facts which are obvious to the expert, such as the uniqueness of Euclidean geometry, require detailed discussion.

Since the time allotted to geometry is now shorter than formerly, much material had to be omitted. The previous remarks explain why many special results and the whole synthetic approach were sacrificed.[1] By using coordinates from the beginning and alternating between geometric and algebraic or analytic arguments we hoped to produce the often sadly lacking ability to pass from the algebraic to the geometric language and conversely. Rather than avoiding any admixture of fields, methods from other branches were used whenever they seemed more effective or provided a natural approach. In the same spirit, variety in proofs was deliberate. The overall aim was to counteract the impression of geometry as an isolated and static subject, and to present its methods and essential content as part of modern mathematics.

The first five chapters of the book were planned as a year course. From the outset, a knowledge of the theory of equations is presupposed along with elementary matrix theory. In Chapters IV and V, intended for the second semester, it is assumed that the student has had some work in rigorous analysis and is familiar with ε, δ methods. The group concept, though used extensively, is not assumed to be known, but is developed as far as needed. In line with most mathematical programs, the course would seem to be most effective for the senior year, though modifications are possible.

In order to emphasize the connectedness of the development, and to de-emphasize the role of problem work, the exercises are collected at the end of the chapter. They were designed to increase familiarity with concepts and methods and are rarely of the routine kind.

The last chapter is in a different spirit since it was written for the more mature student who has mastered the earlier material. Instead of explicit exercises, the proofs of many theorems, and most of the generalizations from two to three dimensions, are left to his initiative. Similarly, there are far fewer figures in this chapter since we feel that the student gains considerably in understanding from the creation of his own sketches. Both the material and the treatment in Chapter VI make it well adapted for a seminar.

We believe that the present book will give the reader new insight into his geometric past and prepare him well for his geometric future in classical differential, as well as Riemannian, and some other branches of modern geometry. He will then be aware of the intrinsic value of projective methods and may feel the urge to acquaint himself with the axiomatic approach, for instance Veblen and Young's work, or even to enjoy some old-fashioned books like Reye's "Geometrie der Lage" or Darboux's "Principes de géométrie analytique."

HERBERT BUSEMANN

December 1952   PAUL J. KELLY

---

[1]Any serious student should, at some time, become familiar with the great discovery, made at the end of the last century, that large parts of geometry do not depend upon continuity. The most outstanding results in this direction, however, are now to be found in modern algebraic geometry, and it is there, the authors believe, that continuity in geometry can be most effectively discussed.

TABLE OF CONTENTS

CHAPTER I

The Projective Plane

1. Ideal Points

Historically, the concept of the projective plane was created to eliminate certain case distinctions in plane geometry arising from the possibility of lines being either parallel or intersecting. Later, the geometry of this plane developed into an important, independent discipline, and this will be the subject of our first two chapters.

In this introductory section, the heuristic ideas which led to the projective plane will be discussed. The exact definition is given in the next section.

Since a pair of non-intersecting, or parallel, lines may be obtained as the limit of two lines whose intersection point moves farther and farther away, parallels may be considered to intersect in "infinite" or "ideal" points adjoined to the ordinary plane. However, if this addition of points is not to create more exceptions than it removes, it must be done in such a way that two distinct "points" determine one line and two distinct lines intersect in one "point." Thus two parallel lines must intersect in the same ideal point for either direction that the lines are traversed. Suppose that L_1 and L_2 are two such parallel lines through the ideal point i, and let a be any ordinary point which is not on L_1 or L_2. Since a and i must determine a line L_3, and since L_3 cannot intersect L_1 or L_2 a second time, L_3 must be the parallel to L_1 and L_2 which passes through a. The adjoined point i lies then on all three parallels, and, by the same argument, must lie on all the lines parallel to L_1.

That a different ideal point must be added for a different family of parallels is easily seen. For let M_1 be a line which cuts L_1 in an ordinary point and take M_2 parallel to M_1. Then M_1 and M_2 pass through an ideal point j which must be distinct from i if M_1 is to have but one intersection with L_1.

Next consider two distinct ideal points, i and j. The line L which they determine cannot pass through any ordinary point a. For the line L_1, determined by a and i, and the line L_2, determined by a and j, would be distinct, ordinary lines both of which would be contained in L from the collinearity of a, i, and j. Therefore the line through i and j must be an "ideal line."

Finally, there can only be one such ideal line. For if L_1 and L_2 were distinct ideal lines they would have to intersect in an ideal point i. A line through an ordinary point a and not through i would have to intersect L_1 and L_2 in distinct ideal points, j and k respectively. The line through j and k would then contain an ordinary point, which has been seen to be contradictory.

These considerations lead to the convention of adding one ideal point for each family of parallels (independent of the two directions along any member of the family) with the totality of new points regarded as an ideal line which contains no ordinary point. In this extended plane it appears that two distinct points determine one line and two distinct lines always intersect in one point.

It is conceivable, however, that this way of extending the plane implies some inherent contradiction.[1] Frequently, when such a question arises in geometry, the number system is taken as a basis for comparison. By introducing numbers, or coordinates, for points, lines, etc., the geometric system is arithmetized. The geometric relations then become relations between numbers, and by checking these the geometric structure may be seen to be either inconsistent or else as consistent as the number system.

In the present case, the simplest way to obtain an arithmetic model of the extended plane is to begin with the representation of the ordinary plane in rectangular coordinates, $\bar{x}, \bar{y}$. A line in this plane is given by a linear equation, $u\bar{x} + v\bar{y} + w = 0$, where u and v are not both zero. The three numbers, u, v, w determine the line in the sense that as coefficients they specify the equation of the line. But there are other triples which determine the same line, namely any triple proportional to u, v, w. Thus one definite line is represented by all triples $\lambda u, \lambda v, \lambda w$, where $\lambda \neq 0$ and one of the numbers u, v is not zero.[2] If the lines corresponding to the triples (u, v, w) and (u', v', w') are distinct, and non-parallel, Cramer's rule gives their intersection point $(\bar{x}, \bar{y})$:

$$(1.1) \qquad \bar{x} = \frac{\begin{vmatrix} v & w \\ v' & w' \end{vmatrix}}{\begin{vmatrix} u & v \\ u' & v' \end{vmatrix}}, \qquad \bar{y} = \frac{\begin{vmatrix} w & u \\ w' & u' \end{vmatrix}}{\begin{vmatrix} u & v \\ u' & v' \end{vmatrix}}, \qquad \begin{vmatrix} u & v \\ u' & v' \end{vmatrix} \neq 0.$$

---

[1]For example, it is impossible to extend the concept of quotient to include fractions with a zero denominator without contradicting some of the usual properties of fractions.

[2]One might be tempted to divide $u\bar{x} + v\bar{y} + w = 0$ by w, obtaining

$$(u/w)\bar{x} + (v/w)\bar{y} + 1 = 0,$$

and then take $(u/w, v/w, 1)$ or just $(u/w, v/w)$ for the line. But this excludes lines through the origin.

Therefore,

(1.2) $$\bar{x} : \bar{y} : 1 = \begin{vmatrix} v & w \\ v' & w' \end{vmatrix} : \begin{vmatrix} w & u \\ w' & u' \end{vmatrix} : \begin{vmatrix} u & v \\ u' & v' \end{vmatrix}.$$

If the lines are parallel, but distinct, the triples (u,v,w) and (u',v',w') are not proportional, but the pairs u,v and u',v' are. The third determinant on the right of (1.2) is zero, but at least one of the other two is not. This is the case where the intersection is to be an ideal point. To let (1.1) give the ideal point involves an inadmissable operation with zero. However, the fact that lines are represented by three coefficients, which might be called coordinates, and the form of (1.2), suggest a way around the difficulty, namely that *points be represented by three coordinates* (x,y,z), *with the understanding that proportional triples* (x,y,z) *and* $(\lambda x, \lambda y, \lambda z)$, $\lambda \neq 0$ *represent the same point.* We can change from our previous representation of a point $(\bar{x},\bar{y})$ to a triple, by adopting for (x,y,z) the triple $(\bar{x},\bar{y},1)$ or $(\lambda\bar{x},\lambda\bar{y},\lambda)$ where $\lambda \neq 0$. We can then change back from an ordinary point given by (x,y,z) to $(\bar{x},\bar{y})$ by taking $\bar{x} = x/z$ and $\bar{y} = y/z$, when $z \neq 0$. Ordinary points then correspond with triples where $z \neq 0$. If (u,v,w) and (u',v',w') represent distinct lines, the form of relation (1.2) suggests, then, that the triple

$$\begin{vmatrix} v & w \\ v' & w' \end{vmatrix}, \begin{vmatrix} w & u \\ w' & u' \end{vmatrix}, \begin{vmatrix} u & v \\ u' & v' \end{vmatrix}$$

be taken as the coordinates of the point of intersection whether or not the lines are parallel. For non-parallels this gives the ordinary intersection point, written as a triple. For parallels it is a triple of the form (x,y,z) where $z = 0$, and where x and y are not both zero since the lines are distinct.

The enlarged plane thus contains as points all triples (x,y,z), excluding $(0,0,0)$, with proportional triples representing the same point. The points on the ideal line, and only these, satisfy the linear equation $z = 0$. Since this may be written as $0 \cdot x + 0 \cdot y + 1 \cdot z = 0$, a triple (u,v,w) representing the ideal line is $(0,0,1)$.

2. The Projective Plane

It is well known that the foregoing numerical representation of the extended Cartesian plane does represent a consistent system. As stated at the beginning, however, we are not primarily interested in this system but in the generalization of it which yields the projective plane. For this purpose, and for other reasons, it will be advantageous to change to a subscript notation in representing coordinates. Thus, instead of (x,y,z) for a point, we will use (x_1,x_2,x_3). The point y will mean the point (y_1,y_2,y_3). For lines,

Greek letters will be used, with the same subscript conventions, so that the line ξ means the line with coordinates (ξ_1, ξ_2, ξ_3).[3]

An immediate advantage of this change is that it enables us to make formal use of the concise notation of the vector calculus. For any two real numbers, λ and μ, and two triples, (a_1, a_2, a_3), (b_1, b_2, b_3), $\lambda a + \mu b$ is defined by

$$(2.1) \qquad \lambda a + \mu b = (\lambda a_1 + \mu b_1, \ \lambda a_2 + \mu b_2, \ \lambda a_3 + \mu b_3).$$

As in the scalar product, $a \cdot b$ or simply ab means:

$$(2.2) \qquad a \cdot b = a_1 b_1 + a_2 b_2 + a_3 b_3 = \sum_{i=1}^{3} a_i b_i.$$

Clearly $a \cdot b = b \cdot a$, and

$$(\lambda a + \mu b) \cdot (\lambda' a' + \mu' b') = \lambda \lambda'(a \cdot a') + \lambda \mu'(a \cdot b') + \lambda' \mu (b \cdot a') + \mu \mu'(b \cdot b').$$

As in a vector product, we put

$$(2.3) \qquad a \times b = \left(\begin{vmatrix} a_2 & a_3 \\ b_2 & b_3 \end{vmatrix}, \ \begin{vmatrix} a_3 & a_1 \\ b_3 & b_1 \end{vmatrix}, \ \begin{vmatrix} a_1 & a_2 \\ b_1 & b_2 \end{vmatrix} \right) = -(b \times a).$$

For any numbers, a_0, b_0, c_0, d_0, a_0', b_0', c_0', d_0', the properties of determinants give

$$\begin{vmatrix} \lambda a_0 + \mu b_0 & \lambda' a_0' + \mu' b_0' \\ \lambda c_0 + \mu d_0 & \lambda' c_0' + \mu' d_0' \end{vmatrix} = \lambda \lambda' \begin{vmatrix} a_0 & a_0' \\ c_0 & c_0' \end{vmatrix} + \lambda \mu' \begin{vmatrix} a_0 & b_0' \\ c_0 & d_0' \end{vmatrix}$$
$$+ \mu \lambda' \begin{vmatrix} b_0 & a_0' \\ d_0 & c_0' \end{vmatrix} + \mu \mu' \begin{vmatrix} b_0 & b_0' \\ d_0 & d_0' \end{vmatrix}.$$

It follows that

$$(2.4) \qquad \begin{aligned} (\lambda a + \mu b) \times (\lambda' a' + \mu' b') &= \lambda \lambda'(a \times a') + \lambda \mu'(a \times b') \\ &+ \mu \lambda'(b \times a') + \mu \mu'(b \times b'). \end{aligned}$$

Finally, we introduce the abbreviation:

$$(2.5) \qquad |\, a, b, c \,| = \begin{vmatrix} a_1 & a_2 & a_3 \\ b_1 & b_2 & b_3 \\ c_1 & c_2 & c_3 \end{vmatrix}.$$

Then,

$$(2.6) \qquad a \cdot (b \times c) = (a \times b) \cdot c = |\, a, b, c \,|.$$

If x, that is (x_1, x_2, x_3), is any triple of real numbers, $[x]$ is taken to denote the class of all triples λx, where $\lambda \neq 0$. *Clearly two classes are identical if*

---

[3] Greek letters, usually λ and μ, will also be used for proportionality constants. Though uniformity has been sought, where possible, reserving a type of letter for a single use would have required more alphabets. Distinctions in the use of a letter, however, are apparent in the context.

they have one triple in common. Therefore the class is determined by any one of its triples, which justifies the notation. The fact that x and y belong to the same class is denoted by $y \sim x$ and the negation of this by $y \not\sim x$. All classes contain infinitely many different triples except the zero class which contains only the triple $0 = (0,0,0)$. Therefore $x \sim 0$ means that $x = 0$, that is, $x_i = 0$, $i = 1,2,3$.

The projective plane is defined to be the set of all classes $[x]$ with the exception of the zero class. A class $[x]$ is called a *point x of the projective plane.* It is important to observe that this definition is independent of Section 1 which merely provided a background for it. Thus, familiar concepts in the Cartesian plane, such as distance between points, area of figures, parallelism of lines, etc., are undefined in the projective plane.

A line in the projective plane is defined as the locus (i.e., the totality) of points x satisfying a linear equation of the form $\xi_1 x_1 + \xi_2 x_2 + \xi_3 x_3 = 0$, where the coefficients ξ_i are not all zero, that is,

$$(2.7) \qquad x \cdot \xi = 0, \ \xi \neq 0.$$

That this equation truly represents points can be seen from the fact that if the triple x satisfies it, the triple $y = \lambda x$, $\lambda \neq 0$, also satisfies it, for $x \cdot \xi = 0$ implies $y \cdot \xi = 0$ and conversely. Not all equations in x_i have this property. For instance,

$$x_1^2 + x_2^2 - 8x_3 = 0$$

is satisfied by $(2,2,1)$ but not by $(-2,-2,-1)$ or $(4,4,2)$ though these triples belong to the same class. For an equation

$$f(x) = f(x_1, x_2, x_3) = 0$$

to hold for the class $[x]$, whenever it holds for any member of the class, it is sufficient that an integer k exists such that for any number $\lambda \neq 0$,

$$(2.8) \qquad f(\lambda x) = f(\lambda x_1, \lambda x_2, \lambda x_3) = \lambda^k f(x), \ or \ that \ a \ real \ number \ k \ exists \ such \ that \ f(\lambda x) = |\lambda|^k f(x).$$

An example of the first type, with $k = 1$, is given by (2.7) where $f(x) = \xi_1 x_1 + \xi_2 x_2 + \xi_3 x_3$. The second type, with $k = \frac{1}{2}$, is exemplified by

$$f(x) = |x_1|^{\frac{1}{2}} + |x_2|^{\frac{1}{2}} + |x_3|^{\frac{1}{2}}.$$

In the first case $f(x)$ is called *homogeneous of degree k,* and in the second is called *positive homogeneous of degree k.* We will not encounter the latter type in projective geometry, but we will meet it in the discussion of projective metrics (compare (25.7)).

The two equations

$$x \cdot \xi = 0 \ \text{and} \ x \cdot \eta = 0$$

where ξ and η are specified, and $\xi \neq 0$, $\eta \neq 0$, represent the same locus of points x if $\xi \sim \eta$. When $\xi, \eta \neq 0$ and $\xi \not\sim \eta$ then the algebra of Section 1 shows that the only point, or class, common to the two loci is $[\xi \times \eta]$. Thus, there is a one-to-one correspondence between the classes $[\xi]$ and the lines $x \cdot \xi = 0$. Consequently, the line $x \cdot \xi = 0$ may be referred to as the line ξ, which, as in Section 1, amounts to taking the coefficients of the equation as coordinates of the line. Then $x \cdot \xi = 0$ becomes, for fixed x, the equation of the point $[x]$ in the line coordinates ξ_i.

Through two distinct points, y and z, there is exactly one line, namely $\xi = y \times z$, since, by (2.6),

$$y \cdot \xi = y \cdot (y \times z) = |\, y,y,z \,| = 0 \text{ and } z \cdot \xi = |\, z,y,z \,| = 0.$$

Thus the projective plane has the uniformity which we sought to obtain for the ordinary plane by adding ideal points.

(2.9)   *Through two distinct points, x and y, there is exactly one line, $x \times y$. Two distinct lines, ξ and η, intersect at exactly one point, $\xi \times \eta$.*

Here, however, there are no "ideal" points, and the line $x_3 = 0$, or $(0,0,1)$, is in no way distinguished from the line $x_1 = 0$ nor from any other line, as will be seen presently.

In Section 1, a representation of the extended Cartesian plane C was obtained with points given by classes $[x]$. Since in this plane lines have linear equations, the points and lines of C may be taken for the elements of a projective plane P_C. But nothing in the definition of a general projective plane P indicates from what source the classes $[x]$ are obtained nor what geometrical meaning is to be assigned to point and line. Thus P_C is only an example of a two-dimensional projective space (see Exercise [2.2]). However, it is an extremely useful one. For a theorem about a general projective plane P is, of course, valid in P_C and through P_C may have an interesting interpretation in C involving non-projective concepts. This is what will be meant by the *"Euclidean" interpretation of a projective theorem.*

3. Projective Coordinates

Points which lie on a line are said to be *collinear*, and lines which pass through a common point are called *concurrent*. From (2.9), any two points are collinear and any two lines are concurrent. If x, y, and z are distinct collinear points, x must lie on the line $y \times z$, hence $x \cdot (y \times z) = |\, x,y,z \,| = 0$ and conversely. The relation $|\, x,y,z \,| = 0$ is also true when two or more of the points coincide. A similar argument for line triples yields:

(3.1)   *The three points x,y,z are collinear if and only if $|\, x,y,z \,| = 0$. The three lines, ξ,η,ζ are concurrent if and only if $|\, \xi,\eta,\zeta \,| = 0$.*

If $y \nleftrightarrow z$, then $| x,y,z | = 0$ is the condition for x to lie on the line $y \times z$ and so is the equation of $y \times z$. On the other hand, $| x,y,z | = 0$ implies the existence of two real numbers λ, μ such that

(3.2) $$x = \lambda y + \mu z.$$

An important point arises here. The points $[y]$ and $[z]$ determine the representative members y and z only to within a factor. If y and z are replaced by $y' = \lambda' y$ and $z' = \mu' z$, $\lambda', \mu' \neq 0$, then $x' = \lambda y' + \mu z'$ will in general be a different point from x, though still a point of $y \times z$. To obtain uniqueness, fixed representations must be taken for y and z. An asterisk will be used for this purpose. Thus where $[y]$ represents the class, and y a representative member which could be changed at will within $[y]$, the symbol y^* denotes the same triple throughout a discussion. The foregoing facts may therefore be formulated:

(3.3)   *If $y \nleftrightarrow z$, and x^*, y^*, z^* are given representations of x, y, and z, where x is on $y \times z$, then uniquely determined numbers λ and μ exist such that*

$$x^* = \lambda y^* + \mu z^*.$$

A general member of $[x]$ is σx^*, where σ is arbitrary but not zero. Since $\sigma x^* = \sigma \lambda y^* + \sigma \mu z^*$, the ratio of the coefficients remains λ/μ, hence this ratio determines $[x]$, and may be used as an "*abscissa*" on $\xi = y \times z$.

More explicitly, if y^* and z^* are fixed representations of two distinct points, y and z, then for all choices of λ and μ, other than $(0,0)$, $\lambda y^* + \mu z^*$ is a point x of $y \times z$. When λ and μ vary, with the ratio λ/μ held constant, the representation runs through the class $[x]$. The point $[x]$ therefore determines the ratio λ/μ, and a definite representation x^* of $[x]$ determines not only the ratio but the numbers λ and μ themselves. The pairs (λ, μ), excepting always $(0,0)$, are called *projective coordinates* on $\xi = y \times z$. Clearly the pairs $(\sigma\lambda, \sigma\mu)$, $\sigma \neq 0$, and only these represent the same point as (λ, μ). Analogous to the projective plane, a *projective line* consists of all classes of number pairs, excepting the zero class, where a class $[(a,b)]$ is called a point. Thus we may say that ξ with the coordinates λ and μ is a one-dimensional projective space.

The selection of y and z as the generating, or base, points of the coordinate system is equivalent to specifying that they correspond to $[(1,0)]$ and $[(0,1)]$ respectively. The coordinate system is then fixed by the choice of representations y^* and z^* (or by a common multiple of these, σy^* and σz^*, $\sigma \neq 0$). An indirect way of specifying the representations y^* and z^*, and so determining the system, is to select a third distinct point u on the line to correspond to $[(1,1)]$, the "*unit*" point. For, representations u^*, y^* and z^* exist such that $u^* = y^* + z^*$. If $\sigma_1 u^*, \sigma_2 y^*, \sigma_3 z^*$ are other representations of the same points, such that $(\sigma_1 u^*) = (\sigma_2 y^*) + (\sigma_3 z^*)$, then $y^* + z^* = (\sigma_2/\sigma_1)y^* + (\sigma_3/\sigma_1)z^*$ implies, since $y \not+ z$, that $\sigma_2/\sigma_1 = \sigma_3/\sigma_1 = 1$. Thus $\sigma_2 y^*$ and $\sigma_3 z^*$ are the same multiple of y^* and z^* respectively and yield the same coordinate system. This establishes:

(3.4)   *Given three distinct points y, z, and u, on a line ξ, there is exactly one system of projective coordinates in which y, z, and u are represented by $(1,0)$, $(0,1)$, and $(1,1)$ respectively. The points $(1,0)$, $(0,1)$, and $(1,1)$ (in this order) are called the reference points of the coordinate system.*

Consider, now, two projective coordinate systems on ξ, say (λ,μ) determined by y^* and z^* and $(\bar{\lambda},\bar{\mu})$ determined by r^* and s^*, so that:

(3.5) $x = \lambda y^* + \mu z^*, \quad x = \bar{\lambda} r^* + \bar{\mu} s^*.$

Then, for suitable numbers, a_{ik},

(3.6) $y^* = a_{11} r^* + a_{21} s^*, \quad z^* = a_{12} r^* + a_{22} s^*$

where $a_{11}/a_{21} \neq a_{12}/a_{22}$, since $y \not+ z$. The substitution of (3.6) into (3.5) yields:

$$x = \bar{\lambda} r^* + \bar{\mu} s^* = \lambda(a_{11} r^* + a_{21} s^*) + \mu(a_{12} r^* + a_{22} s^*).$$

Therefore,

(3.7) $\begin{aligned} \bar{\lambda} &= a_{11}\lambda + a_{12}\mu, \\ \bar{\mu} &= a_{21}\lambda + a_{22}\mu, \end{aligned} \qquad \begin{vmatrix} a_{11} & a_{12} \\ a_{21} & a_{22} \end{vmatrix} \neq 0.$

These are the formulas for the change, on a line, from one system of projective coordinates to a second. Conversely, if λ and μ are projective coordinates, then $\bar{\lambda}$ and $\bar{\mu}$, as defined by (3.7), are also projective coordinates. The proof of this is left as an exercise since the entirely analogous, two-dimensional case is demonstrated in (3.17).

The lines of a projective plane which pass through a common point are said to form a *pencil*. If x is the common point, we speak of the pencil x. In the preceding discussion, if the fixed line ξ is replaced by the fixed point x, the fixed points y and z on ξ by the fixed lines η and ζ through x, and the variable points x on ξ by the variable line ξ through x, then the same argument gives the following facts. If η and ζ are distinct lines through x, then

$$\xi = \lambda\eta^* + \mu\zeta^*, \qquad (\lambda,\mu) \neq (0,0)$$

is a general line of the pencil x. The numbers λ and μ are projective coordinates of ξ in the pencil x. Different values of the ratio λ/μ produce different lines, while if λ and μ vary, with λ/μ constant, ξ runs through the class $[\xi]$. There is exactly one coordinate system in the pencil x for which three given distinct lines have the coordinates $(1,0)$, $(0,1)$, and

(1,1) respectively and these are called the *reference lines* of the coordinate system.

The same type of projective coordinates can be introduced in the projective plane. Let p^1, p^2, p^3 be any three non-collinear points, with representations $p^{1^*}, p^{2^*}, p^{3^*}$ so that $\mid p^{1^*}, p^{2^*}, p^{3^*} \mid \neq 0$. From elementary algebra, the non-vanishing of this determinant implies that for any triple of numbers, $\lambda_1, \lambda_2, \lambda_3$, not $(0,0,0)$, $\lambda_1 p^{1^*} + \lambda_2 p^{2^*} + \lambda_3 p^{3^*}$ is not the zero triple and hence represents a point $[x]$. If it yields $[x]$ in the representation x, changing λ_i to $\sigma\lambda_i$ clearly yields the point $[x]$ in the representation σx. Conversely, if $[x]$ is any point and x^* is one of its representations, the equations

(3.8) $$x^* = \lambda_1 p^{1^*} + \lambda_2 p^{2^*} + \lambda_3 p^{3^*},$$

that is

$$x_i^* = \sum_{k=1}^{3} \lambda_k p_i^{k^*}, \qquad i = 1,2,3$$

have a unique, non-zero solution in the variables λ_i, $i = 1,2,3$. This again follows from standard algebra, since the coefficient determinant $\mid p^{1^*}, p^{2^*}, p^{3^*} \mid$ is not zero. If $[x]$ is represented by σx^*, instead of x^*, the solution of the simultaneous equations changes to $\sigma\lambda_i$, $i = 1,2,3$. Therefore the ratios $\lambda_1 : \lambda_2 : \lambda_3$ determine $[x]$ and conversely. The numbers λ_i are called *projective coordinates of* $[x]$. Putting

(3.9) $d_1 = (1,0,0),\ d_2 = (0,1,0),\ d_3 = (0,0,1),\ e = (1,1,1),$[4]

if d_1, d_2, d_3 are taken for $p^{1^*}, p^{2^*}, p^{3^*}$ (the triangle of reference), then

$$x = x_1 d_1 + x_2 d_2 + x_3 d_3,$$

which shows that the *numbers* (x_1, x_2, x_3) *in any representation of* $[x]$ *are special projective coordinates.*

The construction given for the projective coordinates, λ_i, depended on the representations p^{i^*} chosen for the points p^i, $i = 1,2,3$. If g is any fourth point, not collinear with any pair of the points p^1, p^2, p^3, these four points, as points, will completely determine the projective coordinate system if it is required that g have projective coordinates $(\lambda,\lambda,\lambda)$, $\lambda \neq 0$ (that is, if g is required to be the "unit point" $[(1,1,1)]$). For let $p^{1^*}, p^{2^*}, p^{3^*}$ be respectively representations of p^1, p^2, and p^3 such that, referred to them, g has projective coordinates $(\lambda,\lambda,\lambda)$, $\lambda \neq 0$. Then for some representation g^* of $[g]$,

$$g^* = p^{1^*} + p^{2^*} + p^{3^*}$$

---

[4] In consistency, this should be $d_1^* = (1,0,0)$, etc. However, because these four points appear so frequently the asterisk will not be used, it being understood that exactly these numerical coordinates are meant.

If $\bar{p}^{1*}, \bar{p}^{2*}, \bar{p}^{3*}$ are different representations of p^1, p^2 and p^3, with respect to which g is the unit point, then for some representation of $[g]$, say $\bar{g}^*$,

$$\bar{g}^* = \bar{p}^{1*} + \bar{p}^{2*} + \bar{p}^{3*}$$

The representations with bars are multiples of the corresponding ones without bars, hence numbers, $\sigma, \lambda_1, \lambda_2, \lambda_3$, none of which is zero, exist such that

$$\sigma g^* = \lambda_1 p^{1*} + \lambda_2 p^{2*} + \lambda_3 p^{3*}.$$

From $\sum_{i=1}^{3} (\sigma - \lambda_i) p^{i*} = 0$ it follows that $\lambda_i = \sigma$, $i = 1, 2, 3$, hence $\bar{p}^{i*} = \sigma p^{i*}$,

$i = 1, 2, 3$. Thus, the second set of representations for p^1, p^2, p^3 only differ from the first by a common factor, and this difference does not affect the projective coordinate system. Therefore:

> (3.10) Given four points, no three of which are collinear, there is exactly one projective coordinate system for which these points in a given order have the coordinates $(1,0,0)$, $(0,1,0)$, $(0,0,1)$, and $(1,1,1)$ respectively. These are called the reference points of the coordinate system.

Sets of four points, no three of which are collinear, appear frequently and will be referred to as "quadrangular" sets.

If two projective coordinate systems are given, by:

$$(3.11) \qquad \begin{aligned} x &= x_1' p^{1*} + x_2' p^{2*} + x_3' p^{3*}, & |\, p^{1*},\ p^{2*},\ p^{3*} \,| \neq 0 \\ x &= x_1'' q^{1*} + x_2'' q^{2*} + x_3'' q^{3*}, & |\, q^{1*},\ q^{2*},\ q^{3*} \,| \neq 0, \end{aligned}$$

then for suitable numbers, a_{ik},

$$(3.12) \qquad p^{i*} = a_{1i} q^{1*} + a_{2i} q^{2*} + a_{3i} q^{3*}, \qquad i = 1, 2, 3.$$

Moreover,

$$(3.13) \qquad A = |\, a_{ki} \,| = \begin{vmatrix} a_{11} & a_{12} & a_{13} \\ a_{21} & a_{22} & a_{23} \\ a_{31} & a_{32} & a_{33} \end{vmatrix} \neq 0,$$

for otherwise one of the columns, say the last, would be a linear combination of the other two, hence p^{3*} would be a linear combination of p^{1*} and p^{2*} and so collinear with them.[5] From (3.12) and (3.11) it follows that

$$x = \sum_{i=1}^{3} x_i'' q^{i*} = \sum_{i=1}^{3} x_i' p^{i*} = \sum_{i=1}^{3} x_i' (a_{1i} q^{1*} + a_{2i} q^{2*} + a_{3i} q^{3*}).$$

---

[5]The properties of determinants and matrices up to order 3 are here taken for granted. The reader may refresh his mind on this subject by turning to the beginning of the last chapter, where the corresponding facts for matrices of general order are formulated.

Comparing the coefficients of q^{i^*} in the second and fourth members yields:

$$(3.14) \qquad x_i'' = \sum_{k=1}^{3} a_{ik}x_k', \quad i=1,2,3, \quad |a_{ik}| \neq 0.$$

These are equations for the transformation from the x_i' to the system of coordinates x_i''. Conversely, if the numbers x_i' are projective coordinates, then the x_i'', as defined by (3.14), are also projective coordinates. To see this, let A_{ik} denote the quotient when the co-factor of the element a_{ik} in the matrix (a_{ik}) is divided by the determinant A. By standard algebra, then, the inverse of the transformation (3.14) is given by:

$$(3.15) \qquad x_i' = \sum_{k=1}^{3} A_{ki}x_k'', \quad i=1,2,3, \quad |A_{ik}| \neq 0.$$

Since the x_i' are projective coordinates, a general point x is given by the first equation (3.11),

$$x = \sum_{i=1}^{3} x_i' p^{i^*}$$

Substituting in this from (3.15),

$$x = \sum_{i=1}^{3}\left(\sum_{k=1}^{3} A_{ki}x_k'' \right)p^{i^*}$$

$$= x_1''\left(\sum_{i=1}^{3} A_{1i}p^{i^*} \right) + x_2''\left(\sum_{i=1}^{3} A_{2i}p^{i^*} \right) + x_3''\left(\sum_{i=1}^{3} A_{3i}p^{i^*} \right).$$

Setting $q^{k^*} = \sum_{i=1}^{3} A_{ki}p^{i^*}$, $k=1,2,3$, then

$$(3.16) \qquad x = x_1''q^{1^*} + x_2''q^{2^*} + x_3''q^{3^*}$$

The determinant $|q^{1^*}, q^{2^*}, q^{3^*}| \neq 0$, because $|A_{ik}| \neq 0$. Hence (3.16) shows that the numbers x_i'' are projective coordinates. Thus, we obtain:

If x_i' and x_i'' are two systems of projective coordinates, they are related by equations of the form

$$(3.17) \qquad x_i'' = \sum_{k=1}^{3} a_{ik}x_k', \quad i=1,2,3, \quad |a_{ik}| \neq 0.$$

If the numbers x_i' are projective coordinates, and equations of the given form define the numbers x_i'', then they are also projective coordinates.

The projective plane was defined in terms of classes $[x]$, and the numbers (x_1, x_2, x_3) in a representation of x were shown to be projective coordinates. The distinction between these projective coordinates and any other system of projective coordinates will disappear if it can be shown that straight lines are given by linear equations in any coordinate system. This is easily done. For let $\xi'' \cdot x'' = 0$, $\xi'' \neq 0$, be any linear equation in the coordinates x_i'', and let x_i' be any other projective system. By (3.17), the two are related by

$$x_i'' = \sum_{k=1}^{3} a_{ik} x_k', \qquad i = 1, 2, 3, \qquad |a_{ik}| \neq 0.$$

Substituting from this into $\xi'' \cdot x'' = 0$, the line is given in x_i' by:

$$\xi_1'' \left(\sum_{k=1}^{3} a_{1k} x_k' \right) + \xi_2'' \left(\sum_{k=1}^{3} a_{2k} x_k' \right) + \xi_3'' \left(\sum_{k=1}^{3} a_{3k} x_k' \right) = 0.$$

Rearranging terms,

$$x_1' \left(\sum_{k=1}^{3} a_{k1} \xi_k'' \right) + x_2' \left(\sum_{k=1}^{3} a_{k2} \xi_k'' \right) + x_3' \left(\sum_{k=1}^{3} a_{k3} \xi_k'' \right) = 0.$$

If we define,

(3.18) $$\xi_i' = \sum_{k=1}^{3} a_{ki} \xi_k'', \qquad i = 1, 2, 3,$$

the line has the equation $x' \cdot \xi' = 0$. The numbers ξ_i' cannot all vanish, for if they did, the system (3.18) would be homogeneous, with $|a_{ik}| \neq 0$, which would then imply $\xi'' = 0$. Thus:

(3.19)   *In any projective coordinate system, x_i', straight lines are given by linear equations of the form $x' \cdot \xi' = 0$, $\xi' \neq 0$.*

The numbers x_i, introduced in defining the projective plane, will now be used for arbitrary projective coordinates.

Since the original x_i coordinates have lost their special character, (3.10) shows that *any quadrangular set may be taken for the reference points.* This freedom will be frequently used to simplify calculations, just as in ordinary analytic geometry the coordinate system is often chosen in a convenient manner.

In the proof of (3.19) it was shown that the correspondence of points, $x' \leftrightarrows x''$ given by (3.17) implied the correspondence of lines $\xi' \leftrightarrows \xi''$ in (3.18). One correspondence is said to be induced by the other. More completely:

The coordinate transformation

$$x'_i = \sum_{k=1}^{3} a_{ik}x_k, \qquad i = 1,2,3, \qquad |a_{ik}| \neq 0$$

induces the transformation of the line coordinates

$$\xi'_i = \sum_{k=1}^{3} A_{ik}\xi_k, \qquad i = 1,2,3, \qquad |A_{ik}| = |a_{ik}|^{-1},$$

(3.20)

where A_{ik} is the co-factor of a_{ik} divided by $|a_{ik}|$.[6]
The inverse transformations are :

$$x_i = \sum_{k=1}^{3} A_{ki}x'_k, \qquad i = 1,2,3, \qquad |A_{ki}| = |A_{ik}|$$

$$\xi_i = \sum_{k=1}^{3} a_{ki}\xi'_k, \qquad i = 1,2,3, \qquad |a_{ki}| = |a_{ik}|.$$

4. The Content of Projective Geometry
The Duality Principle

With the exceptional character of the original coordinates removed, we are in a position to describe the *content of projective geometry*. The point x and the line ξ are said to be *incident* if $x \cdot \xi = 0$, and *plane projective geometry is the totality of facts which can be expressed solely in terms of this incidence relation* between points and lines of the projective plane. The term "projective", for describing this geometry, comes from the fact that if a plane P is projected on a second plane P' from a point w (see Figure 2), a point and a line incident with it in P are sent into a point and a line in P' which are incident. Hence a theorem, which is projective in the sense just defined, remains true after a projection (Poncelet, 1788-1867).

Pythagoras' theorem, for example, is not a projective theorem because it involves the concepts of length and angle. Similarly, the statement that the medians of a triangle are concurrent does not belong to projective geometry because distance enters into the definition of a median. Considering how much is ruled out for a theorem to be projective, one might wonder if there are any interesting theorems dealing only with incidence. To show

---

[6] Since projective coordinates are determined only up to a factor, the division by $|a_{ik}|$ could, of course, have been omitted. However, all calculations simplify if the A_{ik} are normalized in this way.

that there are, we discuss two examples before developing the theory systematically. Since the methods so far have been analytic, we choose two theorems which play a basic role when the subject is developed synthetically. The first of these is the Theorem of Desargues (1593-1662).

(4.1)   *If two triangles are such that the lines connecting corresponding vertices are concurrent, then the intersections of corresponding sides are collinear.*

More explicitly this means the following (Figure 1). Let x,y,z and x',y',z' be two triangles (the term triangle implies that the three vertices are

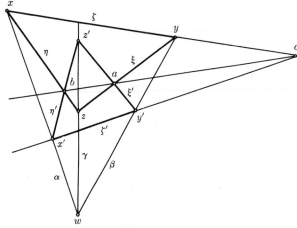

Fig. 1

non-collinear), such that joins of corresponding vertices, $\alpha \sim x \times x'$, $\beta \sim y \times y'$, $\gamma \sim z \times z'$ are three distinct, well defined lines. Then corresponding sides are $\xi \sim y \times z$ and $\xi' \sim y' \times z'$, $\eta \sim z \times x$ and $\eta' \sim z' \times x'$, $\zeta \sim x \times y$ and $\zeta' \sim x' \times y'$, and the intersections of corresponding sides are $a \sim \xi \times \xi'$, $b \sim \eta \times \eta'$, $c \sim \zeta \times \zeta'$. Desargues' theorem states that the concurrence of α,β,γ implies the collinearity of a,b,c or that $|\,\alpha,\beta,\gamma\,| = 0$ implies $|\,a,b,c\,| = 0$. By substituting for the quantities in these determinants, the theorem becomes equivalent to the statement:

(4.2)
$$|x \times x', \quad y \times y', \quad z \times z'| = 0 \; implies$$
$$|(y \times z) \times (y' \times z'), \quad (z \times x) \times (z' \times x'), \quad (x \times y) \times (x' \times y')| = 0.$$

Though (4.2) can be established directly from the algebra of determinants, a simpler proof results from using the fact that any member of its class may be chosen to represent a point. Let w denote the common point of α,β,γ. If w coincides with a vertex, say x, then the proof is immediate. For then y' is on both $x \times y$ and $x' \times y'$ and hence $y' \sim c$. Similarly z' is on $x \times z$ and $x' \times z'$, so $z' \sim b$. But a, by definition, lies on $\xi' \sim y' \times z' \sim b \times c$, hence is collinear with b and c.

We may therefore suppose that w is different from all vertices. Then the line α cannot contain other vertices than x and x', etc. Since $x' \dotplus x$, $y' \dotplus y$, $z' \dotplus z$ (α, β, γ are well defined), then, for a given representation w^* of w, representations of the vertices may be chosen so that

(4.3) $$w^* = x^* - x'^* = y^* - y'^* = z^* - z'^*$$

The relation $x^* - x'^* = y^* - y'^*$ implies $x^* - y^* = x'^* - y'^*$. But $x^* - y^*$ is a point of ζ, and $x'^* - y'^*$ is a point of ζ', so the last given equality implies

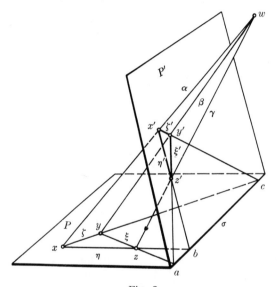

Fig. 2

$x^* - y^* \sim \zeta \times \zeta' \sim c$. By a similar argument (4.3) yields $y^* - z^* \sim \xi \times \xi' \sim a$, and $z^* - x^* \sim \eta \times \eta' \sim b$. Since $1 \cdot (x^* - y^*) + 1 \cdot (y^* - z^*) + 1 \cdot (z^* - x^*) = 0$, it follows that a, b and c are collinear.

An intuitively simple way of seeing Desargues' theorem, though not yet rigorously justified, is to first consider the triangles x,y,z and x',y',z' as lying *in two different planes P and P'*, which intersect in a line σ (Figure 2). If, as before, w denotes the common intersection point of α, β and γ, then since α and γ intersect at w they determine a plane. The lines η and η' lie in this plane and so intersect at a point b. Because it is on η and η', b is on both P and P', hence is on σ. By the same argument, a and c are on σ, so a, b and c are collinear. The plane theorem is now seen to be a limiting case as the plane P approaches P'. Tacitly, it is assumed, of course, that all the intersections exist, in other words that we have a three-dimensional analogue to the projective plane. In a later chapter it will be shown how such a space may be rigorously defined.

If, in (4.2), the algebraic form of Desargues' theorem, the Greek letters ξ,η,ζ and ξ',η',ζ' are substitued respectively for x,y,z and x',y',z', it is still true that

$$(4.4) \quad \begin{aligned} &|\xi \times \xi', \quad \eta \times \eta', \quad \zeta \times \zeta'| = 0 \; implies \\ &|(\eta \times \zeta) \times (\eta' \times \zeta'), (\zeta \times \xi) \times (\zeta' \times \xi'), (\xi \times \eta) \times (\xi' \times \eta')| = 0. \end{aligned}$$

Interpreting these letters as lines and putting $\eta \times \zeta \sim x$, $\eta' \times \zeta' \sim x'$, etc., establishes the *converse of Desargues' theorem*:

(4.5)   *If two triangles have the property that the intersections of correspond-ing pairs of sides are collinear then the lines connecting correspond-ing pairs of vertices are concurrent.*

In a synthetic, or purely geometric, development of projective geometry, the essential axioms are (a) the existence axioms, which guarantee points and lines to work with, (b) the existence and uniqueness of a line through two distinct points and of a point on two distinct lines, and (c) Desargues' theorem.[7] Examples exist to show that (a) and (b) alone do not imply (c). But though non-Desarguean systems are possible, without (c) no interest-ing geometric theory can be developed. For this reason it is added as a basic axiom in the studies referred to.

Our second example is a consequence of Desargues' theorem and concerns the uniqueness of a point v, called the *fourth harmonic point*, corresponding to three given, distinct points x, y and u on a line ξ (Figure 3). The con-struction of v is as follows. Take any point w, not on ξ, and any point z on $x \times w$, distinct from x and w. Let $y \times z$ and $u \times w$ intersect at t, and let $x \times t$ and $y \times w$ intersect at s. Then v is defined as the intersection of ξ and $z \times s$. Though v would seem to depend on z and w, this is not the case.

The construction, for any choice of w and z, yields the same point v on the line ξ. For let the construction be repeated with w' and z' as initial choices de-termining t', s' and v'. By construction, the corresponding sides of the tri-angles w,s,t and w',s',t' intersect in the points x, y and u. By (4.5), the con-verse of Desargues' theorem, the lines $w \times w'$, $s \times s'$, and $t \times t'$ are therefore concurrent at a point g. Considering the triangles w,z,t and w',z',t', we find, in the same way, that the lines $w \times w'$, $t \times t'$, and $z \times z'$ are con-current at a point, which must be g, since g is determined by $w \times w'$ and $t \times t'$. Now the triangles z,w,s and $z',w's'$ satisfy the hypothesis of De-

---

[7] See, for instance, Coxeter's *The Real Projective Plane.*

sargues' theorem, (4.1). Points x and y are the intersections of two pairs of corresponding sides, hence the last pair of corresponding sides, $z \times s$ and $z' \times s'$, must intersect on $x \times y \sim \xi$. Since $z \times s$ already cuts ξ at v, it follows that $v' \sim \xi \times (z' \times s') \sim v$.

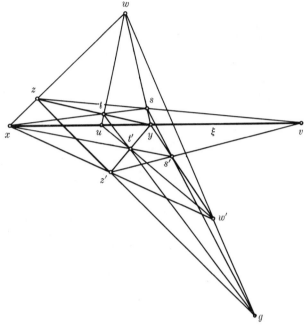

Fig. 3

An analytic proof of this fact is instructive, for a comparison of methods, and will also establish some facts for later use. We first state some conventions that will frequently occur. In any coordinate system the points d_1, d_2, d_3, and e, defined in (3.9), were called the reference points. Similarly the lines

(4.6) $\delta_1 = (1,0,0), \quad \delta_2 = (0,1,0), \quad \delta_3 = (0,0,1), \quad \varepsilon = (1,1,1),$ [8]

denote the *reference lines* of the system. The two reference sets are connected by the relations:

(4.7) $d_1 \times d_2 = \delta_3, \quad d_2 \times d_3 = \delta_1, \quad d_3 \times d_1 = \delta_2$
                    $\delta_1 \times \delta_2 = d_3, \quad \delta_2 \times \delta_3 = d_1, \quad \delta_3 \times \delta_1 = d_2.$

---

[8]As with the points in (3.9), asterisks are not used for these lines and δ_1, for example, means always (1,0,0).

Clearly, from its coordinates:

(4.8)   *The equation of the line δ_i is $x_i = 0$, $i = 1,2,3$.*
The equation of the point d_i (or the pencil d_i) is $\xi_i = 0$, $i = 1,2,3$.

If x is any point distinct from d_1, the line $d_1 \times x$ is

$$(1,0,0) \times (x_1,x_2x_3) = (0,- x_3,x_2),$$

and the intersection point of this line with δ_1 is

$$(d_1 \times x) \times \delta_1 = (0,x_2,x_3) = x_2 d_2 + x_3 d_3.$$

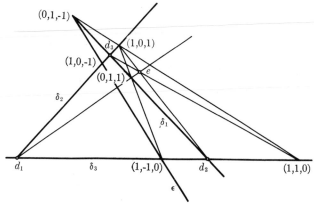

Fig. 4

(Figure 4 illustrates this for $x = e$.) From the definition of projective coordinates on a line, then:

(4.9)   *If $x \not+ d_1$, then $d_1 \times x$ intersects δ_1 at $(0,x_2,x_3)$ and x_2,x_3 are projective coordinates on the line δ_1.*

Similarly :

(4.10)   *If $\xi \not+ \delta_1$, then the line connecting $\delta_1 \times \xi$ and d_1 is $(0,\xi_2,\xi_3)$ and ξ_2,ξ_3 are projective coordinates in the pencil d_1.*

To prove analytically that the harmonic construction yields the fourth harmonic point uniquely, let the coordinate system be chosen so that $w = d_1$, $x = d_2$, $y = d_3$ and $t = e = (1,1,1)$. From (4.9), $u = (0,1,1)$, $z = (1,1,0)$, $s = (1,0,1)$, hence the line $z \times s = (1,-1,-1)$ intersects $\xi = \delta_1$ in $v = (1,0,0) \times (1,-1,-1) = (0,1,-1)$. Now x_2,x_3 are projective coordinates on $\xi = \delta_1$, with x, y and u as reference points. Since this uniquely determines the projective coordinate system on ξ, the point v, with coordinates $(1,-1)$, is uniquely determined by x, y and u. It is therefore independent of z and w. As a corollary of this proof:

(4.11)   *If x, y and u are three distinct points on a line ξ, then in the projective coordinate system on ξ having x, y and u as reference points, the fourth harmonic point to x, y and u has the coordinates $(1,-1)$.*

In purely algebraic terms the uniqueness of the fourth harmonic point v to x, y and u, that is the independence of v from z and w, may be stated as follows. Let x, y and u be distinct points for which $|\,x,y,u\,| = 0$. Take w^i, $i = 1,2$, so that $|\,w^i,x,y\,| \neq 0$, and z^i, $i = 1,2$, so that z^i is distinct from w and from x but satisfies $|\,w^i,z^i,x\,| = 0$, $i = 1,2$. Put $t^i = (w^i \times u) \times (z^i \times y)$ and $s^i = (t^i \times x) \times (w^i \times y)$, $i = 1,2$. Then $|\,x \times y,\ z^1 \times s^1,\ z^2 \times s^2\,| = 0$.

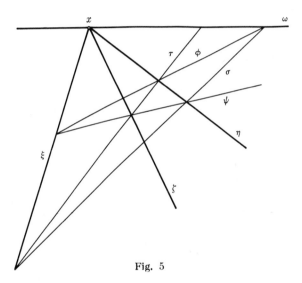

Fig. 5

The truth of this algebraic formulation is not affected if x,y,u,w^i,z^i,t^i and s^i are replaced respectively by $\xi,\eta,\zeta,\psi^i,\varphi^i,\tau^i$ and σ^i. Interpreting the new symbols as lines, the geometric meaning of the algebra is this (Figure 5): let ξ, η, ζ be distinct lines of a pencil x. Take ψ as any line not in the pencil x, and φ a line concurrent with, and distinct from, ξ and ψ. The points $\eta \times \varphi$ and $\psi \times \zeta$ determine a line τ, and the points $\xi \times \tau$ and $\psi \times \eta$ determine a line σ. Then the line ω joining x and $\varphi \times \sigma$ is independent of ψ and φ. It is called *the fourth harmonic line to ξ,η,ζ.*

In two instances, now, the purely formal interchange of Greek and Roman letters has yielded a new geometric theorem. In view of the common algebraic formalism for point and line, it is not difficult to see that a general principle underlies this phenomenon. The classes $[x]$ and $[\xi]$ for point and line both range over the classes of real number triples, with $(0,0,0)$ excluded. On the other hand, projective theorems can be expressed solely in terms of incidence relations of the form $x \cdot \xi = 0$ and this form is symmetric in x and ξ. That is, $x \cdot \xi = 0$ is equivalent to $\xi \cdot x = 0$. Therefore interchanging the concepts of point and line throughout a projective theorem does not impair its validity, but, in general, changes the geometric content and thus gives rise to the so called "dual theorem."

> DUALITY PRINCIPLE: *A projective theorem remains valid if the concepts of point and line are interchanged.*

This fact is only surprising because our intuition and geometric terminology are not conditioned to it since it does not hold in Euclidean, that is ordinary geometry.[9] Instead of automatically seeing the dual of a theorem we have therefore to think through analogues, for instance:

collinear points (points incident with a line)	concurrent lines (lines incident with a point)
the line $x \times y$ connecting x and y (i. e., the line incident with x and y)	the point, $\xi \times \eta$, of intersection of the lines ξ and η (i.e., the point incident with ξ and η)
straight line as the locus of its points.	pencil of lines through a point.

It should be observed here that although the duality principle does not hold in Euclidean geometry, the dual of a Euclidean theorem is frequently correct, in which case it must, of course, be proved independently. Thus even *outside of projective geometry the duality concept is a strong exploratory tool*. Often too, where a theorem is difficult to grasp, its dual may be readily accessible to the intuition. The construction of the fourth harmonic line, for instance, is much harder to visualize than that for the fourth harmonic point.

5. Groups of Transformations. Projectivities

A *mapping*[10] Φ of a set S on a set S' is a correspondence which associates with every element x of S one element $x' = x\Phi$ of S', and which covers S', that is, every element of S' has at least one corresponding element in S. The association is denoted by $x \to x'$. When each element of S' has exactly

---

[9]Ordinary geometry is called Euclidean because it was first represented systematically in Euclid's Elements (3rd century B. C.), which served as *the* textbook for over 2000 years.
[10]The word "transformation" is often used instead of "mapping".

one corresponding point in S the mapping or transformation is said to be *one-to-one*. In that case the association $x' \to x$ is a mapping of S' on S which is called the *inverse* transformation, denoted by Φ^{-1}.

To have some examples, let S be the real x-axis and S' the real x'-axis. Then $x' = \sin x$ is not a mapping of S on S' because only the interval $-1 \leqslant x' \leqslant 1$ on S' is covered. If S' is taken as this interval, instead of the whole line, $x' = \sin x$ is a mapping of S on S', but is clearly not one-to-one. An example that is one-to-one, where S' is again the line, is $x' = ax + b,\ a \neq 0$. The inverse transformation is clearly $x = (x' - b)/a$.

If $x' = x\Phi$ is a mapping of the set S on S' and $x'' = x'\Phi'$ is a mapping of S' on S'', then $x'' = (x\Phi)\Phi'$ is a mapping of S on S''. It is denoted by $\Phi\Phi'$ and is called *the product* of Φ and Φ'. Formally, then:

(5.1) $\Phi'' = \Phi\Phi'$ *means that* $x\Phi'' = x(\Phi\Phi') = (x\Phi)\Phi'.$

From its definition there is no reason for $\Phi\Phi'$ to be the same as $\Phi'\Phi$. In fact, when S and S'' are different sets, $\Phi'\Phi$ is not even defined. If S and S'' are the same set ($S = S''$) then $\Phi'\Phi$ is a mapping of S' on itself while $\Phi\Phi'$ is a mapping of S on itself. Even if $S = S' = S''$, $\Phi\Phi'$ need not be the same transformation as $\Phi'\Phi$. For instance, if S, S' and S'' are all three the real axis, and Φ and Φ' are given by

$$\Phi : x' = ax + b, \quad a \neq 0, \quad b \neq 0; \qquad \Phi' : x' = x^3,$$

then $\Phi\Phi'$ and $\Phi'\Phi$ are the mappings

$$\Phi\Phi' : x'' = (ax + b)^3, \qquad \Phi'\Phi : x'' = ax^3 + b,$$

which are clearly different.

The mapping of a set on itself in which each point is associated with itself is called the *identity mapping*, or simply the *identity*, and is denoted by 1 without indicating the set. It is obviously its own inverse. When a transformation Φ, of S on S', has an inverse, then both $\Phi\Phi^{-1}$ and $\Phi^{-1}\Phi$ are the identity, but the first is the identity on S and the second on S'. For $S = S'$ it is always true that $\Phi\Phi^{-1} = \Phi^{-1}\Phi = 1$.

A non-empty aggregate Γ, whose elements, Φ, are one-to-one mappings of a set S on itself, is said to form a *group* if Φ being in Γ implies that Φ^{-1} is also in Γ, and Φ and Φ' being in Γ implies that $\Phi\Phi'$ is in Γ.

A group Γ *always contains the identity mapping on* S. For being non-empty Γ has at least one mapping Φ. It therefore contains Φ^{-1} and hence $\Phi\Phi^{-1}$, which is the identity. The elements of Γ also satisfy the *associative law*, namely:

(5.2) $\Phi(\Phi'\Phi'') = (\Phi\Phi')\Phi''.$

By hypothesis, $\Phi'\Phi''$ and $\Phi\Phi'$ are in Γ, since Φ, Φ' and Φ'' are. Therefore $\Phi(\Phi'\Phi'')$ and $(\Phi\Phi')\Phi''$ are in Γ also. That they are equal follows directly from the definition, (5.1), of a product, since

and
$$x[\Phi(\Phi'\Phi'')] = (x\Phi)(\Phi'\Phi'') = [(x\Phi)\Phi']\Phi'',$$
$$x[(\Phi\Phi')\Phi''] = [x(\Phi\Phi')]\Phi'' = [(x\Phi)\Phi']\Phi''.$$

A simple example of a group, Γ, is given by the collection of mappings, $x' = ax + b$, of the real axis on itself, where a and b are real numbers and $a \neq 0$. If Φ is $x' = ax + b$ then Φ^{-1} is $x = a^{-1}x' - a^{-1}b$ which is also in Γ. If Φ' is the element $x' = cx + d$, $c \neq 0$, $\Phi\Phi'$ is the mapping

$$x' = (ca)x + (cb + d)$$

and so is in Γ. The reverse product, $\Phi'\Phi$, is $x' = (ac)x + (ad + b)$. Thus, $\Phi\Phi' = \Phi'\Phi$ only if $cb + d = ad + b$.

Two mappings Φ and Φ' are said to *commute* if $\Phi\Phi' = \Phi'\Phi$. When all pairs in a group Γ commute, Γ is said to be a *commutative or an Abelian group*.

We apply these general ideas to some special cases whose importance will soon be evident. Let ξ and ξ' be lines in the same plane, or even in different planes.[11] A mapping of the points of ξ on the points of ξ' is called *projectivity* of ξ on ξ' if projective coordinates $\bar{\lambda}_1, \bar{\lambda}_2$ on ξ, and $\bar{\lambda}'_1, \bar{\lambda}'_2$ on ξ' exist so that in terms of them the mapping takes the form

(5.3) $$\bar{\lambda}'_1 = \bar{\lambda}_1, \qquad \bar{\lambda}'_2 = \bar{\lambda}_2.[12]$$

Clearly *a projectivity is one-to-one, and its inverse is also a projectivity* (of ξ' on ξ). Because any three points of a line may be taken for the reference points of a coordinate system on the line, it follows that if y, z and u are three arbitrary, distinct points of ξ, and y',z',u' are three arbitrary, distinct points of ξ', then a projectivity Φ exists such that $y' = y\Phi$, $z' = z\Phi$, and $u' = u\Phi$.

Suppose now that (5.3) is a projectivity between ξ and ξ' in the indicated coordinates and that λ_1,λ_2 and λ'_1,λ'_2 denote arbitrary projective

---

[11]A "different" projective plane is simply again the collection of all classes of number triples except 0. The only logical difference is that we do not consider the triples as representing the same point set. The distinction is important in space, where many different planes exist, just as there are many different lines in the plane.

[12]This definition may seem strange at first sight. However, it simply is the projective analogue to the idea of defining an ordinary distance preserving mapping of a line ξ on a line ξ' as a mapping of the form $x' = x$, where x and x' are properly chosen abscissas on ξ and ξ'. Similarly a distance preserving transformation of an ordinary plane P on a plane P' may be written as $x' = x, y' = y$, where x,y and x',y' are properly chosen rectangular coordinates on P and P'.

coordinates on ξ and ξ' respectively. It has been shown that for two systems on ξ and two on ξ' relations of the following form exist:

(5.4)
$$\begin{aligned}\bar{\lambda}_1 &= a_{11}\lambda_1 + a_{12}\lambda_2 \\ \bar{\lambda}_2 &= a_{21}\lambda_1 + a_{22}\lambda_2\end{aligned} \qquad |\,a_{ik}\,| \neq 0,$$
$$\begin{aligned}\lambda_1' &= b_{11}\bar{\lambda}_1' + b_{12}\bar{\lambda}_2' \\ \lambda_2' &= b_{21}\bar{\lambda}_1' + b_{22}\bar{\lambda}_2'\end{aligned} \qquad |\,b_{ik}\,| \neq 0.$$

Using (5.3) yields

$$\begin{aligned}\lambda_1' &= b_{11}(a_{11}\lambda_1 + a_{12}\lambda_2) + b_{12}(a_{21}\lambda_1 + a_{22}\lambda_2) \\ \lambda_2' &= b_{21}(a_{11}\lambda_1 + a_{12}\lambda_2) + b_{22}(a_{21}\lambda_1 + a_{22}\lambda_2),\end{aligned}$$

or,

(5.5)
$$\begin{aligned}\lambda_1' &= c_{11}\lambda_1 + c_{12}\lambda_2 \\ \lambda_2' &= c_{21}\lambda_1 + c_{22}\lambda_2\end{aligned}, \qquad |\,c_{ik}\,| \neq 0,$$

where

(5.6)
$$\begin{pmatrix} c_{11}c_{12} \\ c_{21}c_{22} \end{pmatrix} = \begin{pmatrix} b_{11}a_{11} + b_{12}a_{21} & b_{11}a_{12} + b_{12}a_{22} \\ b_{21}a_{11} + b_{22}a_{21} & b_{21}a_{22} + b_{22}a_{22} \end{pmatrix}.$$

Thus the projectivity (5.3) has the same form, in arbitrary coordinates, as the transformation between two arbitrary coordinate systems on ξ or ξ'. That $|\,c_{ik}\,| \neq 0$ follows from the fact that (5.3), and hence (5.5), is a one-to-one mapping of ξ on ξ'. The fact can also be referred to matrix theory, since, from (5.6),

$$(c_{ik}) = (b_{ik}) \cdot (a_{ik}) \text{ which implies } |\,c_{ik}\,| = |\,a_{ik}\,| \cdot |\,b_{ik}\,| \neq 0.$$

The converse is also true, namely that every one-to-one transformation of ξ on ξ' of the form (5.5) is a projectivity. For new coordinates on ξ need only be defined by $\bar{\lambda}_1 = c_{11}\lambda_1 + c_{12}\lambda_2$, $\bar{\lambda}_2 = c_{21}\lambda_1 + c_{22}\lambda_2$ and the transformation takes the form $\lambda_1' = \bar{\lambda}_1$, $\lambda_2' = \bar{\lambda}_2$.

This also shows that there is only one projectivity, Φ, which carries three given, distinct points of ξ into a specified, distinct triple on ξ' in a given order. For let y, z, u and y', z', u' be given, distinct triples on ξ and ξ' respectively. Then $\lambda_1' = \lambda_1$, $\lambda_2' = \lambda_2$ is a projectivity of ξ on ξ' mapping the first triple on the second, where (λ_1, λ_2) and (λ_1', λ_2') are coordinates in ξ and ξ' referred to the respective triples as base points. By definition, coordinates $(\bar{\lambda}_1, \bar{\lambda}_2)$ and $(\bar{\lambda}_1', \bar{\lambda}_2')$ exist on ξ and ξ' such that a second projectivity is expressed by $\bar{\lambda}_1' = \bar{\lambda}_1$, $\bar{\lambda}_2' = \bar{\lambda}_2$. The two coordinate systems on ξ, and the two on ξ', are related by (5.4), hence the second projectivity is expressed by (5.5) with $(1,0)$, $(0,1)$ and $(1,1)$ going into points with the same, or proportional, coordinates. This implies $c_{21} = 0$, $c_{12} = 0$

and $c_{11} = c_{22} \neq 0$. But then (5.5) reduces to $\lambda_1' = \lambda_1$, $\lambda_2' = \lambda_2$, which is the original projectivity. Thus we have the theorem:

> *The general form for a projectivity between two lines, ξ and ξ', with respective coordinates λ_1, λ_2 and λ_1', λ_2', is*

(5.7)
$$\begin{aligned} \lambda_1' &= a_{11}\lambda_1 + a_{12}\lambda_2 \\ \lambda_2' &= a_{21}\lambda_1 + a_{22}\lambda_2 \end{aligned} \qquad |a_{ik}| \neq 0.$$

> *There is exactly one projectivity mapping three given, distinct points of ξ on three given, distinct points of ξ' in a specified order.*

It is now easily shown that:

(5.8)
> *If Φ is a projectivity of ξ on $\bar{\xi}$ and Φ' is a projectivity of $\bar{\xi}$ on ξ' then $\Phi\Phi'$ is a projectivity of ξ on ξ'.*

For, because of (5.7), Φ and Φ' may be written in the forms of (5.4) so that $\Phi\Phi'$ is given by (5.5) whose determinant was shown not to be zero. Because of (5.7), then, $\Phi\Phi'$ is a projectivity.

In particular, if $\xi \sim \xi'$ in (5.8) then $\Phi\Phi'$ is a projectivity of ξ on itself. Since we have already observed that the inverse of a projectivity is a projectivity, it follows:

(5.9)   *The projectivities of a line on itself form a (non-Abelian) group.*

That the group is non-Abelian follows from the previous example of the mapping $x' = ax + b$ if it is written in homogeneous form (see also Exercise [5.2]).

With the duality principle established, the facts for projectivities of lines immediately yield dual facts for *projectivities of pencils*. Since it is cumbersome to always state the dual of a situation under discussion this will, in general, be left to the reader. This is not to underestimate the importance of dualizing theorems, both verbally and pictorially, for added insight.

Two-dimensional projectivities may be treated in the same manner as those for lines. If P and P' are two planes (which may coincide), a mapping of the points of P on the points of P' is called *a collineation (or projectivity) of P on P'* if coordinates $\bar{x}_i$ and $\bar{x}_i'$ exist in P and P' respectively such that the mapping is expressed by

$$\bar{x}_i' = \bar{x}_i, \qquad i = 1,2,3, \text{ or symbolically by } \bar{x}' = \bar{x}.$$

Clearly the inverse of a collineation is also a collineation. Since any quadrangular set may be selected for the reference points of a coordinate system, there is always a collineation of P on P' taking a given, quadrangular set of P into a second, given, quadrangular set of P' in a definite order. The points in P which satisfy $\bar{x} \cdot \bar{\xi} = 0$ go into points which satisfy $\bar{x}' \cdot \bar{\xi}' = 0$, hence the line $\bar{\xi}$ in P goes into the line $\bar{\xi}'$ in P' for which $\bar{\xi}' = \bar{\xi}$. Thus *a projectivity preserves incidence.*

Consider, now, arbitrary coordinate systems, x_i and x_i', $i = 1,2,3$ on P and P' respectively. If $\bar{x}' = \bar{x}$ denotes a collineation of P on P', then, from (3.20), in each plane the two coordinate systems satisfy relations of the form

(5.10)
$$\bar{x}_i = \sum_{k=1}^{3} a_{ik}x_k, \; i = 1,2,3, \; |a_{ik}| \neq 0$$

$$x_i' = \sum_{k=1}^{3} b_{ik}\bar{x}_k', \; i = 1,2,3, \; |b_{ik}| \neq 0.$$

From the collineation, then

(5.11)
$$x_i' = \sum_{j=1}^{3} b_{ij} \left(\sum_{k=1}^{3} a_{jk}x_k \right) = \sum_{j=1}^{3} c_{ij}x_j, \; i = 1,2,3,$$

$$\text{where } c_{ij} = \sum_{k=1}^{3} b_{ik}a_{kj} \text{ and } |c_{ij}| \neq 0.$$

That $|c_{ij}| \neq 0$ again follows either from the fact that the mappings in (5.10) are one-to-one, or from the matrix relation

$$(c_{ij}) = \left(\sum_{k=1}^{3} b_{ik}a_{kj} \right) = (b_{ik})(a_{kj})$$

which implies $|c_{ij}| = |b_{ik}| \cdot |a_{kj}| \neq 0$. Conversely, a mapping of P on P' in the form (5.11) always represents a projectivity. For, new coordinates on P need only be defined by $\bar{x}_i = \sum_{j=1}^{3} c_{ij}x_j$, and the mapping takes the form $x_i' = \bar{x}_i$, $i = 1,2,3$.

Again, this shows that for two quadrangular sets in P and P' respectively there is only one projectivity Φ which maps the first set on the second in a given order. For if x and x' are coordinates in P and P' respectively, with the given sets as reference points, then $x' = x$ is one projectivity with the desired property. For a second projectivity, coordinates $\bar{x}$ in P and $\bar{x}'$ in P' exist such that the mapping is expressed by $\bar{x}' = \bar{x}$. Since (5.10) relates x with $\bar{x}$ and x' with $\bar{x}'$, this second projectivity is expressed by (5.11). Since it sends $(1,0,0)$ into $(1,0,0)$ it follows that $c_{21} = c_{31} = 0$. In the same way, $(0,1,0) \to (0,1,0)$ implies $c_{12} = c_{32} = 0$ and $(0,0,1) \to (0,0,1)$ gives $c_{13} = c_{23} = 0$. The collineation is therefore reduced to $x_i' = c_{ii}x_i$, $i = 1,2,3$, and $(1,1,1) \to (1,1,1)$ implies that the c_{ii}

are equal. Since the x_i are determined only up to a factor, the transformation may be expressed by $x' = x$, which is the original projectivity. Thus we have proved:

> *The general form for a projectivity between the planes P and P', with respective coordinate systems x_i and x_i', is*

(5.12)
$$x_i' = \sum_{k=1}^{3} a_{ik} x_k, \quad i = 1,2,3, \quad |a_{ik}| \neq 0.^{13}$$

> *There is exactly one projectivity which maps in a specified order a given, quadrangular set of P on a given, quadrangular set in P'.*

As in (5.8):

(5.13)
> *If Φ is a projectivity of P on $\overline{P}$ and Φ' is a projectivity of $\overline{P}$ on P', then $\Phi\Phi'$ is a projectivity of P on P'.*

For Φ and Φ' may be taken in the forms (5.10) whence $\Phi\Phi'$ is given by (5.11) which is a projectivity.

If, in (5.13), $P = \overline{P} = P'$ then Φ and Φ', and hence $\Phi\Phi'$, are projectivities of P on itself. Since the inverse of a collineation has been already seen to be a collineation, this establishes:

(5.14)
> *The projectivities (or collineations) of a plane on itself form a (non-Abelian) group.*

Since the algebraic expression for a projectivity is exactly the same as for a coordinate transformation, (3.20) may be reinterpreted as:

> *The projectivity*

$$x_i' = \sum_{k=1}^{3} a_{ik} x_k, \quad i = 1,2,3, \quad |a_{ik}| \neq 0$$

> *induces the line transformation*

(5.15)
$$\xi_i' = \sum_{k=1}^{3} A_{ik} \xi_k, \quad i = 1,2,3, \quad |A_{ik}| \neq 0.$$

> *The inverse projectivity is*

$$x_i = \sum_{k=1}^{3} A_{ki} x_k', \quad i = 1,2,3, \quad |A_{ki}| \neq 0,$$

---

[13] If $P = P'$ the same coordinate system may be used both times. Then this relation gives the coordinates of the image x' of x in the same coordinate system. For $P = P'$ this interpretation will always be adopted without explicit statement.

which, in line coordinates, is

(5.15)
$$\xi_i = \sum_{k=1}^{3} a_{ki}\xi_k', \ i = 1,2,3, \ |a_{ki}| \neq 0.$$

A projectivity Φ of a plane P on P' sets up a correspondence between the points of a line ξ and those of its image ξ' which is called the *mapping of ξ on ξ' induced by* Φ. One and two-dimensional projectivities are related by the following facts.

(5.16)   *If a projectivity Φ of P on P' maps the line ξ (pencil x) on the line ξ' (pencil x'), this induced transformation is a one-dimensional projectivity. Any projectivity of ξ and ξ' (x and x') may be induced by a projectivity of P on P' (which is not unique).*

PROOF: Let y, z and u be three distinct points of ξ with the (neccessarily distinct) images y', z' and u' on ξ'. Let v be any point of P not on ξ and w any point of $u \times v$ which is distinct from u and v. Then $v' = v\Phi$ and $w' = w\Phi$ are distinct points of P' which are collinear with u' but are not on ξ'. If, now, v,y,z,w and v',y',z',w' are taken for the reference points in P and P' respectively, then the form of Φ becomes $x' = x$. In particular the point $(0,x_2,x_3)$ on ξ goes into $(0,x_2',x_3')$ on ξ'. By (4.9), x_2,x_3 and x_2',x_3' are projective coordinates on ξ and ξ' respectively. hence, by definition, the induced mapping of ξ on ξ' is a projectivity.

Conversely let Ψ be a projectivity between ξ in P and ξ' in P'. Let y, z and u be any three distinct points on ξ with y', z' and u' their images under Ψ. Choose v and w as before and let v' be any point of P', not on ξ', and w' any point collinear with u' and v' but distinct from them. There is then, by (5.12), a projectivity Φ of P on P' which maps v,y,z,w on v',y',z',w' in that order. By the first part of the proof, the mapping of ξ on ξ' induced by Φ is a projectivity Ψ'. Since a projectivity of two lines is determined by three points, and y',z',u' are the images of y,z,u under both Ψ and Ψ', it follows that $\Psi' = \Psi$.

This purely analytical approach to projectivities does not reveal their geometric significance. A geometric construction was given for the fourth harmonic point to three given points: the one-dimensional projectivities are the only one-to-one mappings of one line on another in which every harmonic quadruple goes into a harmonic quadruple (in the same order). The projectivities between planes are the only one-to-one mappings which preserve incidence. A proof for the first of these two important facts will be given in the next section, and the second will be proved in Section 9.

6. Cross Ratio. Intervals

Since any distinct triple of points on a line ξ can be sent by a projectivity into any assigned, distinct triple on a line ξ', there is no inherent, projective property to distinguish one collinear triple from another. (In Euclidean geometry, for example, distance distinguishes pairs and hence triples of points from other pairs or triples.) However, *sets of four collinear points have a special property* to which we are led by the following considerations.

In the ordinary (x_1, x_2)-plane, the point $(\bar{x}_1, \bar{x}_2)$, which divides the segment from (x_1, x_2) to (y_1, y_2) in the ratio $\bar{\lambda}/\bar{\mu}$, has the coordinates

$$\bar{x}_1 = \frac{\bar{\mu}x_1 + \bar{\lambda}y_1}{\bar{\lambda} + \bar{\mu}}, \qquad \bar{x}_2 = \frac{\bar{\mu}x_2 + \bar{\lambda}y_2}{\bar{\lambda} + \bar{\mu}}.$$

Setting $\lambda = \bar{\lambda}/(\bar{\lambda} + \bar{\mu})$ and $\mu = \bar{\mu}/(\bar{\lambda} + \bar{\mu})$, it follows that the same point written as $(\mu x_1 + \lambda y_1, \mu x_2 + \lambda y_2)$ divides the segment in the ratio λ/μ. If, now, $[y]$, $[z]$ and $[u]$ are three distinct points on a line ξ in the projective plane with y, z and u as representations, then λ and μ exist, neither zero, such that

$$u = \lambda y + \mu z.$$

However, the ratio λ/μ has no geometric meaning here. For, if $\bar{y} = \sigma y$ and $\bar{z} = \delta z$ are new representations of $[y]$ and $[z]$ then

$$u = \bar{\lambda}\bar{y} + \bar{\mu}\bar{z}, \text{ where } \bar{\lambda} = \lambda/\sigma \text{ and } \bar{\mu} = \mu/\delta,$$

and λ/μ is not, in general, the same as $\bar{\lambda}/\bar{\mu}$. If $[v]$, represented by v, is a fourth point on ξ, then λ', μ' exist such that

$$v = \lambda' y + \mu' z = \bar{\lambda}'\bar{y} + \bar{\mu}'\bar{z}, \text{ where } \bar{\lambda}' = \lambda'/\sigma \text{ and } \bar{\mu}' = \mu'/\delta.$$

Though the ratio λ'/μ' is also (in general) not $\bar{\lambda}'/\bar{\mu}'$, the change in representations for $[y]$ and $[z]$ affects the ratio λ'/μ' in the same way as it affected the ratio λ/μ. The ratio of ratios $(\lambda'/\mu')/(\lambda/\mu)$ is therefore independent of the representations of $[y]$ and $[z]$, that is,

$$(\bar{\lambda}'/\bar{\mu}')/(\bar{\lambda}/\bar{\mu}) = (\lambda'\sigma^{-1}/\mu'\delta^{-1})/(\lambda\sigma^{-1}/\mu\delta^{-1}) = (\lambda'/\mu')/(\lambda/\mu).$$

For this ratio of ratios, associated with the points

$$y, z, \qquad \lambda y + \mu z, \qquad \lambda' y + \mu' z,$$

to have a geometric meaning it only remains to be shown that it is independent of the coordinate system. This follows immediately from the fact that if new coordinates are introduced by

$$x_i' = \sum_{k=1}^{3} a_{ik}x_k, \qquad i = 1, 2, 3, \qquad |a_{ik}| \neq 0,$$

and y_i', z_i', u_i', v_i' indicate the new coordinates of $[y]$, $[z]$, $[u]$ and $[v]$, then

$$y_i' = \sum_{k=1}^{3} a_{ik} y_k, \quad z_i' = \sum_{k=1}^{3} a_{ik} z_k$$

(6.1) $\quad u_i' = \sum_{k=1}^{3} a_{ik} u_k = \sum_{k=1}^{3} a_{ik}(\lambda y_k + \mu z_k) = \lambda \sum_{k=1}^{3} a_{ik} y_k + \mu \sum_{k=1}^{3} a_{ik} z_k = \lambda y_i' + \mu z_i'$

$$v_i' = \sum_{k=1}^{3} a_{ik} v_k = \sum_{k=1}^{3} a_{ik}(\lambda' y_k + \mu' z_k) = \lambda' y_i' + \mu' z_i'.$$

Thus the points are

$$y', z', \lambda y' + \mu z', \lambda' y' + \mu' z',$$

which have the same associated ratio of ratios. This quotient of ratios is called the *cross ratio of the four points*, and is denoted by $R(y,z; u,v)$.

(6.2)　　*If y, z, u and v are four distinct collinear points, with $u = \lambda_1 y + \lambda_2 z$ and $v = \mu_1 y + \mu_2 z$, then the cross ratio of the four points in the given order is $R(y,z; u,v) = \lambda_2 \mu_1 / \lambda_1 \mu_2$.*

Dually :

(6.3)　　*If $\eta, \zeta, \varphi, \psi$ are four distinct, concurrent lines, where $\varphi = \lambda_1 \eta + \lambda_2 \zeta$ and $\psi = \mu_1 \eta + \mu_2 \zeta$, the cross ratio of the four lines in the given order is $R(\eta, \zeta, \varphi, \psi) = \lambda_2 \mu_1 / \lambda_1 \mu_2$.*

Since the equations for a projectivity have algebraically the same structure in point and line coordinates and are formally identical with those for coordinate transformations, and because one-dimensional projectivities are induced by two-dimensional ones, it follows from (6.1) that:

(6.4)　　*The cross ratio of four collinear points, or four concurrent lines, is invariant under one and two-dimensional projectivities.*

The cross ratios of points and lines are connected by the fact (Figure 7 in the next section):

(6.5)　　*If $\eta, \zeta, \varphi, \psi$ are four distinct lines which are concurrent at a, and the line ξ intersects them respectively in the four distinct points y, z, u and v, then $R(y,z; u,v) = R(\eta, \zeta; \varphi, \psi)$.*

PROOF: Let $u = \lambda_1 y + \lambda_2 z$ and $v = \mu_1 y + \mu_2 z$. Since $R(\eta, \zeta, \varphi, \psi)$ is independent of the choice of the representation of η and ζ, we may put $\eta = a \times y$, $\zeta = a \times z$. Then

$$\varphi \sim a \times u = a \times (\lambda_1 y + \lambda_2 z) = \lambda_1 (a \times y) + \lambda_2 (a \times z) = \lambda_1 \eta + \lambda_2 \zeta,$$
$$\psi \sim a \times v = a \times (\mu_1 y + \mu_2 z) = \mu_1 (a \times y) + \mu_2 (a \times z) = \mu_1 \eta + \mu_2 \zeta,$$

hence

$$R(y,z; u,v) = \lambda_2\mu_1/\lambda_1\mu_2 = R(\eta,\zeta; \varphi,\psi).$$

For the distinct points $y,z,u = \lambda_1 y + \lambda_2 z,\ v = \mu_1 y + \mu_2 z$, the definition (6.2) gives $R(y,z; v,u) = \mu_2\lambda_1/\mu_1\lambda_2$, hence:

(6.6) $$R(y,z; v,u) = 1/R(y,z; u,v).$$

Also, since u and v are distinct, $\delta = \begin{vmatrix} \lambda_1 & \lambda_2 \\ \mu_1 & \mu_2 \end{vmatrix} \neq 0$, so $u = \lambda_1 y + \lambda_2 z$ and $v = \mu_1 y + \mu_2 z$ imply that $y = (\mu_2/\delta)u - (\lambda_2/\delta)v$ and $z = (-\mu_1/\delta)u + (\lambda_1/\delta)v$. Therefore:

(6.7) $$R(u,v; y,z) = (-\lambda_2/\delta)(-\mu_1/\delta)/(\mu_2/\delta)(\lambda_1/\delta) = R(y,z; u,v).$$

From (6.6) and (6.7):

(6.8)   *The cross ratio $R(y,z; u,v)$ is not altered by the interchange of the pairs y,z and u,v in that order, or by the interchange of the order in both pairs. Interchange of the order in only one pair produces the reciprocal.*

If y, z and u are the reference points on a line ξ and v is a fourth point with coordinates (λ_1,λ_2), then $u = y + z$ and $v = \lambda_1 y + \lambda_2 z$, hence:

(6.9) $$R(y,z; u,v) = \lambda_1/\lambda_2.$$

Therefore, the value of $R(y,z; u,v)$ determines v. A consequence of this is:

(6.10)   *A one-to-one mapping of a line ξ on a line ξ' which preserves cross ratio is a projectivity.*

The proof, which is simple, is left as an exercise since a much stronger theorem will be established in (6.15). The theorem (6.10) was stated explicitly because projectivities are often defined as one-to-one mappings which preserve cross ratio.

Because of (6.9), the cross ratio for four points is extended in the following way to the case where two of the points coincide. Since u has coordinate $(1,1)$, we define

$$R(y,z; u,u) = R(y,y; u,v) = 1.$$

Because z has coordinates $(0,1)$, we set

$$R(y,z; u,z) = R(z,y; z,u) = 0,$$

and since y is $(1,0)$ we take

$$R(y,z; u,y) = R(u,y; y,z) = \infty.$$

The relation (6.9) expresses the cross ratio in special coordinates. To obtain an expression for general coordinates, let the transformation from the special λ_1, λ_2 coordinates to general λ'_1, λ'_2 coordinates be given by

$$\lambda'_1 = a_{11}\lambda_1 + a_{12}\lambda_2 \qquad A = \mid a_{ik} \mid \neq 0.$$
$$\lambda'_2 = a_{21}\lambda_1 + a_{22}\lambda_2$$

If r and s are an arbitrary pair of points on ξ, with coordinates $(r_1,r_2),(s_1,s_2)$ in the special system and $(r'_1,r'_2),(s'_1,s'_2)$ in the second, then:

(6.11)
$$\mid r',s' \mid = \begin{vmatrix} r'_1 & r'_2 \\ s'_1 & s'_2 \end{vmatrix} = \begin{vmatrix} a_{11}r_1 + a_{12}r_2, & a_{21}r_1 + a_{22}r_2 \\ a_{11}s_1 + a_{12}s_2, & a_{21}s_1 + a_{22}s_2 \end{vmatrix}$$
$$= \begin{vmatrix} r_1 & r_2 \\ s_1 & s_2 \end{vmatrix} \cdot \begin{vmatrix} a_{11} & a_{12} \\ a_{21} & a_{22} \end{vmatrix} = A \cdot \mid r,s \mid.$$

In the special coordinates, with (v_1,v_2) for the former (λ_1,λ_2)

$$R(y,z; u,v) = v_1/v_2 = \frac{\begin{vmatrix} 1 & 0 \\ 1 & 1 \end{vmatrix} \cdot \begin{vmatrix} 0 & 1 \\ v_1 & v_2 \end{vmatrix}}{\begin{vmatrix} 0 & 1 \\ 1 & 1 \end{vmatrix} \cdot \begin{vmatrix} 1 & 0 \\ v_1 & v_2 \end{vmatrix}} = \frac{\begin{vmatrix} y_1 & y_2 \\ u_1 & u_2 \end{vmatrix} \cdot \begin{vmatrix} z_1 & z_2 \\ v_1 & v_2 \end{vmatrix}}{\begin{vmatrix} z_1 & z_2 \\ u_1 & u_2 \end{vmatrix} \cdot \begin{vmatrix} y_1 & y_2 \\ v_1 & v_2 \end{vmatrix}}$$
$$= \frac{\mid y,u \mid \cdot \mid z,v \mid}{\mid z,u \mid \cdot \mid y,v \mid}.$$

From (6.11), then, in the general coordinates

(6.12) $$R(y,z; u,v) = \frac{\begin{vmatrix} y'_1 & y'_2 \\ u'_1 & u'_2 \end{vmatrix} \cdot \begin{vmatrix} z'_1 & z'_2 \\ v'_1 & v'_2 \end{vmatrix}}{\begin{vmatrix} z'_1 & z'_2 \\ u'_1 & u'_2 \end{vmatrix} \cdot \begin{vmatrix} y'_1 & y'_2 \\ v'_1 & v'_2 \end{vmatrix}} = \frac{\mid y',u' \mid \cdot \mid z',v' \mid}{\mid z',u' \mid \cdot \mid y',v' \mid}.$$ [14]

If the ratio $\bar{y} = y'_1/y'_2$, which determines a point y, is taken as a non-homogeneous coordinate, (6.12) simplifies to:

(6.13) $$R(y,z; u,v) = \frac{(\bar{u} - \bar{y})(\bar{v} - \bar{z})}{(\bar{u} - \bar{z})(\bar{v} - \bar{y})}.$$

The particular character of the fourth harmonic point is also indicated by cross ratio, since from (4.11) and (6.9) we obtain:

(6.14)   *The point v is the fourth harmonic point to y, z and u if and only if* $R(y,z; u,v) = -1.$

When v is the fourth harmonic point to y, z and u, the four points are said to form a *harmonic set or quadruple* (in the order y,z,u,v). The *pair u and v is also said to divide the pair y and z harmonically*, and, finally, *u and v are*

---

[14]The proof for this relation may seem artificial, but is was chosen because it is a simple example for a method frequently used to derive an expression in general coordinates from an expression in special coordinates.

said to be harmonic conjugates with respect to y and z. These different terminologies stem, in part, from the many forms for the basic cross ratio. For if $R(y,z; u,v) = -1$, then (6.8) yields

$$R(y,z; v,u) = R(z,y; u,v) = R(z,y; v,u) = R(u,v; y,z) = R(u,v; z,y)$$
$$= R(v,u; z,y) = -1.$$

Since (6.4) shows that cross ratio is invariant under a projectivity, and (6.14) relates the harmonic property to cross ratio, we are now in a position to prove the first of the two fundamental facts stated at the end of the last section.

(6.15)   *A one-to-one mapping, Ψ, of the line ξ on the line ξ', which sends harmonic quadruples into harmonic quadruples, is a projectivity.*

PROOF: (Due to Darboux, 1842-1917). Let the images under Ψ of the reference points, y, z and u on ξ be indicated by y', z' and u'. The points y', z' and u' are distinct, since Ψ is one-to-one, therefore a projectivity Ψ_1 between ξ and ξ' exists which also maps y,z,u into y',z',u'. Because Ψ_1^{-1} is a projectivity, and hence preserves the harmonic relationship, $\Phi = \Psi\Psi_1^{-1}$ is a one-to-one mapping of ξ on itself in which the image of a harmonic set is again harmonie. Also Φ leaves y, z and u fixed. If it can be shown that Φ is the identity the theorem will follow since $1 = \Psi\Psi_1^{-1}$ implies $\Psi = \Psi_1$.

To see that Φ is the identity, since a general point x of ξ and the cross ratio $\lambda = R(y,z; u,x)$ uniquely determine each other, the transformation is determined by the function

$$\Phi(\lambda) = R(y,z; u,x\Phi) \text{ where } \lambda = R(y,z; u,x).$$

Since y, z and u are fixed, $\Phi(0) = 0$, $\Phi(1) = 1$ and $\Phi(\infty) = \infty$. It must be shown that $\Phi(\lambda) = \lambda$ for all other λ.

If λ_1, λ_2 are arbitrary (finite) values of λ, corresponding to the points x^1, x^2, and if x^3 is the point corresponding to $\lambda_3 = (\lambda_1 + \lambda_2)/2$, then from (6.12) rearranging terms

$$1/R(y,x^3; x^1,x^2) = \frac{\begin{vmatrix} 1 & \lambda_2 \\ 0 & 1 \end{vmatrix} \cdot \begin{vmatrix} (\lambda_1 + \lambda_2)/2 & \lambda_1 \\ 1 & 1 \end{vmatrix}}{\begin{vmatrix} 1 & \lambda_1 \\ 0 & 1 \end{vmatrix} \cdot \begin{vmatrix} (\lambda_1 + \lambda_2)/2 & \lambda_2 \\ 1 & 1 \end{vmatrix}} = -1.$$

Therefore $y, x^3\Phi, x^1\Phi, x^2\Phi$ must also be a harmonic quadruple, or

$$-1 = \frac{\begin{vmatrix} 1 & \Phi(\lambda_2) \\ 0 & 1 \end{vmatrix} \cdot \begin{vmatrix} \Phi(\lambda_3) & \Phi(\lambda_1) \\ 1 & 1 \end{vmatrix}}{\begin{vmatrix} 1 & \Phi(\lambda_1) \\ 0 & 1 \end{vmatrix} \cdot \begin{vmatrix} \Phi(\lambda_3) & \Phi(\lambda_2) \\ 1 & 1 \end{vmatrix}} = \frac{\Phi(\lambda_3) - \Phi(\lambda_1)}{\Phi(\lambda_3) - \Phi(\lambda_2)}.$$

Hence,

(1)
$$\Phi\left(\frac{\lambda_1 + \lambda_2}{2}\right) = \Phi(\lambda_3) = \frac{\Phi(\lambda_1) + \Phi(\lambda_2)}{2}.$$

Applying (1), when $\lambda_1 = 2\lambda$, $\lambda_2 = 0$, yields

(2)
$$2\Phi(\lambda) = \Phi(2\lambda).$$

Applying (1), when $\lambda_1 = 2\lambda$, $\lambda_2 = 2\mu$, and using (2), gives

(3)
$$\Phi(\lambda + \mu) = \Phi(\lambda) + \Phi(\mu).$$

This implies $\Phi(m\lambda) = \Phi[(m-1)\lambda] + \Phi(\lambda)$, hence induction shows that

(4)
$$\Phi(m\lambda) = m\Phi(\lambda) \text{ for all positive, integral } m.$$

Since $\Phi(1) = 1$, for integral m,n this gives $\Phi(m) = m$, and $\Phi(1) = n\Phi(1/n)$, or $\Phi(1/n) = 1/n$, and $\Phi\left(\dfrac{m}{n}\right) = m\Phi\left(\dfrac{1}{n}\right) = \dfrac{m}{n}$. Finally, (1) yields

$$[\Phi(\lambda) + \Phi(-\lambda)]/2 = \Phi(0) = 0,$$

or

(5)
$$\Phi(-\lambda) = -\Phi(\lambda), \text{ hence } \Phi\left(-\frac{m}{n}\right) = -\Phi\left(\frac{m}{n}\right) = -\frac{m}{n}.$$

Thus,

(6)
$$\Phi(\lambda) = \lambda \text{ for all rational } \lambda.$$

Next, the points whose abscissas are λ^2, 1, λ and $-\lambda$ form a harmonic set because

$$\frac{\lambda^2 - \lambda}{\lambda^2 + \lambda} \cdot \frac{1 + \lambda}{1 - \lambda} = -1.$$

The images of these points then form a harmonic set, so

$$\frac{\Phi(\lambda^2) - \Phi(\lambda)}{\Phi(\lambda^2) - \Phi(-\lambda)} \cdot \frac{1 - \Phi(-\lambda)}{1 - \Phi(\lambda)} = -1.$$

Using (5) this reduces to $\Phi(\lambda^2) = [\Phi(\lambda)]^2$. If now h is any positive number it can be written in the form λ^2, therefore

(7)
$$\Phi(h) = \Phi(\lambda^2) = [\Phi(\lambda)]^2 > 0, \text{ for } h > 0.$$

(Because $\lambda \rightarrow \Phi(\lambda)$ is one-to-one and $\Phi(0) = 0$, $\lambda \neq 0$ prevents $\Phi(\lambda) = 0$ in (7).) Next, (3) and (7) imply

(8)
$$\Phi(\lambda + h) - \Phi(\lambda) = \Phi(h) > 0, \text{ for } h > 0.$$

That is, $\Phi(\lambda)$ is an increasing function of λ.

For any real number λ there exists an increasing sequence of rationals $\{\lambda_i\}$ and a decreasing sequence of rationals $\{\lambda_i'\}$ both of which converge to λ. From (6) and (8), $\lambda_i < \lambda < \lambda_i'$ implies

$$\lambda_i = \Phi(\lambda_i) < \Phi(\lambda) < \Phi(\lambda_i') = \lambda_i',$$

for all i, hence $\lim\limits_{i \to \infty} \lambda_i = \lim\limits_{i \to \infty} \lambda_i' = \lambda = \Phi(\lambda)$. Thus $\Phi(\lambda) = \lambda$, for all λ, which establishes the theorem.

As previously indicated, the importance of this theorem lies in two implications: first, it shows that the concept of a projectivity between lines belongs inherently to projective geometry. Second, since the harmonic relationship can be defined entirely in terms of incidence, it follows from (6.15) that *in projective geometry the only important one-to-one transformations of a line on a line (or a pencil on a pencil) are the projectivities.* There are, however, projectively significant mappings of lines on lines which are not one-to-one. These play an important role in algebraic geometry.

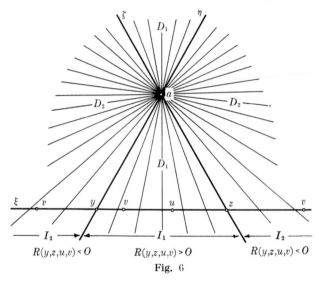

Fig. 6

Cross ratio may also be used in another problem. Up to now, nothing has been said about the *arrangement of points on a projective line*. In the model of Section 1 a projective line was obtained by adding to an ordinary line a "point at infinity." Intuitively, this amounts to "closing" the line, and it is natural therefore to expect the projective line to have some of the characteristics of a closed curve such as a circle. In particular, two points on a circle decompose it into arcs, and we seek some analogue of this on the projective line.

If a is an ordinary point in the model plane (Section 1), the pencil of ordinary lines does not change its appearance if points at infinity are added to the lines of the pencil. If η and ζ are two lines through a they divide the lines of the pencil into two classes according as they lie in the angular domain D_1 or D_2 (Figure 6). If η and ζ are excluded from the two domains, and ξ is a line not through a, the line ξ is composed of $\eta \times \xi$, $\zeta \times \xi$, and two sets of points I_1 and I_2, the first being the intersections of ξ with lines in D_1 and the second the intersections of ξ with the lines in D_2. Though the sets I_1 and I_2 are intervals, and connected from the projective point of view, the Euclidean intuition on which this construction is based makes it unsatisfactory. A truly projective definition may be based on the following observation.

(6.16)
$$\textit{If } y,z,u,v,w \textit{ are five, distinct, collinear points, then}$$
$$R(y,z; u,v)R(y,z; v,w) = R(y,z; u,w).$$

This is an immediate consequence of (6.13), since

$$R(y,z; u,v)R(y,z; v,w) = \frac{(\bar{u} - \bar{y})(\bar{v} - \bar{z})}{(\bar{u} - \bar{z})(\bar{v} - \bar{y})} \cdot \frac{(\bar{v} - \bar{y})(\bar{w} - \bar{z})}{(\bar{v} - \bar{z})(\bar{w} - \bar{y})} = \frac{(\bar{u} - \bar{y})(\bar{w} - \bar{z})}{(\bar{u} - \bar{z})(\bar{w} - \bar{y})}$$
$$= R(y,z; u,w).$$

Two distinct points, y and z, on a projective line ξ determine *intervals* defined in the following way. If u and v are points of ξ, neither of which is y or z, they belong to different intervals if $R(y,z; u,v) < 0$ and to the same interval if $R(y,z; u,v) > 0$. Because neither u nor v can be y or z, $R(y,z,u,v)$ cannot be zero or infinity and so has a definite sign, while $R(y,z,u,u) = 1 > 0$ shows that u and u always belong to the same interval. This relationship of two points being on the same interval is symmetric since $R(y,z,u,v) > 0$ implies $R(y,z,v,u) = R(y,z,u,v)^{-1} > 0$. It is also transitive, for if u and v are on the same interval, and so are v and w, then, because of (6.16), u and w are on the same interval.

In applying this definition it is convenient to take u, distinct from y and z, as a third fixed point. Then one interval determined by y and z is that containing u, and consisting of all points v for which $R(y,z; u,v) > 0$. The other is composed of points v for which $R(y,z; u,v) < 0$.

7. Perspectivities

Projectivities between lines (or pencils) in the same plane may be expressed in terms of very simple projectivities called perspectivities. If ξ and ξ' are distinct lines in the plane P and a is a point on neither ξ or ξ', each line of the pencil through a cuts ξ and ξ' respectively in a pair of points y and y'. The association of y and y' is called the *perspectivity of ξ and ξ' from the center a* (Figure 7). Dually, if p and p' are distinct pencils in the plane and α is a line in neither pencil, each point of α determines with p and p' respectively a pair of lines η and η'. The association of η and η' is called *a perspectivity of the pencils p and p' from the axis α* (Figure 8).

If η,ζ,φ,ψ are four lines concurrent at a and if ξ and ξ' are lines not through a which respectively cut the given four lines in the points y,z,u,v and y',z',u',v' (see figure), then by (6.5),

$$R(y,z; u,v) = R(\eta,\zeta; \varphi,\psi) = R(y',z'; u',v').$$

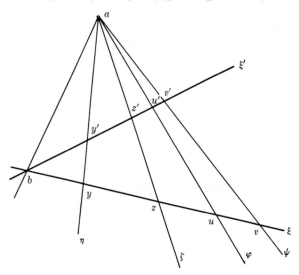

Fig. 7

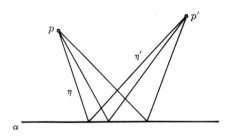

Fig. 8

Therefore a perspectivity preserves cross ratio. Since it is also a one-to-one transformation, (6.15) shows that it is a projectivity. Clearly the perspectivity maps $\xi \times \xi'$ on itself. The converse of this is also true, that is:

(7.1) *A perspectivity of ξ on ξ' is a projectivity which maps $\xi \times \xi'$ on itself. Any projectivity of ξ on ξ' which maps $\xi \times \xi'$ on itself is a perspectivity.*

To see the latter part, suppose $b \sim \xi \times \xi'$ to be self-corresponding under a projectivity Φ (Figure 7). Let y and z be any other two points of ξ with images y' and z' on ξ', and define $a \sim (y \times y') \times (z \times z')$. The perspectivity of ξ on ξ' from a is also a projectivity Φ'. In both ξ and ξ' the images of b, y and z are respectively b, y', and z'. Since, by (5.7), three points uniquely determine a projectivity, $\Phi = \Phi'$, hence Φ is a perspectivity.

As mentioned at the beginning of this section, projectivities may be broken up into perspectivities. More precisely:

(7.2)   *A projectivity between two distinct lines, which is not a perspectivity, is the product of two perspectivities.*

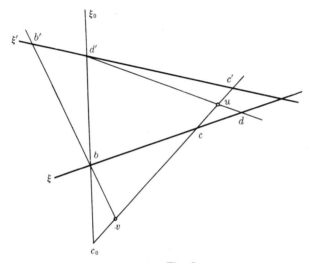

Fig. 9

For let Φ be such a projectivity between the lines ξ and ξ' (Figure 9). On ξ select a distinct triple of points, b,c,d, none of which is $\xi \times \xi'$, and let b',c',d' denote their images under Φ. Take ξ_0 as the line $b \times d'$ and let u indicate $(c \times c') \times (d \times d')$. The perspectivity, Φ_1, of ξ on ξ_0 from u sends $b \to b$, $d \to d'$, and sends c into a point c_0 on ξ_0. If v denotes $(c \times c') \times (b \times b')$, then the perspectivity, Φ_2, of ξ_0 on ξ' from v maps $b \to b'$, $c_0 \to c'$, $d' \to d'$. Under $\Phi_1 \Phi_2$, then, b, c and d go into b', c' and d'. Because Φ_1 and Φ_2 are projectivities, $\Phi_1 \Phi_2$ is a projectivity and since it coincides with Φ on three points it is identical with Φ.

The counterpart of (7.2) for a single line is:

(7.3)   *A projectivity of a line on itself is always expressible as the product of three (or fewer) perspectivities.*

For if Φ is a projectivity of ξ on itself, let ξ' be any line distinct from ξ and let a be any point which is not on ξ or ξ'. If Φ_1 denotes the perspectivity of ξ on ξ' from a, then Φ_1^{-1} is also a perspectivity. The product $\Phi\Phi_1$ is a projectivity of ξ on ξ', hence, by (7.2), it is the product of two perspectivities Φ_2, Φ_3. But $\Phi\Phi_1 = \Phi_2\Phi_3$ implies $\Phi = \Phi_2\Phi_3\Phi_1^{-1}$, q.e.d.

Both (7.2) and (7.3) exhibit again the projective character of projectivities. These ideas may also be used to obtain a classic theorem due to Pappus (3rd century A. D.).

(7.4)   *If x,y,z are distinct points of a line ζ, and x',y',z' are distinct points of another line ζ', then the points $a \sim (y \times z') \times (y' \times z)$, $b \sim (z \times x') \times (z' \times x)$, and $c \sim (x \times y') \times (x' \times y)$ are collinear.*

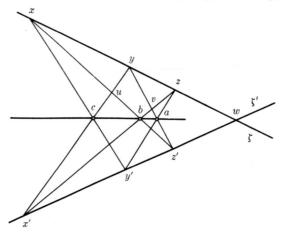

Fig. 10

PROOF: (Figure 10). It may be supposed that none of the six given points is $w \sim \zeta \times \zeta'$, since in that case the theorem is trivial. Let Φ_1 indicate the perspectivity, from center x, of $x' \times y$ on ζ'. Under this mapping, $x' \to x'$, $c \to y'$, $y \to w$ and $u \sim (x' \times y) \times (x \times z')$ goes into z'. The perspectivity, Φ_2, of ζ' on $y \times z'$ from the center z maps $y' \to a$, $w \to y$, $z' \to z'$ and sends x' into $v \sim (x' \times z) \times (y \times z')$. The product $\Phi_1\Phi_2$ is a projectivity of $(x' \times y)$ on $(y \times z')$ in which $x' \to v$, $c \to a$, $u \to z'$, and $y \to y$. Since y is self-corresponding, $\Phi_1\Phi_2$ is a perspectivity (see (7.1)). Its center is then at $b \sim (x' \times v) \times (z' \times u)$. Since c and a are an associated pair in the perspectivity, they are collinear with its center b.

8. The Projectivities of a Line on Itself

If Γ_1 is a subset of a group of transformations Γ, and Γ_1 is also a group, it is called a *sub-group* of Γ. A trivial sub-group of any group is formed by the identity element 1 by itself. If Γ_1 is any other sub-group of Γ, it must contain at least one element $\Phi \neq 1$. By repeated application of the product property, Γ_1 must also contain all powers of Φ, that is $\Phi\Phi = \Phi^2$, $\Phi^2 \cdot \Phi = \Phi^3$, etc. For Γ_1 to consist of just two distinct elements it is necessary that Φ^2 be either Φ or 1. It cannot be Φ since $\Phi^2 = \Phi$ leads to $\Phi^2\Phi^{-1} = \Phi\Phi^{-1}$ or $\Phi = 1$, contrary to assumption. If $\Phi^2 = 1$, the two elements form a group.

A transformation, which is not the identity, but whose square is the identity, is called an *involution*. The reflection of the Euclidean plane about a line is an example. The condition $\Phi^2 = 1$ is equivalent to $\Phi = \Phi^{-1}$ hence an involution is its own inverse.

In (5.9) it was shown that the projectivities of a line ξ on itself form a group Γ_ξ, and we wish to determine whether any elements in this group are involutions. If u_1, u_2 are coordinates on ξ, then by (5.7) a general projectivity Φ of ξ on itself has the form:

$$(8.1) \qquad \Phi : \begin{array}{l} u_1' = a_{11}u_1 + a_{12}u_2 \\ u_2' = a_{21}u_1 + a_{22}u_2, \end{array} \qquad \begin{vmatrix} a_{11} & a_{12} \\ a_{21} & a_{22} \end{vmatrix} \neq 0,$$

where the same coordinate system is used for u and u'. Since the coefficients determine the mapping, Φ can be referred to by its matrix (a_{ik}), $i,k = 1,2$. Because of (5.6), Φ^2 corresponds to the matrix

$$\begin{pmatrix} a_{11}^2 + a_{12}a_{21} & a_{11}a_{12} + a_{12}a_{22} \\ a_{21}a_{11} + a_{22}a_{21} & a_{21}a_{12} + a_{22}^2 \end{pmatrix}$$

which will represent the identity if it reduces to $\begin{pmatrix} b & 0 \\ 0 & b \end{pmatrix}$, $b \neq 0$, the matrix for the identity. By inspection, this will be the case if $a_{11} = -a_{22}$. Since this also implies $\Phi \neq 1$, it is a sufficient condition for Φ to be an involution. Conversely, if $\Phi^2 = 1$, then

$$a_{11}^2 + a_{12}a_{21} = a_{21}a_{12} + a_{22}^2,$$

and

$$a_{21}a_{11} + a_{22}a_{21} = a_{11}a_{12} + a_{12}a_{22} = 0.$$

These may also be written as:

$$(a_{11} - a_{22})(a_{11} + a_{22}) = 0 \text{ and } a_{21}(a_{11} + a_{22}) = a_{12}(a_{11} + a_{22}) = 0.$$

If Φ is not the identity, the case $a_{11} + a_{22} \neq 0$ is ruled out, since with it the foregoing equations imply $a_{11} = a_{22}$ and $a_{12} = a_{21} = 0$, the conditions for $\Phi = 1$. Hence $a_{11} + a_{22} = 0$ is also a necessary condition for Φ to be an involution.

(8.2)   *The projectivity (8.1) is an involution if and only if $a_{11} + a_{22} = 0$.*

If Φ is an involution, it is its own inverse hence $u' = u\Phi$ implies $u = u'\Phi$. Moreover, if this relation holds for a single pair of distinct points, a projectivity Φ must be an involution. That is:

(8.3)   *If Φ is a projectivity of ξ on itself and if two distinct points u and v on ξ exist such that $u = v\Phi$ and $v = u\Phi$, then Φ is an involution.*

For the fact that Φ interchanges u and v shows it is not the identity. On the other hand, if w is any point of ξ, and $w' = w\Phi$, the points u,v,w,w' go into the points $v,u,w',w'\Phi$. Since Φ preserves cross ratio, this with (6.8) implies

$$R(u,v; w,w') = R(v,u; w',w'\Phi) = R(u,v; w'\Phi,w'),$$

hence $w = w'\Phi = w\Phi^2$, or $\Phi^2 = 1$. This result, in turn, implies:

(8.4)   *If x, y, u and v are four distinct points of ξ there is exactly one involution Φ, of ξ on itself, such that $v = u\Phi$ and $y = x\Phi$.*

There is exactly one projectivity Φ, of ξ on itself, which maps u, v and x into v, u and y. By (8.3) it is an involution.

The involutions in Γ_ξ have a basic character in the following sense.

(8.5)   *A projectivity Φ of ξ on itself, which is not an involution, is expressible as the product of two involutions.*

First, if $\Phi = 1$, then for any involution Φ_1 in Γ_ξ, $\Phi = \Phi_1^2 = 1$. If Φ is neither the identity nor an involution, there exists at least one point u which is not left fixed or interchanged with its image, that is, $v = u\Phi \not\sim u$ and $w = v\Phi \not\sim u,v$. There exists then a projectivity Φ_1 which maps v,u,w into v,w,u, and (8.3) shows that Φ_1 is an involution. Because $\Phi_2 = \Phi\Phi_1$ interchanges u and v it is also an involution. Therefore

$$\Phi = \Phi \cdot 1 = \Phi\Phi_1^2 = (\Phi\Phi_1)\Phi_1 = \Phi_2\Phi_1,$$

which is the product of two involutions.

We now consider the problem of determining the *fixed points*, that is the points mapped on themselves, in a projectivity of a line on itself. If Φ in (8.1) represents a general projectivity of ξ on itself, its fixed points are clearly determined by the simultaneous equations:

(8.6) $$\begin{array}{ll} \lambda u_1 = a_{11}u_1 + a_{12}u_2 \\ \lambda u_2 = a_{21}u_1 + a_{22}u_2 \end{array} \qquad \lambda \neq 0.$$

Substituting from these in the relation $u_1u_2 - u_2u_1 = 0$ yields

$$u_1(a_{21}u_1 + a_{22}u_2) - u_2(a_{11}u_1 + a_{12}u_2) = 0,$$

which may be written

(8.7) $$a_{21}u_1^2 + (a_{22} - a_{11})u_1u_2 - a_{12}u_2^2 = 0.$$

This is a quadratic equation for the quantity (u_1/u_2) with the discriminant

(8.8) $$\Delta = (a_{11} - a_{22})^2 + 4a_{12}a_{21}.$$

When Φ is the identity, $a_{11} = a_{22}$ and $a_{12} = a_{21} = 0$, so (8.7) is identically zero, confirming the fact that every point is fixed. If $\Phi \neq 1$, then it is called:

(8.9)
elliptic   if $\Delta < 0$; there are no (real) fixed points,
parabolic   if $\Delta = 0$; there is one fixed point,
hyperbolic if $\Delta > 0$; there are two fixed points.

As an example, consider the distance preserving translation $x' = x + a$ of the real axis on itself. The mapping has no fixed elements at a finite distance, but if the axis is interpreted as a projective line the point at ∞ remains fixed. It is thus a parabolic projectivity. For the mapping $x' = -x + a$, the point $x = \frac{a}{2}$ remains fixed, as well as the point at ∞, so the projectivity is hyperbolic. It is also an involution. There are no examples for the remaining case quite so elementary and familiar since elliptic mappings are not distance preserving. In the translations, $x' = x + a$, we have also the instance of a sub-group, Γ', of the projective group Γ. Though no element of the sub-group Γ' is an involution, any member of it, $x' = x + a$, is the product of the involutions $x' = -x$ and $x'' = -x + a$.

In a hyperbolic projectivity, Φ, there are two fixed points, v,w. For any pair of non-fixed points, u,z with images u',z', the invariance of cross ratio under Φ implies $R(v,w; u,z) = R(v,w; u',z')$. Expressing this, for simplicity, in non-homogeneous coordinates by means of (6.13) gives it the form:

$$\frac{(\bar{v} - \bar{u})(\bar{w} - \bar{z})}{(\bar{v} - \bar{z})(\bar{w} - \bar{u})} = \frac{(\bar{v} - \bar{u}')(\bar{w} - \bar{z}')}{(\bar{v} - \bar{z}')(\bar{w} - \bar{u}')}.$$

Merely rearranging the terms in this equality yields:

$$R(v,w; u,u') = \frac{(\bar{v} - \bar{u})(\bar{w} - \bar{u}')}{(\bar{v} - \bar{u}')(\bar{w} - \bar{u})} = \frac{(\bar{v} - \bar{z})(\bar{w} - \bar{z}')}{(\bar{v} - \bar{z}')(\bar{w} - \bar{z})} = R(v,w; z,z').$$

Since u and z were arbitrary, this implies:

(8.10) *If v,w are the fixed points of a hyperbolic projectivity, the cross ratio $R(v,w; u,u\Phi)$ is constant for all u distinct from v and w.*

If Φ is an involution, it was shown in (8.2) that $a_{11} = -a_{22}$. It follows then from (8.8) that $\Delta = 4(a_{11}^2 + a_{12}a_{21}) = 4(-a_{11}a_{22} + a_{12}a_{21}) = -4\,|\,a_{ik}\,| \neq 0$. This shows:

(8.11)   *There are no parabolic involutions.*

When Φ is hyperbolic it has two fixed elements, v,w. Using the fact that $u = u\Phi^2$, and applying (8.10), gives $R(v,w;\ u,u\Phi) = R(v,w;\ u\Phi,u)$. By (6.8), the second of these is the reciprocal of the first, so $[R(v,w;\ u,u\Phi)]^2 = 1$. Hence, for $u \neq v,w$, $R(v,w;\ u,u\Phi) = \pm 1$. But the value $+1$ would imply $u = u\Phi$ (compare Section 6) making u a fixed point and hence not distinct from v and w. Therefore the cross ratio is -1 and the set $v,w,u,u\Phi$ is harmonic. That is:

(8.12)   *A hyperbolic involution maps each point into its harmonic conjugate with respect to the two fixed points.*

It follows that such an involution is completely determined by the two fixed points.

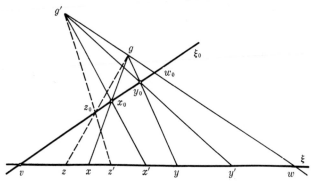

Fig. 11

More generally, since a projectivity is determined by the images of any three points, a hyperbolic projectivity will be known if the two fixed points v,w and one pair of corresponding points, x and x', are given. For the same reason, both a hyperbolic and a parabolic projectivity are determined if one fixed point v, and two pairs of corresponding points, x and x', y and y', are given. To better illustrate the geometric side of these results, we give a construction for the last two cases (Figure 11).

The fixed point v, and the corresponding points, x,x' and y,y', of the projectivity Φ are given on the line ξ. Through v draw any line $\xi_0 \neq \xi$, and select on ξ_0 any distinct pair of points x_0,y_0 different from v. Let g be the intersection of $x \times x_0$ and $y \times y_0$ and let g' be the intersection of $x' \times x_0$ and $y' \times y_0$. Consider the perspectivity Φ_1 of ξ on ξ_0 from the center g, and Φ_2, the perspectivity of ξ_0 on ξ from the center g'. The product $\Phi_1\Phi_2$ sends v,x,y, into v,x',y', via v,x_0,y_0, and is therefore the desired projectivity Φ. The image $z' = z\Phi$, of any point z, can be obtained in the following way. First z_0 is found as the intersection of ξ_0 with $g \times z$, and z' is then the intersection of ξ with $g' \times z_0$.

In this construction, a point w of ξ, $w \neq v$, will be mapped on itself if and only if $g \times w$ and $g' \times w$ intersect ξ_0 in the same point w_0. Therefore, the second fixed point is $w = (g \times g') \times \xi$. The projectivity is parabolic if $g \times g'$ passes through v, otherwise it is hyperbolic. A consequence of this construction is the following addition to (7.3).

(8.13)   *A hyperbolic or a parabolic projectivity of a line on itself is express-ible as the product of two perspectivities.*

However,

(8.14)   *An elliptic projectivity of a line ξ on itself is never the product of fewer than three perspectivities.*

For suppose Φ to be expressible as $\Phi = \Phi_1\Phi_2$, where Φ_1 and Φ_2 are perspectivities of ξ on ξ_0 and ξ_0 on ξ respectively. By the definition of a perspectivity, the point $\xi \times \xi_0$ is fixed under both Φ_1 and Φ_2. It is thus a fixed point for $\Phi_1\Phi_2$, hence Φ is not elliptic.

9. Collineations

A projective theorem was defined to be a fact expressible solely in terms of the incidence relation of point and line. Consequently, a one-to-one mapping of a plane P on a plane P' which preserves incidence will preserve all projective properties. It is therefore important to know the class of all one-to-one incidence preserving transformations of one plane on a second. This problem was mentioned at the end of Section 5 and is answered in the theorem:

(9.1)   *A one-to-one incidence preserving mapping of the projective plane P on the projective plane P' is a collineation (and conversely).*

PROOF: Let $x \to x' = x\Phi$ be such a mapping. Choose p_1, p_2, p_3, p_4 to be any quadrangular set in P. Then the image set p_i', $i = 1,2,3,4$ must also be quadrangular. For suppose that p_1', p_2', p_3' lie on a line ζ'. Let y be any point of P, other than p_3, which is not on $p_1 \times p_2$, and let x denote the intersection of $y \times p_3$ with $p_1 \times p_2$. Since x is on $p_1 \times p_2$, the point x' is on $p_1' \times p_2'$, hence $x' \times p_3' \frown \zeta'$. Because p_3 and all points on $p_1 \times p_2$ also map onto ζ', it follows that Φ carries all of P into ζ' and hence does not cover P'. The set p_i', as stated, is therefore quadrangular.

Now from (5.12) there is a projectivity Φ_1, of P on P', which maps p_i into p_i', $i = 1,2,3,4$. Therefore $\Phi\Phi_1^{-1}$ is a one-to-one incidence preserving transformation of P on itself which leaves p_i fixed, $i = 1,2,3,4$. If it can be shown that $\Phi\Phi_1^{-1}$ is the identity then $\Phi = \Phi_1$ establishes the theorem.

As in the proof of (6. 15), the problem can thus be transposed to showing that if $x \to x\Phi'$ is a one-to-one incidence preserving mapping of P on itself, under which the points of a quadrangular set, p_1,p_2,p_3,p_4, are left fixed, then Φ' is the identity. Since the harmonic relationship, as shown by the construction for the fourth harmonic point, is defined entirely in terms of incidence, it follows that it is invariant under Φ'. Now consider $g \sim (p_1 \times p_2) \times (p_3 \times p_4)$ (Figure 12). Because Φ' preserves incidence, and leaves p_i fixed, $i = 1,2,3,4$, it leaves g fixed. The induced mapping of $p_1 \times p_2$ on itself is one-to-one, with the harmonic relationship invariant, and so, by (6.15), is a projectivity. Having three fixed points, the projectivity is the identity (see (5.7)), hence every point of $p_1 \times p_2$ is fixed under Φ'. By the same reasoning all of the points on the lines $p_i \times p_j$,

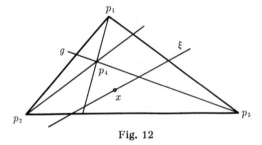

Fig. 12

$i \neq j$, $i,j = 1,2,3,4$, are fixed points of Φ'. If x is a point not on any of these six lines, an arbitrary line ξ through x cuts the six lines in at least three distinct points. The induced mapping of ξ on itself is therefore the identity, hence x is a fixed point of Φ'. Every point of P is therefore a fixed point of Φ' and $\Phi' = 1$.

In (5.14) it was shown that the projectivities of a plane on itself form a group Γ. In many respects the elements of Γ differ from the one-dimensional projectivities in the group Γ_ξ. For instance, one important difference is that, whereas Γ_ξ contains elliptic elements [that is, projectivities without (real) fixed points], there are no such mappings in Γ.

(9.2)   *A collineation of the plane P on itself has at least one fixed point and one fixed line.*[15]

Let Φ be the collineation

$$x'_i = \sum_{k=1}^{3} a_{ik}x_k, \quad i = 1,2,3, \quad |a_{ik}| \neq 0.$$

---

[15]To say that a line is fixed is not to say that the individual points on the line are fixed. If this also is the case the line is said to be pointwise invariant. The same distinction exists between a pencil being invariant and linewise invariant.

It must be shown that there is at least one point x for which $x' \sim x$, or, algebraically, that the equations

$$(9.3) \qquad \lambda x_i = \sum_k a_{ik} x_k, \quad i = 1,2,3$$

have, for a suitable $\lambda \neq 0$, a solution $(x_1, x_2, x_3) \neq (0,0,0)$. If the equations are written in the usual homogeneous form

$$(9.4) \qquad \sum_k a_{ik} x_k - \lambda x_i = 0, \quad i = 1,2,3,$$

the determinant of the system is

$$\Delta(\lambda) = \begin{vmatrix} a_{11} - \lambda & a_{12} & a_{13} \\ a_{21} & a_{22} - \lambda & a_{23} \\ a_{31} & a_{32} & a_{33} - \lambda \end{vmatrix}.$$

For fixed λ, (9.4) will have a non-trivial solution if and only if $\Delta(\lambda) = 0$. But $\Delta(\lambda) = 0$ is a cubic equation with real coefficients, and hence has at least one real root, which is not 0 since $\Delta(0) = |a_{ik}| \neq 0$. Corresponding to this root, then, there is at least one fixed point.[16] The existence of a fixed line follows, of course, from the same argument applied to the collineation in the form $\xi_i' = \sum_k A_{ik} \xi_k$, $i = 1,2,3$.

A consequence of (9.2) is that there is only one type of involutary collineation of a plane on itself. It belongs to the special collineations, called homologies, described as follows. A *homology* Φ is a collineation of the plane on itself, other than the identity, which leaves fixed every point of a line α and every line through a point a not on α. The line α and the point a are called respectively the *axis* and the *center of the homology*. From the preservation of incidence, a is fixed under Φ and a general point z goes into z' on the line $a \times z$.

$$(9.5) \qquad \begin{array}{l} \textit{Given a line } \alpha \textit{ and a collinear triple of distinct points, } a, z \textit{ and } z', \\ \textit{none of which is on } \alpha, \textit{ there is exactly one homology } \Phi, \textit{ with center} \\ a \textit{ and axis } \alpha, \textit{ which maps } z \textit{ on } z'. \end{array}$$

PROOF: If b and c are any two points of α, not on $\zeta \sim a \times z \sim a \times z'$, then neither of the quadruples a,b,c,z or a,b,c,z' contains a collinear triple, hence there is exactly one collineation Φ carrying a,b,c,z into a,b,c,z' in that order. The projectivity, which Φ induces in the pencil at a, leaves

---

[16]Although $\Delta(\lambda) = 0$ cannot have more than 3 roots, (9.3) may have infinitely many roots x, because for certain values of a_{ik} and λ the three equations may reduce to only one essential condition (or even none in case of the identity).

$a \times b$, $a \times c$, and ζ fixed and is therefore the identity. Because every line through a is invariant, every point on α is fixed, so Φ is a homology.

It is now easy to construct the image $x' = x\Phi$ of any point x (Figure 13). If x is not on ζ, let $z \times x$ cut α in y. Then x' must lie on both the lines $a \times x$ and $z' \times y$ and hence is their intersection. For x on ζ, let $b \times x$ cut $c \times z$ in r. Then r' is given by $(a \times r) \times (c \times z')$ and x' by $\zeta \times (b \times r')$. This construction implies, of course, the uniqueness of Φ.

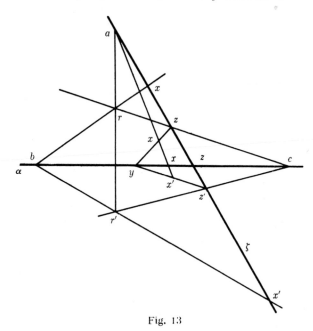

Fig. 13

A *homology* is called *harmonic* if it has the property that a general pair of correspondent points, z and z', are separated harmonically by the center a and the point in which $z \times z'$ intersects the axis α. An immediate consequence of the construction just given is the lemma:

(9.6)   *If Φ is a homology with center a and axis α, and if for one pair of correspondent points, z and z', the quadruple z, z', a and $\bar{z} = \alpha \times (z \times z')$ is harmonic, then Φ is a harmonic homology.*

For Φ induces on $a \times z$ a projectivity with a and $\bar{z}$ as fixed points, hence the projectivity is hyperbolic. Since z, z', a and $\bar{z}$ form a harmonic set then, by (8.10), a and $\bar{z}$ separate harmonically every corresponding pair on $a \times z$. If, now, x is any point not on $a \times z$ or α, and $\bar{x} = \alpha \times (a \times x)$, then the construction shows that $a,\bar{z},z,z'$ and $a,\bar{x},x,x\Phi = x'$ are perspective from a point y. Hence the latter quadruple is also harmonic, proving (9.6).

Clearly a harmonic homology is an involution of the plane. What is more:

(9.7)   *Harmonic homologies are the only collineations which are involutions.*

For let Φ be an involutary collineation, and choose two pairs, a,a' and b,b', of corresponding points such that no three of the four are collinear (Figure 14). Then, by (5.12), Φ is the only collineation which maps a,a',b,b' into a',a,b',b. On the other hand, if $c = (a \times b') \times (a' \times b)$, $d = (a \times b) \times (a' \times b')$, $f = (a \times a') \times (b \times b')$, $g = (a \times a') \times (d \times c)$ and $h = (b \times b') \times (d \times c)$, then, by the harmonic construction, a,a',f,g and b,b',f,h are harmonic sets. The harmonic homology with center f and

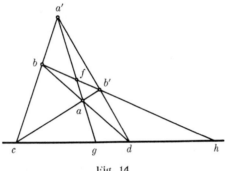

Fig. 14

axis $d \times c$ must, then, send $a \to a'$ and $b \to b'$, and so coincides with Φ.

If the definition used for a homology is changed only in requiring that the center be on, rather than off, the axis, the resulting transformation is called an *elation*.

(9.8)   *Given a line α and a pair of points z,z' not on α, there is exactly one elation Φ, having α as an axis and z,z' as corresponding points.*

For let $\zeta = z \times z'$ (Figure 15). If the desired elation Φ exists, it is clear that its center must be $a = \zeta \times \alpha$. But in the manner argued for homologies, the image of any point x not on α or ζ is now determined. For if $y = \alpha \times (z \times x)$, then x' must be the intersection point of $y \times z'$ and $a \times x$. This shows there could not be two elations Φ, and also indicates how to construct Φ. For if b is any point of α distinct from y and a, then Φ has to be the unique collineation sending a,b,z,x into a,b,z',x' in that order. If Φ is taken as this collineation, then, since a, b and y are fixed under Φ, the entire line α is pointwise invariant. Similarly, since Φ leaves fixed the lines α, $z \times z'$ and $x \times x'$ in the pencil on a it carries every line of the pencil into itself, hence is an elation.

(9.9) *A collineation Φ, not the identity, which leaves all points of a line α invariant, is an elation if there are no other fixed points, and is a homology otherwise.*

Let Φ have a fixed point a not on α. Since every line η of the pencil on a has two fixed points, a and $\eta \times \alpha$, the pencil is linewise invariant and Φ is a homology. If for every z not on α, $z' \not= z$, set $\zeta = z \times z'$ for some pair z,z' and $a = \alpha \times \zeta$ (Figure 15). Since $\zeta' = \zeta \Phi$ contains a and z', which

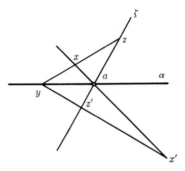

Fig. 15

are on ζ, Φ carries ζ into itself. For any point x not on ζ or α, the same argument shows that Φ takes $x \times x'$ into itself. The point $\zeta \times (x \times x')$, being fixed, is then by assumption on α and hence is a. Therefore Φ is an elation with a as its center.

(9.10) *The product of two harmonic homologies Φ_1, Φ_2, with the same axis α, is an elation (which may be the identity). Conversely, every elation can be expressed as the product of two harmonic homologies having the same axis as the elation.*

To show this, we need the following fact.

(9.11) *If a line α, and two points z and z', not on α, are given, there is exactly one harmonic homology with axis α and z,z' as corresponding points.*

Take $a' = \alpha \times (z \times z')$ and define a to be the fourth harmonic point to z, z' and a'. The harmonic homology, with center a and axis α, pairs z and z' and is clearly the only harmonic homology with this property.

Returning to (9.10), let Φ_1 and Φ_2 be two harmonic homologies with the same line α as axis. If a point z, not on α, is fixed under $\Phi_1 \Phi_2$, then

$z = z\Phi_1\Phi_2$ implies $z\Phi_2 = z\Phi_2^{-1} = z\Phi_1\Phi_2\Phi_2^{-1} = z\Phi_1$. If z is fixed under Φ_1 it is also fixed under Φ_2, and must then be the center of both homologies, so $\Phi_1 = \Phi_2$. For $z \not+ z\Phi_1$, z and $z\Phi_1$ determine the center of Φ_1. Since $z \not+ z\Phi_2, z$ and $z\Phi_2$ determine the same center for Φ_2, and again $\Phi_1 = \Phi_2$. Thus in either case $\Phi_1 = \Phi_2$, and $\Phi_1\Phi_2 = \Phi_1^2$ is the identity, since Φ_1 is an involution. Should there be no fixed points of $\Phi_1\Phi_2$ which are not on α, then (9.9) shows the mapping to be an elation.

To see the converse, suppose Φ to be an elation, not the identity, with α as its axis. Take any point z which is not on α and let $z' = z\Phi$. There is, by (9.11), a harmonic homology Φ_1, with axis α, such that $z' = z\Phi_1$. Take Φ_2 a second harmonic homology with the same axis and with center z'. Then $z' = z'\Phi_2 = z\Phi_1\Phi_2$, which shows that $\Phi_1\Phi_2$ is not the identity. By the preceding argument, this implies that points of α, and only these, are fixed under $\Phi_1\Phi_2$, which is thus an elation. Since Φ and $\Phi_1\Phi_2$ have α and z,z' in common, they are the same elation.

To obtain *algebraic expressions for homologies and elations* in forms that will be needed later, use can be made of the fact that in both types the points of the axis are fixed. Take the axis as δ_3. If the collineation $x_i' = \sum_k a_{ik}x_k$, $|a_{ik}| \neq 0$, $i = 1,2,3$ leaves δ_3 pointwise invariant, it leaves

(1,0,0) and (0,1,0) fixed which implies $a_{21} = a_{31} = 0$ and $a_{12} = a_{32} = 0$. To insure $|a_{ik}| \neq 0$, the term a_{33} cannot vanish, hence can be taken as unity. Finally, because (1,1,0) goes into itself, $a_{11} = a_{22}$. We conclude, then, that:

A collineation Φ which leaves every point of $x_3 = 0$ fixed has a representation of the form

(9.12)
$$x_1' = bx_1 + a_1x_3$$
$$x_2' = bx_2 + a_2x_3$$
$$x_3' = x_3.$$

This mapping Φ will be a homology or an elation according as there is or is not a fixed point off the axis δ_3. In the former case a point $y = (y_1,y_2,1)$ exists satisfying $y_1 = by_1 + a_1$ and $y_2 = by_2 + a_2$. For $b = 1$, these equations have a solution only if $a_1 = a_2 = 0$, and the mapping is the identity. If either a_1 or a_2 is not zero, there will be a solution $y_1 = a_1/(1-b)$ and $y_2 = a_2/(1-b)$ if and only if $b \neq 1$. When $(a_1,a_2) \neq (0,0)$ and $b = 1$, Φ has no fixed points off the axis and it is an elation. Its center is the intersection of the axis δ_3 with the line joining any pair of corresponding points. One such pair is (0,0,1) and $(a_1,a_2,1)$, whose join cuts δ_3 in $(a_1,a_2,0)$. Summing up these results:

(9.13) When the collineation (9.12) is not the identity, it represents a homology with center $(a_1,a_2,1-b)$, if $b \neq 1$, and an elation with center $(a_1,a_2,0)$, if $b = 1$. The axis in both cases is δ_3.

The homology Φ of (9.12) will be harmonic if it is not the identity but its square is the identity. Since Φ^2 is given by

$$x_1'' = b^2 x_1 + b a_1 x_3 + a_1 x_3$$
$$x_2'' = b^2 x_2 + b a_2 x_3 + a_2 x_3$$
$$x_3'' = x_3,$$

$\Phi^2 = 1$ implies $b^2 = 1$ and $a_1(b + 1) = a_2(b + 1) = 0$. If $b = 1$, then $a_1 = a_2 = 0$ and Φ is the identity. Hence $\Phi \neq 1$ implies $b = -1$, that is:

(9.14)   *The collineation (9.12) is a harmonic homology if and only if* $b = -1$.

An application of (9.13) is the result:

(9.15)   *The elations with a common axis α form an Abelian (or commutative) group.*[17]

For if δ_3 is chosen as α, then any two elations, Φ_a and Φ_b, with this axis can be represented as:

$$\Phi_a : x_i' = x_i + a_i x_3, \quad i = 1,2, \qquad \Phi_b : x_i' = x_i + b_i x_3, \quad i = 1,2,$$
$$x_3' = x_3. \qquad\qquad\qquad\qquad x_3' = x_3.$$

Then $\Phi_a \Phi_b$ is given by

$$\Phi_a \Phi_b : x_i'' = x_i + (a_i + b_i)x_3, \ x_3'' = x_3, \ i = 1,2,$$

and the symmetry in a_i and b_i shows $\Phi_a \Phi_b = \Phi_b \Phi_a$. That Φ_a^{-1} and $\Phi_a \Phi_b$ are elations, with axis α, follows by inspection.

Exercises[18]

[2.1]   Verify in detail the relation following (2.2). Do the same for (2.4) and (2.6).

[2.2]   In Cartesian three-space, the direction numbers of a line through the origin form a class of triples. The coefficients in the equation of a plane through the origin also form a class of triples. Show that if these classes are taken for "point" and "line" respectively, they define a projective two-space.

[3.1]   Show that the points

$$a^* = (2,3,-2), \ b^* = (1,2,-4) \text{ and } c^* = (0,1,-6)$$

are collinear. Find λ and μ such that $a^* = \lambda b^* + \mu c^*$. Also find representations b'^* and c'^* of b and c such that $a^* = b'^* - c'^*$.

---

[17] In such statements, it is understood that the identity is included as an element.

[18] The notation [2.4], for example, means the fourth exercise concerning material in Section 2. It is not necessarily related to theorem (2.4). An asterisk on a problem number indicates a more difficult exercise.

[3.2] If ξ, η, ζ, φ are respectively the lines $x_1 - x_3 = 0$, $x_2 + x_3 = 0$, $2x_1 + x_2 - x_3 = 0$, $x_1 + x_2 + 2x_3 = 0$, use the triple notation to find the line $(\xi \times \eta) \times (\zeta \times \varphi)$, and write its equation.

[3.3] If $(1,-1,2)$, $(3,2,1)$ and $(0,-1,1)$ are taken for the reference points on a line, find the projective coordinates of $(5,2,3)$. If $(1,1,0)$, $(-1,2,-3)$ and $(1,-3,4)$ are the reference points of a second system, find the equations relating the two systems of coordinates.

[3.4] Show that if λ and μ are projective coordinates on a line, and $\bar{\lambda}$, $\bar{\mu}$ are defined by (3.7), then $\bar{\lambda}, \bar{\mu}$ are projective coordinates.

[3.5] Show that if $d_1 \times a$, $d_2 \times b$ and $d_3 \times c$ are concurrent at e, then $(d_1 \times d_2) \times (a \times b)$, $(d_2 \times d_3) \times (b \times c)$ and $(d_3 \times d_1) \times (c \times a)$ are collinear (Theorem of Desargues).

[3.6] Show: if a_1, a_2, a_3, b is a quadrangular set and $c_i = (b \times a_i) \times (a_j \times a_k)$, where i, j, k take the values $(1,2,3)$, $(2,3,1)$ and $(3,1,2)$, then the three points $(c_i \times c_j) \times (a_i \times a_j)$ are collinear. (Hint: choose $a_i = d_i$ and $b = e$.)

[3.7] If $(1,2,1)$, $(1,1,0)$, $(2,1,1)$ and $(0,1,7)$ are taken for the reference points of a coordinate system, find the coordinates of $(1,1,1)$ in this system.

[3.8] Find the transformation in point coordinates induced by
$$\xi'_1 = 2\xi_1 - 3\xi_2 + \xi_3, \; \xi'_2 = 4\xi_1 + 2\xi_2 - 6\xi_3, \; \xi'_3 = \xi_1 + \xi_2 + 3\xi_3.$$
Find the inverse transformations.

[4.1] State Desargues' theorem entirely in terms of the basic incidence relation $x \cdot \xi = 0$ and dualize it mechanically.

[4.2] Prove the uniqueness of the fourth harmonic line without making explicit use of the duality principle.

[4.3] State the dual to (7.4), the theorem of Pappus.

[4.4] Give a proof of (4.5) which uses Desargues' theorem, but which is not the dual of the proof given for (4.1).

[4.5] Obtain the result of problem [3.5] from the theorem of Desargues.

[4.6] Verify the values of the coordinates in Figure 4.

[4.7] If, in the pencil x, φ is the fourth harmonic line to ξ, η, ζ, and a line σ, not in the pencil, cuts these four lines in the points a, b, c, d, show that d is the fourth harmonic point to a, b and c.

[5.1] Show that if Φ is a one-to-one transformation of a set S on itself, and $\Phi^n = 1$, then the mappings $1, \Phi, \Phi^2, \cdots, \Phi^{n-1}$ form an Abelian group.

[5.2] Find two definite projectivities of a line on itself which do not commute (compare (5.9)).

[5.3]   Find the transformations $\Phi : x' = ax + b$, $a \neq 0$, for which $\Phi^2 = 1$.
Do the same for $\Phi^3 = 1$.

[5.4]   Find the collineation which sends the points $(1,0,1)$, $(2,0,1)$, $(0,1,1)$
and $(0,2,1)$ into d_1, d_2, d_3 and e respectively.

[5.5]   For a given integer $n > 1$, find a collineation $\Phi \neq 1$ of the plane
P on itself for which $\Phi^n = 1$.

[5.6]   The transformation $\Phi : \xi_1' = \xi_1 + \xi_2$, $\xi_2' = 2\xi_1 - \xi_3$, $\xi_3' = \xi_2 + \xi_3$
sends the line $(1,1,1)$ into the line $(2,1,2)$. Obtain this fact from the
equations of the lines and the point transformation form of Φ.

[6.1]   Show that the points $x = (1,4,1)$, $y = (0,1,1)$ and $z = (2,3,-3)$
lie on a line ξ and find w on ξ such that $R(x,y,z,w) = -4$.

[6.2]   Let x, y, u and v be four distinct points of the Euclidean plane on ξ
and let p be a point of this plane not on ξ. If (x,y) denotes the
smaller of the two angles between $p \times x$ and $p \times y$ show that
$|R(x,y,u,v)| = \sin(x,u) \sin(y,v) \csc(x,v) \csc(y,u)$.

[6.3]   Let x,y,u,v be four distinct, collinear points, with $\lambda = R(x,y,u,v)$.
Show that under the 24 permutations of x, y, u and v, the cross
ratio assumes four times each of the values λ, λ^{-1}, $1 - \lambda$, $(1 - \lambda)^{-1}$,
$(\lambda - 1)\lambda^{-1}$, $\lambda(\lambda - 1)^{-1}$.

[6.4]   With the same assumptions and notation as in [6.3], show that
there are less than 6 distinct values for the cross ratio, if and only
if in some order the points form a harmonic set.

[6.5]* Let p_1,p_2,p_3 be the vertices of a triangle, and g_i,g_i' be points on the
side opposite p_i, $i = 1,2,3$. Let $k_i = R(g_i,g_i',p_j,p_k)$ where i,j,k take
the respective values $(1,2,3)$, $(2,3,1)$, $(3,1,2)$. Show that g_1', g_2' and
g_3' are collinear if and only if $k_1k_2k_3 = 1$.

[7.1]   In the proof of (7.2) where is the hypothesis used that Φ is not a
perspectivity?

[7.2]   Carry through the details of (7.2) analytically for the projectivity
in problem [5.6].

[7.3]   Dualize the construction in the proof of (7.2).

[7.4]   Dualize the Theorem of Pappus and its proof.

[8.1]* Let p and g be two different pencils and Φ be a projectivity which is
not a perspectivity of p on g. If $a = [(p \times g)\Phi] \times [(p \times g)\Phi^{-1}]$,
show that the pairs of points in which a line through a intersects
corresponding lines in p and g are the pairs of corresponding points
in an involution.

[8.2]   Prove that there is a projectivity of ξ on itself carrying the distinct
points a,b,c,d into b,a,c,d respectively if and only if the points
form a harmonic quadruple.

[8.3]   Find the involutions of a line on itself which have (1,1) and (2,0) as fixed points.

[8.4]   If Φ is a mapping of the set S on itself and $\Phi \neq 1$, but $\Phi^n = 1$, where n is the smallest number with these two properties ($n > 1$, $\Phi^n = 1$), then n is called the period of Φ.
Show that a parabolic projectivity is never periodic. Prove there is always a projectivity with a given period.

[8.5]* If at least one of two distinct involutions on a line is elliptic show that there is exactly one pair of points which are corresponding in both involutions.

[8.6]* Give an example of two hyperbolic involutions of a line on itself which have no pair of corresponding points in common.

[8.7]   Show that the value of $R(v,w,u,u\Phi)$ in (8.10) is a_{22}/a_{11}.

[8.8]   Find the parabolic projectivities of ξ on itself which have (2,1) as its fixed point and sends (2,3) into (1,0).

[8.9]   Dualize the construction referring to Figure 11.

[8.10] The two fixed points, v and w, of a hyperbolic projectivity are given, as well as one pair, x and x', of corresponding points. Give a construction for the image, y', of an arbitrary point y.

[9.1]   Find the fixed points and lines of $x_1' = x_1 + x_2$, $x_2' = 8x_1 + 3x_2$, $x_3' = x_1 + x_2 + 2x_3$.

[9.2]   Prove: An elation induces a parabolic projectivity on every line through the center distinct from the axis. Dualize.

[9.3]   If a and α are respectively the center and axis of a homology Φ, show that for y, distinct from a and not on α, the cross ratio $R(y, y\Phi, a, \alpha \times (a \times y))$ is constant.

[9.4]   Show that a homology induces a hyperbolic projectivity on every line through its center. Dualize.

[9.5]   Let a,b,c and a',b',c' be two triangles such that $a \times a'$, $b \times b'$ and $c \times c'$ are concurrent. Show that a homology exists which maps a,b,c into a',b',c' respectively.

[9.6]   Find the values of α for which
$$x_1' = x_1 \cos \alpha + x_2 \sin \alpha, \; x_2' = -x_1 \sin \alpha + x_2 \cos \alpha, \; x_3' = x_3$$
is a homology.

[9.7]   Show that a harmonic homology with d_i as center and δ_i as axis carries the locus $\alpha \, |x_1|^k + \beta \, |x_2|^k = |x_3|^k$ into itself. Give a Euclidean interpretation.

[9.8]   Find the elation with axis δ_1 which carries (2,1,1) into (3,2,0).

[9.9]   Find the collineations which leave the line $(1,1,1) = \varepsilon$ pointwise invariant.

[9.10] Show that two distinct, harmonic homologies commute if and only if the center of each lies on the axis of the other.

[9.11] Use the collineation (9.12) to verify the footnote statement concerning the roots in (9.3).

[9.12] Dualize the statements and proofs of (9.8) and (9.9).

CHAPTER II

Polarities and Conic Sections

10. Polarities

From an algebraic point of view, the collineation $x_i' = \sum\limits_{k=1}^{3} a_{ik}x_k$ is a linear, homogeneous transformation of number triples into other number triples. If the original triples are interpreted as points and the image triples as lines, the mapping may be written as

$$(10.1) \qquad \xi_i' = \sum_{k=1}^{3} a_{ik}x_k, \qquad |a_{ik}| \neq 0, \qquad i = 1,2,3.$$

Such a transformation of P on P', which associates points with lines, is called a *correlation*. While mappings of this type are less important, in general, than collineations, certain special correlations are of the greatest importance and are intimately related to conic sections.

The transformation (10.1) induces a correlation of the lines of P into the points of P'. From (5.15) this mapping is

$$(10.2) \qquad x_i' = \sum_{k=1}^{3} A_{ik}\xi_k, \qquad |A_{ik}| \neq 0, \qquad i = 1,2,3,$$

while the respective inverses of (10.1) and (10.2) are

$$(10.3) \qquad x_i = \sum_{k=1}^{3} A_{ki}\xi_k', \qquad |A_{ki}| \neq 0, \qquad i = 1,2,3,$$

and

$$(10.4) \qquad \xi_i = \sum_{k=1}^{3} a_{ki}x_k', \qquad |a_{ki}| \neq 0, \qquad i = 1,2,3.$$

That the correlation (10.1) preserves incidence follows exactly as in the proof of (3.19), save that formally $x \cdot \xi = 0$ now implies $\xi' \cdot x' = 0$ and conversely. Collinear points, then, map into concurrent lines and conversely. Making use of this fact, (5.12) and (9.1) yield:

*If a,b,c,d are points of P, no three of which are collinear, and
α',β',γ',δ' are lines of P', no three of which are concurrent, then
(10.5)   there is exactly one correlation of P on P' in which a → α', b → β',
c → γ', d → δ'. Every one-to-one mapping x → ξ' of the points
of P on the lines of P' which preserves incidence is a correlation.*

If a correlation of P on P' is denoted by γ and a correlation of P'
on P'' is denoted by γ', then $\gamma \cdot \gamma'$ is a collineation of P on P''. More
particularly, when P,P',P'' are all the same plane, then γ and $\gamma \cdot \gamma'$ are
respectively a correlation and a collineation of P on itself. The import-
ant special class of correlations, γ, referred to at the beginning of this
section, are those for which γ^2 is the identity collineation of the plane
on itself.

DEFINITION :   *A correlation of the plane P on itself, whose square is the
identity, is called a polarity of P. The line which is the
image of the point x is called the polar of x and the point
which is the image of line ξ is called the pole of ξ.*

Since, for a polarity, $\gamma^2 = 1$ it follows at once that:

(10.6)   *If ξ is the polar of x, then x is the pole of ξ.*

Because collinear points go into concurrent lines,

(10.7)   *If x traverses a line η, then ξ, the polar of x, traverses the pencil
whose center z is the pole of η.*

To require that $\gamma^2 = 1$ is the same as requiring that $\gamma = \gamma^{-1}$. Therefore
(10.4) must be the same transformation as (10.1). Consequently the set
of coefficients a_{ik} and the set a_{ki} must be proportional, or $a_{ik} = \lambda a_{ki}$, $\lambda \neq 0$,
for all i and k. Repeating the equality gives $a_{ik} = \lambda a_{ki} = \lambda(\lambda a_{ik})$, so
$\lambda = \pm 1$. The case $\lambda = -1$ would imply $a_{kk} = -a_{kk}$ and hence $a_{kk} = 0$.
Then $|a_{ik}|$ would equal

$$\begin{vmatrix} 0 & a_{12} & a_{13} \\ -a_{12} & 0 & a_{23} \\ -a_{13} & -a_{23} & 0 \end{vmatrix} = -a_{12} \begin{vmatrix} -a_{12} & a_{23} \\ -a_{13} & 0 \end{vmatrix} + a_{13} \begin{vmatrix} -a_{12} & 0 \\ -a_{13} & -a_{23} \end{vmatrix} = 0.$$

Since this contradicts $|a_{ik}| \neq 0$, it follows that $\lambda = 1$, hence that $\gamma^2 = 1$
implies $a_{ik} = a_{ki}$. Conversely, if $a_{ik} = a_{ki}$, (10.1) and (10.4) are identical.
Therefore:

(10.8)   *The correlation $\xi_i = \sum_{k=1}^{3} a_{ik}x_k$ is a polarity if and only if $a_{ik} = a_{ki}$.*

Because for a polarity (10.1) and (10.4) are identical, as are (10.2) and

(10.3), the distinction between x and x' and that between ξ and ξ' may be dropped. Henceforth, we will write a polarity in the simpler form

(10.9)
$$\xi_i = \sum_{k=1}^{3} a_{ik}x_k, \qquad |a_{ik}| \neq 0, \qquad a_{ik} = a_{ki}.$$
$$x_i = \sum_{k=1}^{3} A_{ik}\xi_k, \qquad |A_{ik}| \neq 0, \qquad A_{ik} = A_{ki},$$

where A_{ik} is the co-factor of a_{ik} divided by $|a_{ik}|$.

The point x is said to be conjugate to the point y if it lies on the line η which is the polar of y. The polar line η is thus the locus of all points conjugate to y. Since $\eta_i = \sum_k a_{ik}y_k$, the condition for x to be conjugate to y, namely that $x \cdot \eta = \sum_i x_i\eta_i = 0$, can be written as

(10.10)
$$\sum_{i,k} a_{ik}x_iy_k = 0.$$

The symmetry, $a_{ik} = a_{ki}$, shows that y is also conjugate to x, as could be seen, too, from the condition $\gamma^2 = 1$.

According to (10.10) a point x is conjugate to itself, or more briefly is *self-conjugate,* if and only if it satisfies

(10.11)
$$\sum_{i,k} a_{ik}x_ix_k = 0.$$

Since self-conjugate points satisfy a quadratic equation, it is not surprising that they represent a conic section, which explains the importance of polarities.

It will prove useful to have the following explicit dualizations of some of the foregoing concepts. *The line ξ is conjugate to the line η* if it contains the pole y of η. The pencil y is thus the locus of lines conjugate to η. For conjugacy of ξ and η, the condition is

(10.12)
$$\sum_{i,k} A_{ik}\xi_i\eta_k = 0,$$

so ξ is a *self-conjugate line* if and only if

(10.13)
$$\sum_{i,k} A_{ik}\xi_i\xi_k = 0.$$

Consider now a polarity, γ, which associates the point x and the line ξ as pole and polar. Then, point by point, ξ maps into the pencil x. If

y_i, $i = 1,2,3,4$, are any four points of ξ, and η_i, $i = 1,2,3,4$, are their respective polars, then the same argument as in (6.4) shows that

$$R(y_1,y_2;\ y_3,y_4) = R(\eta_1,\eta_2;\ \eta_3,\eta_4).$$

If ξ is not self-conjugate, it does not pass through x, hence it intersects η_i at y_i', $i = 1,2,3,4$. Then, from the preceding result and (6.5) it follows that

$$R(y_1,y_2,y_3,y_4) = R(y_1',y_2',y_3',y_4').$$

The mapping $y \to y'$, where y is on ξ and y' is the intersection of ξ and the polar of y, is thus a projectivity of ξ on itself. Each point of ξ goes into its conjugate point on ξ, hence the mapping interchanges conjugates in pairs, and is an involution $\gamma(\xi)$. Similarly, γ induces an involution $\gamma(x)$ in the pencil whose center x is the pole of ξ. Lines through x, which correspond under $\gamma(x)$, pass through points which correspond under $\gamma(\xi)$. Hence $\gamma(x)$ has fixed lines, or is hyperbolic, if and only if $\gamma(\xi)$ is hyperbolic. Summed up:

(10.14)   *A polarity γ induces on any non-self-conjugate line (pencil x) an involution $\gamma(\xi)$ of ξ ($\gamma(x)$ of x) which interchanges the conjugate pairs of ξ (of x). If ξ is the polar to x, then $\gamma(\xi)$ and $\gamma(x)$ are both hyperbolic or both are elliptic.*

If a line ξ contains no self-conjugate points it is clearly not a self-conjugate line. On the other hand, if ξ is not self-conjugate then its self-conjugate points are the fixed points of $\gamma(\xi)$. Because there are no parabolic involutions, it follows that ξ has either two self-conjugate points or none at all. The former case also implies ξ is not self-conjugate, that is:

(10.15)   *A line ξ, which contains at least two self-conjugate points, is not a self-conjugate line (and therefore contains exactly two self-conjugate points).*

PROOF: Suppose ξ to be self-conjugate. Then its pole x lies on ξ and is a self-conjugate point. By assumption ξ contains another self-conjugate point y. The line η, polar to y, therefore passes through y. But η also passes through x since the polar line to x contains y. Hence $\eta \sim x \times y \sim \xi$. But $x \not\cap y$, which contradicts the fact that the correlation γ is a one-to-one mapping.

Combining these results yields:

(10.16)   *A self-conjugate line contains exactly one self-conjugate point (namely its pole). A line which is not self-conjugate contains two self-conjugate points or none at all.*

If x and y are a distinct pair of conjugate points, and the line $x \times y$ is not self-conjugate, then its pole z is the intersection of the polar lines of x and y. The triangle x,y,z, which is then called *self-polar*, has the property that each vertex is the pole of the opposite side. As might be expected, a polarity takes its simplest form when the triangle of reference is self-polar. To follow up this idea in greater generality, suppose first that a correlation, $\xi_i = \sum_k a_{ik}x_k$, carries d_i into δ_i, $i = 1,2,3$. That d_1 goes into δ_1 implies

$$0 = a_{21}\cdot 1 + a_{22}\cdot 0 + a_{23}\cdot 0$$
$$0 = a_{31}\cdot 1 + a_{32}\cdot 0 + a_{33}\cdot 0,$$

or that $a_{21} = a_{31} = 0$. Similarly, from δ_2 and δ_3 being the images of d_2 and d_3, we obtain $a_{12} = a_{32} = 0$ and $a_{13} = a_{23} = 0$. The condition $a_{ik} = a_{ki}$, which characterizes polarities, is trivially satisfied, so the correlation is a polarity. Since d_1,d_2,d_3 may be chosen arbitrarily this shows that:

(10.17) *A correlation of P on itself which maps the vertices of any one triangle into the opposite sides of the triangle is a polarity.*

(10.18) *A polarity in which the coordinate triangle is self-polar has the form*

$$\xi_i = b_i x_i \qquad b_i \neq 0 \qquad i = 1,2,3.$$
$$x_i = \xi_i/b_i$$

This follows from setting $b_i = a_{ii}$ in the above computation, whence $|a_{ik}| = b_1 b_2 b_3 \neq 0$ yields $b_i \neq 0$, $i = 1,2,3$.

For a polarity in the form (10.18), the conditions (10.10) and (10.11) for conjugacy of points x,y and lines ξ,η become

(10.19) $$\sum_i b_i x_i y_i = 0 \text{ and } \sum_i \xi_i \eta_i/b_i = 0,$$

while the loci of self-conjugate points and self-conjugate lines are given by

(10.20) $$\sum_i b_i x_i^2 = 0 \text{ and } \sum_i \xi_i^2/b_i = 0.$$

In standard terminology these simplifications may be described by saying that choosing the coordinate triangle to be self-polar transforms the equation of the self-conjugate loci into a sum of squares. As a theorem this becomes:

(10.21) *If $\sum_{i,k} a_{ik}x_i x_k$, where $a_{ik} = a_{ki}$, is a non-degenerate quadratic form (i. e., $|a_{ik}| \neq 0$), then a transformation $x_j = \sum_k c_{jk}y_k$, $|c_{jk}| \neq 0$, $j = 1,2,3$, exists which reduces the quadratic form to the type $\sum_i b_i y_i^2$, $b_j \neq 0$, $j = 1,2,3$.*

The coefficients, b_i, in the reduced form are either all of the same sign, or else two are alike and the third different. However, replacing b_i by $-b_i$ in $\xi_i = b_i x_i$ does not change the polarity, so we may suppose all the b_i to be positive, or else two to be positive and one negative. In the first of these cases, since (10.20) has no real solutions there are no self-conjugate points or lines and the *polarity is called elliptic*. In the second case, the polarity is said to be *hyperbolic*, and in this type there are infinitely many self-conjugate points and lines.

Returning to the involutions induced on lines and pencils by a polarity γ, these must be elliptic if γ is, since under γ no self-conjugate points or lines exist. But much less information than this suffices to tell that a polarity is elliptic.

(10.22)   *If $\eta \cdot a = 0$ and the involutions $\gamma(\eta)$ and $\gamma(a)$, induced by a polarity γ, are both elliptic, then γ is elliptic.*

PROOF: Let a be the point d_2. Its conjugate point on η is different from a, since $\gamma(\eta)$ is elliptic, and so may be chosen as d_3. The polar lines to d_2 and d_3 intersect at the pole of η and this point may be taken as d_1. By construction, the triangle d_1, d_2, d_3 is self-polar hence the locus of self-conjugate points takes the form $\sum b_i x_i^2 = 0$. Because the coefficients b_i are not zero, there is no loss of generality in supposing $b_3 > 0$. Since $\gamma(\eta)$ is elliptic, $\eta \sim \delta_1$ contains no self-conjugate points, hence $b_2 x_2^2 + b_3 x_3^2 = 0$ has no (real) non-zero solution. Because $b_3 > 0$ it follows that $b_2 > 0$. The fact that $\gamma(d_2)$ is elliptic implies that $\gamma(\delta_2)$ is elliptic, hence δ_2, that is $x_2 = 0$, contains no self-conjugate points. Therefore $b_1 x_1^2 + b_3 x_3^2 = 0$ has no non-zero solutions, and $b_3 > 0$ implies $b_1 > 0$. Since $b_i > 0$, $i = 1, 2, 3$, $\sum b_i x_i^2 = 0$ contains no real points, q.e.d.

This theorem provides a simple analytical criterion for deciding whether the general polarity γ, $\xi_i' = \sum_k a_{ik} x_i$, $a_{ik} = a_{ki}$, is hyperbolic or elliptic. For, from (10.22), γ will be elliptic if and only if $\gamma(\delta_3)$ and $\gamma(d_1)$ are elliptic. The pair $(x_1, x_2, 0)$ and $(y_1, y_2, 0)$, associated under $\gamma(\delta_3)$, are conjugates and so are related by

$$a_{11} x_1 y_1 + a_{12}(x_1 y_2 + x_2 y_1) + a_{22} x_2 y_2 = 0.$$

The fixed points of $\gamma(\delta_3)$ are then those for which

$$a_{11} x_1^2 + 2 a_{12} x_1 x_2 + a_{22} x_2^2 = 0.$$

This equation will have no real solutions, or $\gamma(\delta_3)$ will be elliptic, when $a_{11} a_{22} - a_{12}^2 > 0$, that is, when $A_{33} \mid a_{ik} \mid > 0$. Dually, there will be no real fixed elements in $\gamma(d_1)$, or $\gamma(d_1)$ will be elliptic, when $A_{22} A_{33} - A_{23}^2 > 0$.

Since the inverse of the matrix (A_{ki}) is (a_{ik}) and for a polarity these matrices are symmetric,

$$a_{11} = (A_{22}A_{33} - A_{23}^2)/|A_{ik}| = (A_{22}A_{33} - A_{23}^2)|a_{ik}|.$$

Therefore:

The polarity $\xi_i' = \sum_k a_{ik}x_i$ *is elliptic if and only if*

(10.23) $a_{11}a_{22} - a_{12}^2 > 0$ *and* $a_{11} \cdot |a_{ik}| > 0$.

11. Conic Sections

In a hyperbolic polarity, $\xi_i = \sum_k a_{ik}x_k$, the locus of self-conjugate points

satisfies the non-degenerate, quadratic equation

(11.1) $\sum_{i,k} a_{ik}x_ix_k = 0, \qquad a_{ik} = a_{ki}, \qquad |a_{ik}| \neq 0.$

This locus, which may also be written in the form

(11.2) $a_{11}x_1^2 + a_{22}x_2^2 + a_{33}x_3^2 + 2a_{12}x_1x_2 + 2a_{13}x_1x_3 + 2a_{23}x_2x_3 = 0, \quad |a_{i,k}| \neq 0,$

is satisfied by infinitely many points. It is natural, then, to define with von Staudt (1798-1867), the creator of pure projective geometry:

A point conic is the locus of self-conjugate points in a hyperbolic polarity.[1]
A line conic is the locus of self-conjugate lines in a hyperbolic polarity.

If C is a point conic, determined by a hyperbolic polarity γ, the self-conjugate lines of γ are defined to be the *tangents* of C. This is justified by (10.16), which shows that the self-conjugate lines are those which have exactly one point in common with C. It follows, then, that the locus of the lines tangent to a point conic C is a line conic. If C is given by (11.1), the corresponding line conic, that is C in line coordinates, is

(11.3) $\sum_{i,k=1}^{3} A_{ik}\xi_i\xi_k = 0, \qquad A_{ik} = A_{ki}, \qquad |A_{ik}| \neq 0.$

The distinction between point and line conic has therefore more of a logical than a practical significance. *But it is important to be aware of the distinc-*

---

[1]Whereas the projective treatment of conics as developed here is a product of the last century, the theory of conics itself is almost as old as Euclid's Elements. The most famous ancient treatise on conics is due to Apollonius who died about 200 B.C.

tion because curves of higher order are in general not self-dual. For simplicity, we will use the word conic, by itself, for point conic.

With reference to a conic C lines fall into three classes: the *tangents,* the *secants* which cut C twice, and *lines which do not intersect* C. Dually, a line conic C provides three catagories for points: those on C, those on two tangents, and those on no tangent. We have no name in our vocabulary for points of the last two classes because we are not accustomed to thinking of a curve as a line locus. We will use, for them, the terms *two-tangent point* and *no-tangent point.*

(11.4) *A point x, not on a conic, is a no-tangent or two-tangent point according as $\gamma(x)$, the involution induced on the pencil x, is elliptic or hyperbolic. Dually, if ξ is not tangent to a conic, it is non-intersecting or a secant according as $\gamma(\xi)$, the involution on ξ, is elliptic or hyperbolic.*

PROOF: If $\gamma(x)$ is elliptic (hyperbolic), it has no (two) fixed elements, hence there are no (two) self-conjugate lines through x, hence no (two) tangents through x. Conversely, if x is a no-tangent (two-tangent) point, there are no (two) self-conjugate lines through x, hence no (two) fixed lines under $\gamma(x)$, so $\gamma(x)$ is elliptic (hyperbolic).

(11.5) *Every line through a no-tangent point is a secant. Every point on a non-intersecting line is a two-tangent point.*

It has already been shown that the induced involution, $\gamma(x)$, on a no-tangent point x is elliptic. If the induced involution on any line η through x were also elliptic, then, contrary to assumption, the polarity γ would be elliptic (see (10.22)). Therefore $\gamma(\eta)$ is hyperbolic for all η through x, and each of these lines has two self-conjugate points.

(11.6) *The polar to a no-tangent point x is a non-intersecting line ξ.*

For if $\gamma(x)$ is elliptic, then $\gamma(\xi)$ is elliptic, hence ξ is non-intersecting.

(11.7) *If x and x' are conjugate points, not on C, and $x \times x'$ intersects C in u and v, then the points u, v, x and x' form a harmonic set.*

This follows from the fact (see (8.12)) that in a hyperbolic involution a point and its image are harmonic conjugates with respect to the fixed points. As a corollary of this and (11.6):

(11.8) *If a point x is not on C, and a line η through x cuts C in u and v, then x', the fourth harmonic point to u, v and x, is on ξ the polar line to x. When x is a no-tangent point, x' traverses all of ξ as η traverses the pencil at x.*

The polar line to a point x, not on C, may be constructed in the following way (Figure 16). Draw two secants η, ζ through x, cutting C in b, b' and c, c' respectively. Put $u = (b \times c) \times (b' \times c')$ and $v = (b' \times c) \times (b \times c')$, and let $b'' = \eta \times (u \times v)$ and $c'' = \zeta \times (u \times v)$. By the construction of Section 4, b, b', x, b'' and c, c', x, c'' are harmonic quadruples. Then, from (11.8), $u \times v$ is the polar of x. Clearly u is also the pole of $v \times x$, and v is the pole of $u \times x$, hence u, v and x in pairs are conjugates.

(11.9)   *The polar of a two-tangent point x is the line connecting the contact points of the two tangents through x.*

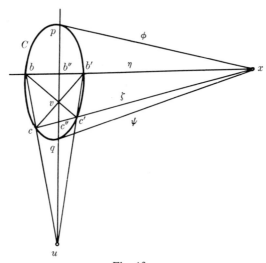

Fig. 16

Let φ and ψ be the tangents with contact points p and q. Since x is on the polar of p and the polar of q, the polar of x passes through p and q.

If a conic is given by $\sum\limits_{i,k} a_{ik} x_i x_k = 0$, then the locus of points conjugate to a given point y, namely the polar of y, is expressed by

(11.10) $$\sum_{i,k} a_{ik} x_i y_k = 0.$$

In particular, if y is on C, (11.10) gives the equation of the tangent through y.

It was shown that the conic in (11.1) takes the simple form $\sum\limits_{i} b_i x_i^2 = 0$ if the coordinate triangle is self-polar, and we will have occasion to make use of this. Frequently, however, another form is convenient which corresponds to the representation $xy = k$ of a rectangular hyperbola in ordinary analytic geometry. This form of the hyperbola arises from selecting the axes, $x = 0$ and $y = 0$, for the asymptotes, that is, the lines contacting C at infinity. To derive an analogue, suppose for the general case, (11.2), that the points d_1 and d_3 are on the conic and that the lines δ_3 and δ_1 are tangent to C at d_1 and d_3 respectively (Figure 17). The fact that C contains $d_1 = (1,0,0)$ yields directly that $a_{11} = 0$. Similarly d_3 on C implies $a_{33} = 0$. Using (11.10), the tangent at d_1 has then the form $2(a_{12}x_2 + a_{31}x_3) = 0$ and since this must be δ_3, that is $x_3 = 0$, it follows

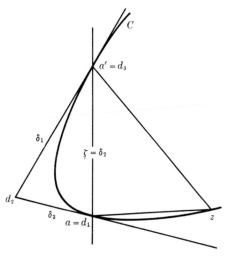

Fig. 17

that $a_{12} = 0$. In the same way, the fact that δ_1 is a tangent at d_3 implies that $a_{23} = 0$. Therefore (11.2) takes the form $a_{22}x_2^2 + 2a_{31}x_3x_1 = 0$. Since now $|a_{ik}| = -a_{22}a_{31}^2$, neither a_{22} or a_{31} vanishes. Thus:

(11.11)   *If the coordinate triangle is such that δ_3 and δ_1 are tangent to the conic at d_1 and d_3 respectively, then the equation of C has the form $x_2^2 - kx_1x_3 = 0$, where $k \neq 0$.*

Here k will be 1 if C contains one of the points $(1,1,1)$ or $(-1,1,-1)$, in which case it will contain both.

The following is an application of (11.11):

(11.12)   *If the line ζ intersects the conic C at a and a', then one of the intervals determined by a and a' consists of no-tangent points, and the other of two-tangent points.*[2]

---

[2]For the definition of interval compare the end of Section 6.

For let $a = d_1$, $a' = d_3$, and take d_2 to be the intersection point of the tangents at a and a'. Finally, assume that $(1,1,1)$ is on C so that the equation of the conic is $x_2^2 - x_1 x_3 = 0$. For this form it is easily verified that $a_{ik} = A_{ik}$, so the corresponding line conic has the form $\xi_2^2 - \xi_1 \xi_3 = 0$. Since ζ joins d_1, d_3, it is the line δ_2, and any point y on it has coordinates of the form $(y_1, 0, y_3)$. A general line η through y satisfies $\eta \cdot y = \eta_1 y_1 + \eta_3 y_3 = 0$, so $\eta_1 = -\lambda y_3$ and $\eta_3 = \lambda y_1$ for a suitable λ. To be a tangent, η must satisfy $\eta_2^2 - \eta_1 \eta_3 = 0$, that is $\eta_2^2 + \lambda^2 y_1 y_3 = 0$. This is possible when $y_1 y_3 < 0$ and impossible when $y_1 y_3 > 0$. Since $y = y_1 d_1 + y_3 d_3$, this means there are no tangents through y if $R(d_1, d_3; d_1 + d_3, y) = y_1/y_3 > 0$ and there is a tangent through y when $R(d_1, d_3; d_1 + d_3, y) = y_1/y_3 < 0$.

12. The Theorems of Steiner and Pascal

With the results of the last section a famous theorem due to Steiner (1796-1863) may be verified.

(12.1)   THEOREM OF STEINER: *If a and a' are two distinct points of a conic C, and z is a variable point on C, then the mapping of the pencil a on the pencil a' given by $a \times z \to a' \times z$ is a projectivity, but not a perspectivity.*

For let a be chosen as d_1 and a' as d_3 and take δ_3 and δ_1 as the tangents to C at a and a' respectively (Figure 17). In addition, let $e = (1,1,1)$ be on C so that the equation of the conic is $x_2^2 - x_1 x_3 = 0$. Then the lines $\lambda x_2 - \mu x_1 = 0$ and $\mu x_2 - \lambda x_3 = 0$ intersect in a point z of C. The first of these is the line $a' \times z \sim (-\mu \delta_1 + \lambda \delta_2)$ in the pencil at a, and the second is $a \times z \sim (\mu \delta_2 - \lambda \delta_3)$ in the pencil at a. Then

$$(12.2) \quad R(\delta_2, \delta_3, \delta_2 + \delta_3, \mu \delta_2 - \lambda \delta_3) = -\mu/\lambda = R(\delta_1, \delta_2, \delta_1 + \delta_2, -\mu \delta_1 + \lambda \delta_2)$$

shows the transformation to be a projectivity. It is not a perspectivity since δ_2, common to both pencils, is not its own image. It is also clear that a perspectivity would require the conic to be degenerate.

The Steiner construction also defines a conic, that is:

(12.3)   *The locus C of intersections of corresponding lines in two, distinct, projective, non-perspective, pencils is a conic.*

To see this, the steps above need only be retraced. Let Φ be the given projectivity of the pencils a and a'. Put $d_1 = a$ and $d_3 = a'$. Since Φ is not a perspectivity, the common element $\delta_2 \sim d_1 \times d_3$ is not mapped on itself. Therefore δ_3 and δ_1 may be chosen so that $\delta_3 = \delta_2 \Phi^{-1}$ and $\delta_1 = \delta_2 \Phi$. Finally, take $g = (-1, 1, -1)$ to be on C. Then Φ maps

$$d_1 \times g \sim (0,1,1) = \delta_2 + \delta_3 \text{ into } d_3 \times g \sim (1,1,0) = \delta_1 + \delta_2.$$

Because it preserves cross ratio, Φ must map $\mu\delta_2 - \lambda\delta_3$ on $-\mu\delta_1 + \lambda\delta_2$. The intersection of these lines satisfies $x_2^2 - x_1 x_3 = 0$ and is therefore a conic.

An interesting illustration of Steiner's theorem in the Euclidean plane appears in connection with the inscribed angles of a circle which have a common chord. Let a and a' on a circle C define a chord, and take x_1, x_2, x_3, x_4 to be any four distinct points of C. Setting $\xi_i = a \times x_i$, $\xi_i' = a' \times x_i$, $i = 1,2,3,4$, it follows that the angle between ξ_i and ξ_j is either equal to that between ξ_i' and ξ_j' or is its supplement. In any case the sine of the two angles is the same. The exercise, given in [6.2], then shows that

$$R(\xi_1, \xi_2, \xi_3, \xi_4) = R(\xi_1', \xi_2', \xi_3', \xi_4'),$$

so the mapping $\xi \rightarrow \xi'$ is a projectivity. This can also be interpreted as showing that the cross ratio of the lines joining four given points on a circle to a variable fifth point of the circle is constant.

We now consider some applications of Steiner's theorem:

(12.4)   *Five points, no three of which are collinear, lie on exactly one conic.*

For if a, a', b_1, b_2, b_3 are the given points, then the association

$$a \times b_i \rightarrow a' \times b_i, \qquad i = 1,2,3,$$

uniquely determines a projectivity Φ of the pencil a on the pencil a'. It is not a perspectivity because b_1, b_2, b_3 are not collinear. The locus of the intersection of corresponding lines is thus a conic through the five points. It is unique since Φ is.

(12.5)   *Four points and the tangent through one of them, or three points and the tangents at two of them, determine a conic uniquely.*

The first part means, more precisely, that given a quadrangular set of points a, a', b_1, b_2, and also a specified line α through a, but not through a', b_1 or b_2, then there is one and only one conic through the four points which has α as its tangent at a. To see this, put $\alpha' \sim a \times a'$. Then the projectivity determined by $\alpha \rightarrow \alpha'$, $a \times b_1 \rightarrow a' \times b_1$, and $a \times b_2 \rightarrow a' \times b_2$ yields the desired, unique conic. To show the second part, let a, a', b be the given non-collinear triple, and α and α' be the given lines through a and a' respectively, where neither line intersects the triple again. The unique projectivity, fixed by the correspondence $\alpha \rightarrow a \times a'$, $a \times a' \rightarrow \alpha'$, and $a \times b \rightarrow a' \times b$, determines the only conic through the triple which has α and α' as its tangents at a and a'.

Another consequence of Steiner's theorem is the classic result due to Pascal (1623-1662):

(12.6)   THEOREM OF PASCAL: *Pairs of opposite sides of a hexagon inscribed in a conic intersect on a straight line.*

Explicitly, if x,y,z,x',y',z' lie on a conic C, then the points

$$a \sim (y \times z') \times (y' \times z), b \sim (z \times x') \times (z' \times x), \text{and } c \sim (x \times y') \times (x' \times y)$$

are collinear (Figure 18). (Note: the theorem of Pappus is the theorem of Pascal for the case when C degenerates into two distinct straight lines (compare (7.4)).)

By Steiner's theorem, the projectivity which the conic determines, between the pencils at x and z, gives the association $x \times x' \to z \times x'$, $x \times y' \to z \times y'$, $x \times z' \to z \times z'$, and $x \times y \to z \times y$. In the order given, these four lines of the pencil x intersect the line $x' \times y$ in the points $x', c, p \sim (x \times z') \times (x' \times y)$ and y. The corresponding lines of the pencil z cut the line $z' \times y$ in the points $q \sim (z \times x') \times (z' \times y)$, a, z' and y. The projectivity between the pencils at x and z thus induces a projectivity between the lines $x' \times y$ and $z' \times y$ in which x', c, p and y are

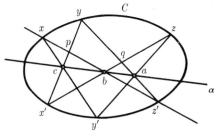

Fig. 18

sent into q, a, z' and y, respectively. Since y, the intersection of the lines, is self-corresponding, the mapping is a perspectivity. Its center is $b \sim (p \times z') \times (x' \times q)$. The join of the corresponding points a and c, then, must pass through b.

The dual of Pascal's theorem is that of Brianchon (1785-1864):

(12.7)   THEOREM OF BRIANCHON: *The lines joining opposite vertices of a hexagon circumscribed about a conic are concurrent.*[3]

Pascal's theorem gives a very simple construction for the second point z' in which a conic through five given points x,y,z,x',y' again cuts a given secant ξ through x. Determine points c, b and a by means of $c \sim (x \times y') \times (x' \times y)$, $b \sim \xi \times (z \times x')$, and $a \sim (b \times c) \times (y' \times z)$. The desired intersection is $z' \sim \xi \times (y \times a)$. This construction by means of incidence exhibits clearly the projective character of a conic.

---

[3] This was one of the first important theorems to be actually discovered by the duality principle.

13. Collineations of Conics

Our next aim is to study the mapping of one conic on a second, or on itself, which are induced by collineations. In particular, the collineations which take a conic into itself will prove to be fundamental in the study of non-Euclidean geometry. In that connection, it is unfortunate that the case where the conic is "imaginary", that is, where it is defined by an elliptic polarity, is just as important as the real case. For the imaginary case it is meaningless to speak of the collineations carrying the conic into itself unless we admit imaginary points. However, the polarity, whether elliptic or hyperbolic, is defined in terms of real points and real lines. Therefore many properties of imaginary conics may be obtained without using imaginary elements, and some of the following results are so phrased as to apply to both the real and imaginary case.

Let Φ be a collineation of the plane P on the plane P' (the two planes may coincide). If γ is a polarity of P on itself which maps the point x into the line ξ, we write, as before, $\xi = x\gamma$. The mappings γ and Φ together define a correlation of P' on itself under the association $(\xi\gamma)\Phi = x' \to \xi' = \xi\Phi$. That is, under γ the line ξ maps into a point x of P which is sent by Φ into x' on P', while the image of ξ under Φ is a line ξ' in P'. More concisely, $x' = x\Phi = (\xi\gamma)\Phi = \xi(\gamma\Phi) = \xi'\Phi^{-1}(\gamma\Phi) = \xi'(\Phi^{-1}\gamma\Phi)$, which shows that the correlation $\xi' \to x'$ is the mapping $\Phi^{-1}\gamma\Phi$.

This is again a polarity. For, by the definition of polarity, γ^2 is the identity, and from this and the associative property it follows that

$$(\Phi^{-1}\gamma\Phi)^2 = (\Phi^{-1}\gamma\Phi)(\Phi^{-1}\gamma\Phi) = (\Phi^{-1}\gamma)(\Phi\Phi^{-1})(\gamma\Phi)$$
$$= \Phi^{-1}(\gamma1\gamma)\Phi = \Phi^{-1}\gamma^2\Phi = \Phi^{-1}1\Phi = \Phi^{-1}\Phi = 1.$$

If γ is hyperbolic then self-conjugate points exist, that is, points x which are incident with their images $\xi = x\gamma$. Since γ, Φ and Φ^{-1} preserve incidence, such points and lines go into pairs of the same type. The induced polarity, $\Phi^{-1}\gamma\Phi$, then contains self-conjugate elements and is therefore hyperbolic. Using the same argument for the elliptic case, it follows that:

(13.1)   *A collineation Φ of a plane P on a plane P' induces, for any polarity γ of P, the polarity $\gamma' = \Phi^{-1}\gamma\Phi$ of P'. Both polarities are hyperbolic or else both are elliptic.*

The following theorem gives exact information about the totality of collineations which send a given conic into itself.

(13.2)   *If $a,\bar{a},b$ are three distinct points of P on a conic C, and $a',\bar{a}',b'$ are three distinct points of P' on a conic C', then there exists exactly one collineation of $P \to P'$ in which $C \to C'$ and $a \to a',\bar{a} \to \bar{a}'$ and $b \to b'$.*

Let α and $\bar{\alpha}$ be the tangents to C at a and $\bar{a}$, and take α' and $\bar{\alpha}'$ to denote the tangents to C' at a' and $\bar{a}'$. Put $c = \alpha \times \bar{\alpha}$ and $c' = \alpha' \times \bar{\alpha}'$. Since no three points in either the quadruple $a,\bar{a},b,c$, or $a',\bar{a}',b',c'$ are collinear, there is just one collineation Φ which maps the first set on the second in the order given. Under Φ, C goes into a conic C'' through $a',\bar{a}',b'$ with $\alpha',\bar{\alpha}'$ as its tangents at a' and $\bar{a}'$. But the conic through three given points, with tangents specified at two of them, is unique (see (12.5)), so $C'' = C'$. On the other hand, any collineation which sends C into C' in the manner specified has to send c into c' to preserve incidence, hence it pairs the indicated quadruples and so coincides with Φ.

This theorem implies that from the projective point of view we cannot distinguish conics as being ellipses, hyperbolas or parabolas. For there exists a collineation carrying one conic into a second whatever their types. This was already apparent in (11.11) and is not surprising in view of the fact that the distinct cases in ordinary geometry correspond to the different types of intersection between a plane and a cone. There is then an actual perspectivity of an ellipse, for example, into a parabola. However, the degree of freedom we have in transforming one conic into a second is remarkable.

Our interest will center later on collineations carrying a conic into itself. To obtain a formulation, which may also serve for imaginary conics, we show:

(13.3)   *If the conic C on P is defined by the hyperbolic polarity γ, then the collineation Φ of P on itself carries C into itself if and only if Φ and γ commute, that is $\Phi\gamma = \gamma\Phi$.*

By (13.1), the polarity induced by Φ with respect to γ is $\Phi^{-1}\gamma\Phi$. If $\Phi\gamma = \gamma\Phi$, then $\gamma = \Phi^{-1}\gamma\Phi$, hence the induced polarity also defines C, so $C\Phi = C$. Or, to follow the argument through for a point x on C, let $y = x\Phi$. Take $\xi = x\gamma$ and $\eta = y\gamma$. Then $\eta = x\Phi\gamma = x\gamma\Phi = \xi\gamma^2\Phi = \xi\Phi$. Under Φ, $x \to y$, $\xi \to \eta$. Since x is on ξ, y is on η, so y, being self-conjugate, is on C.

For the converse theorem, assume Φ carries C into itself. If x is not on C, the geometric construction of Section 11 for the polar ξ to x shows that $x\Phi$ is the pole of $\xi\Phi$. That is, $(x\Phi)\gamma = \xi\Phi$. But $\xi = x\gamma$, so $x(\Phi\gamma) = x(\gamma\Phi)$ for non-self-conjugate points. If x lies on C, then Φ carries the tangent at x into the tangent at the image of x. The image tangent is thus $\xi\Phi$, and is also $(x\Phi)\gamma$, being the polar of the image contact point. Again $x(\Phi\gamma) = \xi\Phi = x(\gamma\Phi)$. Hence $\Phi\gamma = \gamma\Phi$ for all x.

The collineations of a plane on itself form a group Γ, and those elements of Γ which leave a conic C invariant obviously form a sub-group of Γ. For if C is its own image under Φ and Ψ, it goes into itself under $\Phi\Psi$ and Ψ^{-1}. The following proof is valid for the imaginary case also.

(13.4) *For a fixed polarity γ (elliptic or hyperbolic), all the collineations Φ in Γ which commute with γ form a sub-group Γ_γ of Γ.*

For if both Φ and Ψ commute with γ, then $(\Phi\Psi)\gamma = \Phi(\gamma\Psi) = \gamma(\Phi\Psi)$, hence $\Phi\Psi$ commutes with γ and therefore belongs to Γ_γ. Also if $\Phi\gamma = \gamma\Phi$, then $\Phi\gamma\Phi^{-1} = \gamma\Phi\Phi^{-1} = \gamma$, hence $\gamma\Phi^{-1} = \Phi^{-1}\gamma$, so Φ^{-1} belongs to Γ_γ.

(13.5) *If Φ commutes with the polarity γ, and leaves x fixed, then Φ leaves ξ, the polar of x, fixed also.*

Using $x = \xi\gamma$ and $x\Phi = x$ yields $\xi\Phi = x\gamma\Phi = x\Phi\gamma = x\gamma = \xi$.

Combining this with the fact, (9.2), that all collineations have at least a fixed point and a fixed line, we see that every collineation in Γ_γ leaves fixed at least one point and line which are pole and polar.

Because any point and its conjugate on a secant of a conic are separated harmonically by the intersection points of the secant and the conic, it follows that a conic goes into itself under a harmonic homology whose center a is the pole of the axis α. More generally, and including imaginary conics:

(13.6) *The harmonic homology Φ with center a and axis α commutes with the polarity γ if $\alpha = a\gamma$.*

For suppose $\alpha = a\gamma$. By choosing d_3 as a, and taking any two distinct conjugate points on α as d_1 and d_2, the coordinate triangle is self-polar. Hence, by (10.18), γ has the form $\gamma : \xi_i = b_i x_i$, $b_i \neq 0$, $i = 1,2,3$. Using (9.14) and the fact that d_3 goes into itself, it follows that Φ is expressed by

$$\Phi : x_1' = -x_1, \qquad x_2' = -x_2, \qquad x_3' = x_3.$$

The calculation of $\Phi\gamma$ and $\gamma\Phi$ gives both in the form

$$\xi_1 = -b_1 x_1', \qquad \xi_2 = -b_2 x_2', \qquad \xi_3 = b_3 x_3',$$

hence $\Phi\gamma = \gamma\Phi$.

If Φ is a collineation of P on itself which leaves a conic C invariant, the induced mapping $x \to x'$ of C on itself is called a *projectivity of C on itself.* By theorem (13.2), there is exactly one projectivity of C on itself which carries three given points of C into three other given points of C.

(13.7) *If $x \to x'$ is a projectivity of C and p,q are arbitrary points of C, then the mapping $(p \times x) \to (q \times x')$ is a projectivity of the pencil p on the pencil q.*[4]

Suppose the projectivity $x \to x'$ to be induced by Φ, and take $p' = p\Phi$. Then Φ induces a projectivity of the pencil p on the pencil p' by means of $p \times x \to p' \times x'$. By Steiner's theorem, $p' \times x' \to q \times x'$ maps the pencil p' projectively on the pencil q. The product of these two project-

---

[4]For $x = p$, the symbol $p \times x$ is to be interpreted as the tangent to C at p.

ivities is again a projectivity and is, on the other hand, the given mapping
$p \times x \rightarrow q \times x'$.

(13.8)   If Φ is a projectivity $x \rightarrow x'$ of C, and b is any point of C not fixed under Φ, then the points $(b \times x') \times (b' \times x)$ lie on a straight line α. This line, called the axis of Φ, is the same for all choices of b.

By the previous theorem, $b' \times x \rightarrow b \times x'$ is a projectivity of b' on b. It maps $b' \times b$, the common element of the two pencils, into itself and so is a perspectivity (Figure 19). Hence, the intersections of corresponding lines lie on a line α. To see that α is independent of b, let c be any other initial choice and α' the corresponding line. Let x be any point of C, distinct from b, b', c, c', and not fixed under Φ. Applying Pascal's theorem to

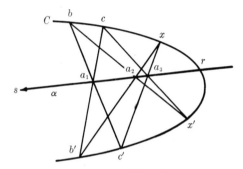

Fig. 19

the inscribed hexagon b, x', c, b', x, c', the points $a_1 = (b \times c') \times (b' \times c)$, $a_2 = (b \times x') \times (b' \times x)$ and $a_3 = (c \times x') \times (c' \times x)$ are collinear. But, by the first part of the theorem, $\alpha \sim a_1 \times a_2$ and $\alpha' \sim a_1 \times a_3$, so $\alpha \sim \alpha'$. If α intersects C at r and s, these are the fixed points of Φ. For $(b' \times r) \times (b \times r')$ on α implies $r' \sim r$. In this case it is obvious that α is independent of b.

The dual to a projectivity of a point conic is the mapping of the lines of a line conic on themselves induced by a collineation of the plane which carries the line conic into itself. If the collineation carries the point x (tangent ξ) into the point x' (tangent ξ') of C, then Φ carries the tangent ξ at x (contact point x of ξ) into the tangent ξ' at x' (contact point x' on ξ'). Therefore any projectivity of C as a point conic may also be regarded as a projectivity of C as a line conic. Hence the dual of (13.8) is:

(13.9)   If Φ is the projectivity $\xi \rightarrow \xi'$ of C, and $\beta \neq \beta'$ is any line of C, then the lines $(\beta \times \xi') \times (\beta' \times \xi)$ all pass through a common point a. This point, called the center of Φ, is independent of the choice of β.

If the axis α of the projectivity Φ of C on itself intersects C at r and s, then, as already observed, these are fixed under Φ. The tangents ζ and η at r and s respectively are then fixed under Φ. Since

$$(\beta \times \zeta') \times (\beta' \times \zeta) \sim (\beta \times \zeta) \times (\beta' \times \zeta) \sim \zeta$$

is through a, and similarly η is through a, it follows that $a \sim \zeta \times \eta$. Because of (11.9), the center a is thus the pole of the axis α. As will be seen in (13.15), this is true whether or not the axis cuts the conic.

There is a close connection between projectivities on a conic and projectivities on a line or pencil. To see this, we first define the *projection* π *of a conic* C *on any line* ξ *from a point* p *which is on the conic but not on* ξ. The image of a point $a \not\mathrel{+} p$ on C is $a' = a\pi = \xi \times (a \times p)$. The point p itself is mapped into p', the intersection point of ξ with the tangent through p. If Φ is any mapping of C on itself, then, for any x on the conic, $x\pi \to x\Phi\pi$

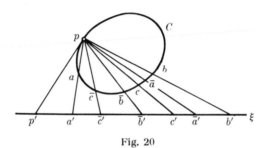

Fig. 20

defines a mapping of ξ on itself. *This mapping is called the projection on* ξ *from* p *of the mapping* Φ. The connection mentioned above is given by the theorem:

(13.10)   *The projection on a line* ξ, *of a mapping* Φ *of* C *on itself, is a projectivity of* ξ *if and only if* Φ *is a projectivity of* C.

PROOF: (Figure 20). Let $a \to \bar{a}$ be a projectivity Φ on C. The mapping of lines, $p \times a \to p \times \bar{a}$ is then, by (13.7), a projectivity of the pencil p on itself. Since a line cuts a pencil in sets of points with the same cross ratios as the intercepted lines, it follows that the projection of Φ on ξ is a projectivity of ξ on itself.

For the converse, suppose that the mapping Φ' on ξ given by $a\pi = a' \to \bar{a}' = a\Phi\pi$, that is, the projection $a \to \bar{a}$ of Φ, is a projectivity on ξ. Let a', b' and c' be any three distinct point on ξ and let $\bar{a}', \bar{b}', \bar{c}'$ be their images under Φ'. Denote by $a, b, c, \bar{a}, \bar{b}$ and $\bar{c}$ the respective images of the six points under π^{-1}. There is a unique projectivity Φ^* of C on itself which carries a, b and c into $\bar{a}, \bar{b}$ and $\bar{c}$ in that order. As has just been shown, the projection Φ'' of Φ^* on ξ is a projectivity on ξ.

But $a' \to \bar{a}'$, $b' \to \bar{b}'$ and $c' \to \bar{c}'$ under both Φ' and Φ'', hence $\Phi' = \Phi''$. Therefore Φ and Φ^* are the same mapping of C on itself.

This connection between the projectivities on a conic and those on a line yields a great variety of theorems, of which we can give only a few examples.

(13.11) *If two triangles a,b,c and a',b',c' are inscribed in a conic C, then they are circumscribed about a second conic C'.*

Projecting the line α from a' on C (see Figure 21) carries the points b,c,a_1,a_2 into b,c,b',c', and projection from a onto α' carries b,c,b',c' into a_1',a_2',b',c'. By (13.10), the association $b \to a_1'$, $c \to a_2'$, $a_1 \to b'$, $a_2 \to c'$ is part of a projectivity of α on α' which is clearly not a perspectivity. By the dual to Steiner's theorem, the connections of corresponding points of the projectivity $\alpha \to \alpha'$ define a line conic C' containing α and α' as elements. For the above quadruples, this gives γ, β, γ' and β' as lines of C'.

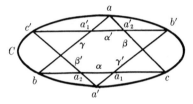

Fig. 21

Therefore α, β, γ and α', β', γ' are lines of C' and hence are tangents of the corresponding point conic.

Using a familiar notion, a projectivity Φ of C, which is not the identity, is called an involution if its square is the identity. Clearly Φ is an involution if and only if its projections on lines are involutions of the lines.

(13.12) *If the collineation Φ induces an involution Ψ on C then Φ is a harmonic homology.*

For if a,a' and b,b' are two pairs of C interchanged by Ψ, then the proof of (9.7) shows that the collineation which carries a,a',b,b' into a',a,b',b is a harmonic homology and must also be Φ.

(13.13) *The projectivity $\Phi : x \to x'$ on C is an involution if and only if all the lines $x \times x'$ are concurrent.*

PROOF: If the lines $x \times x'$ are concurrent at a point a, then Φ interchanges the intersection points on all secants through a and hence is an involution. On the other hand, if $x \to x'$ is an involution, then, by the previous theorem, the collineation Φ inducing it is a harmonic homology. The joins of all pairs which correspond under Φ pass through a the center of the homology, hence the lines $x \times x'$ are concurrent at a (Figure 22). It can also be seen that the center and axis of Φ are respectively the center and axis, defined in (13.8) and (13.9), of the projectivity Φ_0 inducing Φ on C. For, by definition, the axis α of Φ_0 is the locus of fourth harmonic points to x,x',a and is thus the polar of a. If b,b' and c,c' are corresponding points under Φ_0, then using these points in the construction for the polar of a (Section 11) shows that $(c \times b') \times (c' \times b)$ lies on α. By the definition of α', the axis of Φ, it follows that $\alpha = \alpha'$.

The theorem that the center of a projectivity Φ_0 is pole to the axis has been shown for all involutions. It is now easy to extend this to general projectivities. We show first:

(13.14)   *If a projectivity Φ of C, not the identity, is expressible as $\Phi = \Phi_1\Phi_2$, where Φ_1 and Φ_2 are involutions with axes α_1, α_2 and centers a_1, a_2, then $\alpha_1 \times \alpha_2$ is the center of Φ and $a_1 \times a_2$ is its axis.*

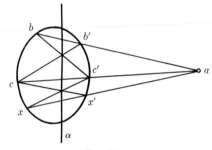

Fig. 22

For any point x on C we use the notation $x\Phi_1 = x_1$ and $x\Phi_2 = x_2$. Also we write $x_1\Phi_2 = x_{12}$ and $x_2\Phi_1 = x_{21}$. Then under $\Phi_1\Phi_2$ (Figure 23), $x \rightarrow x\Phi_1\Phi_2 = x_1\Phi_2 = x_{12}$, $x_1 \rightarrow x_1\Phi_1\Phi_2 = (x\Phi_1)\Phi_1\Phi_2 = x\Phi_2 = x_2$, and $x_{21} \rightarrow x_{21}\Phi_1\Phi_2 = (x_2\Phi_1)\Phi_1\Phi_2 = x_2\Phi_2 = (x\Phi_2)\Phi_2 = x$. That is, $\Phi_1\Phi_2$ sends x, x_1 and x_{21} into x_{12}, x_2 and x. But $(x \times x_2) \times (x_1 \times x_{12}) \sim a_2$ and $(x_1 \times x) \times (x_{21} \times x_2) \sim a_1$, therefore $a_1 \times a_2$ is the axis of $\Phi_1\Phi_2$. The axis α_i of Φ_i is the pole of a_i, $i = 1,2$, so $\alpha_1 \times \alpha_2$ is the pole a of $a_1 \times a_2$. Considering Φ_1 and Φ_2 as projectivities of C as a line conic then, by the dual of the proof just given, $\alpha_1 \times \alpha_2$ is the center of $\Phi_1\Phi_2$, which completes the proof.

Since, by (8.5), every projectivity on a line is either an involution, or the product of two involutions, (13,10) implies that the same fact holds for projectivities on a conic. This, with the theorem just shown, establishes:

(13.15)   *A projectivity Φ on a conic is either an involution or the product of two involutions. The center and axis of Φ are always pole and polar.*

If Φ is not an involution, the collineations inducing the two involutions expressing Φ are, by (13.12), harmonic homologies. Thus we have:

(13.16) *A collineation which leaves a conic fixed is a harmonic homology or else the product of two harmonic homologies which leave C fixed.*

Finally we conclude from (13.13):

(13.17) *An elliptic involution and an arbitrary involution on the same line, or on the same conic, have a pair of corresponding points in common.*

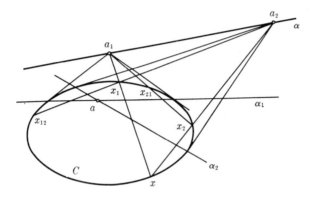

Fig. 23

For if Φ_1 and Φ_2 are involutions on the conic C, with centers a_1, a_2 and Φ_1 is elliptic, then a_1 is a no-tangent point. The line $a_1 \times a_2$ is then a secant, (11.5), and cuts C in two points u and u'. From (13.13), u and u' correspond to each other in both Φ_1 and Φ_2. This implies the same result for projectivities on a line (compare (13.10)).

14. Pencils of Conics

From two lines ξ and η, a pencil of lines is obtained in the form $\lambda\xi + \mu\eta$. This notion can be extended to conics. If $\sum\limits_{i,k} a_{ik}x_ix_k = 0$, $a_{ik} = a_{ki}$ and $\sum\limits_{i,k} b_{ik}x_ix_k = 0$, $b_{ik} = b_{ki}$ are two conics, which we may abbreviate as $g_a(x,x)$ and $g_b(x,x)$, then the form $\lambda g_a(x) + \mu g_b(x) = 0$ is taken to define a *pencil of conics*. Since $g_a(x,x)$ and $g_b(x,x)$ are obtained from polarities the discriminants of these conics, $|a_{ik}|$ and $|b_{ik}|$, are not zero. The discriminant of a general conic in the pencil, however, is $|\lambda a_{ik} + \mu b_{ik}|$, which may vanish for particular values of λ and μ.

To provide for this case, a "*degenerate conic*" is defined to be the locus satisfying a quadratic equation of the form:

(14.1) $\sum_{i,k} a_{ik}x_ix_k = 0$, $a_{ik} = a_{ki}$, $|a_{ik}| = 0$, *where not all* a_{ik} *are zero.*

The nature of this locus can be found by observing that $|a_{ik}| = 0$ implies that the equations

(14.2) $\sum_{k=1}^{3} a_{ik}u_k = 0$, $i = 1,2,3$

have a solution $(u_1,u_2,u_3) \neq (0,0,0)$. If new coordinates are chosen in such a way that $u = d_3$, it follows from (14.2), that $a_{13} = a_{23} = a_{33} = 0$ and, from $a_{ik} = a_{ki}$, that $a_{31} = a_{32} = 0$. Then (14.1) takes the form

(14.3) $a_{11}x_1^2 + 2a_{12}x_1x_2 + a_{22}x_2^2 = 0$.

Let $\Delta = a_{11}a_{22} - a_{12}^2$. If $\Delta > 0$ then the only real values satisfying (14.3) are $x_1 = x_2 = 0$, hence the point $d_3 = (0,0,1)$ is the entire locus. If $\Delta = 0$, the ratio x_1/x_2 in (14.3) has the value $\lambda/\mu = -a_{12}/a_{11}$, and the locus is the line $\mu x_1 - \lambda x_2 = 0$, where $(\lambda,\mu) \neq (0,0)$. If $\Delta < 0$, the ratio x_1/x_2 in (14.3) has the two values $\lambda_1/\mu_1 = -a_{12} + \sqrt{-\Delta}/a_{11}$ and $\lambda_2/\mu_2 = -a_{12} - \sqrt{-\Delta}/a_{11}$. The locus then represents the two lines $\mu_1 x_1 - \lambda_1 x_2 = 0$ and $\mu_2 x_1 - \lambda_2 x_2 = 0$. Corresponding to these cases, then:

(14.4)   *A degenerate conic represents a point, or a line (sometimes called a double line), or a pair of distinct lines.*[5]

In order to have a comprehensive term we will call the locus given by $\sum_{i,k} a_{ik}x_ix_k = 0$, $a_{ik} = a_{ki}$, where not all $a_{ik} = 0$, a quadratic curve, whether or not it is degenerate, and if non-degenerate, whether or not it contains points. For reference we formulate the obvious facts that if $g_a(x,x)$ and $g_b(x,x)$ represent distinct conics C_a and C_b:

(14.5)   *The pencil of conics* $\lambda g_a(x,x) + \mu g_b(x,x) = \sum_{i,k}(\lambda a_{ik} + \mu b_{ik})x_ix_k = 0$
yields a quadratic curve for each pair of values $(\lambda,\mu) \neq (0,0)$. *Proportional pairs give the same curve, and every member of the family can be obtained from a pair* $1,\mu$, *excepting* C_b, *or from a pair* $\lambda,1$, *excepting* C_a.

The condition for degeneracy of the curve corresponding to $1,\mu$ in the pencil is that $\Delta(\mu) = |a_{ik} + \mu b_{ik}| = 0$. This is always a cubic, for the

---

[5] The remainder of this section is not used later in the book.

degree of $\Delta(\mu)$ is no greater than three, and the coefficient of μ^3 in $\Delta(\mu)$ does not vanish since it is given by

$$\lim_{\mu \to \infty} \Delta(\mu)/\mu^3 = \lim_{\mu \to \infty} | a_{ik} + \mu b_{ik} |/\mu^3 = \lim | \mu^{-1} a_{ik} + b_{ik} | = | b_{ik} | \neq 0.^6$$

The equation $\Delta(\mu) = 0$, being a real cubic, has always one real root and may have three. Correspondingly,

(14.6)   *A pencil of conics contains at least one degenerate conic, and at most three.*

The intersection points of C_a and C_b clearly belong to every curve of the pencil. They are the only points with this property and are called the *base points* of the pencil.

(14.7)   *If p is not a base point, there is one and only one curve of the pencil passing through p.*

For at least one of the numbers $g_a(p,p)$, $g_b(p,p)$ is not zero since p is not a base point. Therefore the equation $\lambda g_a(p,p) + \mu g_b(p,p) = 0$ determines the ratio λ/μ uniquely, which selects one curve through p. A consequence of this is that two curves of the pencil intersect each other in the base points and only these.

If there are any base points, then all the non-degenerate conics of a pencil are real since they contain points. This is not necessarily the case when no base points exist. For example, if $g_a = 4x_1^2 - x_2^2 + x_3^2$, and $g_b = -x_1^2 + x_2^2 + x_3^2$, then $g_a + 2g_b = 2x_1^2 + x_2^2 + 3x_3^2 = 0$ is imaginary. If γ_a and γ_b are the polarities which define C_a and C_b, then $\gamma_a\gamma_b$ is a collineation which is not the identity since $C_a \neq C_b$. A point x is fixed under this collineation if and only if it has the same polar line under γ_a and γ_b. For $x\gamma_a = x\gamma_b$ implies $x\gamma_a\gamma_b = x\gamma_b^2 = x$. And if $x\gamma_a\gamma_b = x$, then $x\gamma_a = x\gamma_b^{-1} = x\gamma_b$. Dually, the fixed lines of $\gamma_a\gamma_b$ are those having the same pole in γ_a and γ_b. If C_a and C_b have ζ as a common tangent at a base point z, then ζ is the image of z under both γ_a and γ_b so that z and ζ are fixed under $\gamma_a\gamma_b$. Moreover, ζ is tangent to every non-degenerate conic of the pencil. For let a general conic C of the system be given by $\overline{\lambda}g_a(x,x) + \overline{\mu}g_b(x,x) = 0$. Its tangent at z is the line $\overline{\lambda}g_a(x,z) + \overline{\mu}g_b(x,z) = 0.^7$ Since both $g_a(x,z) = 0$ and $g_b(x,z) = 0$ represent the tangent ζ, a constant ρ exists such that

---

[6] In a polynomial $f(x) = a_n x^n + a_{n-1} x^{n-1} + \cdots + a_1 x + a_0$, the coefficient for the leading term is $\lim_{x \to \infty} f(x)/x^n = \lim_{x \to \infty} \left(a_n + \dfrac{a_{n-1}}{x} + \dfrac{a_{n-2}}{x^2} + \ldots + \dfrac{a_0}{x^n} \right) = a_n.$

[7] $g_a(x,z) = \displaystyle\sum_{i,k} a_{ik} x_i z_k.$

$g_b(x,z) \equiv \rho g_a(x,z)$, $\rho \neq 0$. The tangent to C is then $(\bar{\lambda} + \rho\bar{\mu})g_a(x,z) = 0$, which is ζ.

We illustrate these ideas with some simple types of pencils. We will do so in detail to exemplify the methods that have been developed. Since five points determine a conic, (12.4), there cannot be more than four base points. If there are four, z_1, z_2, z_3, z_4 (Figure 24), then C_a and C_b cannot have a common tangent at any one of these, since a conic is also completely specified by four points and the tangent at one of them. Such pencils, with four base points, are called *quadrangular*. Putting $\beta_{ik} = z_i \times z_k$, $i \neq k$, the degenerate conics of the pencil are the pairs of lines (β_{12}, β_{34}), (β_{13}, β_{24}), (β_{14}, β_{23}), which are opposite sides in the quadrangle of base points. For if p is any point on β_{24}, say, but not a base point, there is a unique curve of the system through p. It is degenerate since it contains the collinear

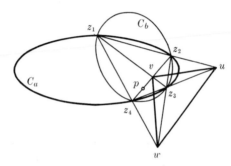

Fig. 24

triple p, z_2, z_4. Because it also passes through z_1 and z_3 it must be the pair (β_{13}, β_{24}). Setting $u = \beta_{12} \times \beta_{34}$, $v = \beta_{13} \times \beta_{24}$, and $w = \beta_{14} \times \beta_{23}$, the construction for the polar of a point shows the triangle u, v, w to be self-polar with respect to both γ_a and γ_b. Under $\gamma_a\gamma_b$, then, u, v, w are the fixed points and $u \times v$, $v \times w$ and $w \times u$ are the fixed lines. In particular, if u, v and w are taken for the vertices of the coordinate triangle, C_a and C_b reduce to the simple form of (10.21) and the pencil becomes

$$\lambda \sum_i a_i x_i^2 + \mu \sum_i b_i x_i^2 = 0.$$

The next simplest case (Figure 25) is that of three base points, z_1, z_2, z_3. For the choice $z_i = d_i$, $i = 1, 2, 3$, C_a and C_b take the forms

$$g_a = a_{12}x_1x_2 + a_{23}x_2x_3 + a_{31}x_3x_1 = 0$$
$$\text{and} \quad g_b = b_{12}x_1x_2 + b_{23}x_2x_3 + b_{31}x_3x_1 = 0.$$

At least one pair of sides in the coordinate triangle is a curve of the pencil. For let p be a point of δ_1 distinct from d_2 and d_3. Then the curve of the pencil through p is degenerate and consists of δ_1 and a line ζ_1 through d_1. If ζ_1 is not δ_2 or δ_3, consider a point p' on δ_2 distinct from d_1 and d_3. Again, the curve of the pencil through p' must consist of δ_2 and a line ζ_2 through d_2. Since the intersection of two curves of the system consists exactly of the base points, $\zeta_1 \times \zeta_2$ must be d_1, d_2 or d_3. The only point of this set on ζ_1 is d_1, so ζ_2 contains d_1. Passing also through d_2, it is the line δ_3. The degenerate conic consisting of δ_2, δ_3, that is $x_2 x_3 = 0$, is thus an element of the pencil. This implies that two non-zero numbers, $\bar{\lambda}, \bar{\mu}$, exist such that $\bar{\lambda} g_a(x,x) + \bar{\mu} g_b(x,x) = x_2 x_3$. Setting $a'_{ik} = \bar{\lambda} a_{ik}$ and $b'_{ik} = - \bar{\mu} b_{ik}$, then g_a and g_b in the new coefficients satisfy $g_a(x,x) - g_b(x,x) = x_2 x_3$. By inspection of this equation, $a'_{12} = b'_{12}$ and $a'_{31} = b'_{31}$. The tangents to C_a and C_b at z_1, expressed in the new coefficients, are $a'_{12}x_2 + a'_{31}x_3 = 0$ and $b'_{12}x_2 + b'_{31}x_3 = 0$, respectively, and hence coincide. This tangent, together with δ_1, forms a degenerate conic of the pencil because

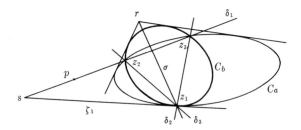

Fig. 25

$b'_{23} g_a(x,x) - a'_{23} g_b(x,x) = (b'_{23} - a'_{23})x_1(a'_{12}x_2 + a'_{31}x_3) = 0$ and $b'_{23} \neq a'_{23}$, since C_a and C_b are dinstinct conics. Therefore this tangent must be the line ζ_1 introduced in the beginning of the discussion.

Altogether it has been shown that *if C_a and C_b intersect in three points z_1, z_2, z_3, then they have a common tangent at one of these points*, say ζ_1 at z_1. The degenerate loci in the pencil are the pair $z_2 \times z_3$ and ζ_1 and the pair $z_1 \times z_2$ and $z_1 \times z_3$. If z_1, z_2, z_3 are the vertices d_1, d_2, d_3 of the coordinate triangle, the pencil takes the form

$$(\lambda + \mu)(c_2 x_1 x_3 + c_3 x_1 x_2) + (\lambda c'_1 + \mu c''_1)x_2 x_3 = 0.$$

Since z_1 has the polar ζ_1 under γ_a and γ_b, z_1 and ζ_1 are fixed under $\gamma_a\gamma_b$. It is easily verified that the other fixed elements of $\gamma_a\gamma_b$ are the points $s = \zeta_1 \times (z_2 \times z_3)$ and, dually, the line σ connecting z_1 to the intersection point r of the two other common tangents of C_a and C_b.

Finally, we consider the case where C_a and C_b have two common points, z_1 and z_2, with a common tangent at each (Figure 26). If z_1 and z_2 are chosen as d_1 and d_2, and the tangents at these points are δ_2 and δ_1, respectively, then all conics of the pencil have δ_2 and δ_1 as tangents at z_1 and z_2. By (11.11) the pencil of conics is given by $\lambda x_1 x_2 + \mu x_3^2 = 0$. This equation differs from previous pencils in the fact that the pairs (0,1) and (1,0) yield degenerate elements of the pencil. This fact is immaterial, for if C_a is given by $x_1 x_2 + k_a x_3^2 = 0$ and C_b by $x_1 x_2 + k_b x_3^2 = 0$, then $\lambda(x_1 x_2 + k_a x_3^2) + \mu(x_1 x_2 + k_b x_3^2) = (\lambda + \mu) x_1 x_2 + (\lambda k_a + \mu k_b) x_3^2 = 0$ represents a general curve in the pencil expressed in terms of C_a and C_b, or else in terms of degenerate loci. For this case, clearly, the fixed points and lines of $\gamma_a \gamma_b$ are the vertices and sides of the coordinate triangle.

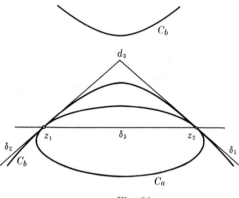

Fig. 26

We conclude this section with some general remarks on pencils. If C_a and C_b define the pencil $\lambda g_a(x,x) + \mu g_b(x,x) = 0$, then the polar of a point y with respect to a general conic C of the pencil is $\lambda g_a(x,y) + \mu g_b(x,y) = 0$. Since $g_a(x,y) = 0$ and $g_b(x,y) = 0$ are the polars of y with respect to C_a and C_b, the polar of y with respect to C is a line of the pencil on $z \sim [(y\gamma_a) \times (y\gamma_b)]$. If y is a fixed element of $\gamma_a \gamma_b$, then $y\gamma_a = y\gamma_b$, so the pencil is undefined, and conversely. When y is not fixed, z is a conjugate of y with respect to every conic of the pencil. The situation is not dual for the locus of poles of a fixed line with respect to varying conics of the pencil. For if z_1 and z_2 are two points of the fixed line ξ, the polar loci of z_1 and z_2 are respectively $\lambda g_a(x,z_1) + \mu g_b(x,z_1) = 0$ and $\lambda g_a(x,z_2) + \mu g_b(x,z_2) = 0$. Eliminating λ, μ from these two equations, gives

$$g_a(x,z_1)g_b(x,z_2) - g_a(x,z_2)g_b(x,z_1) = 0$$

as the locus in x for the poles of ξ. This is quadratic in x and in general represents a conic.

If $\lambda g_a(x,x) + \mu g_b(x,x) = 0$ is a pencil of conics, where C_a and C_b are non-degenerate, the locus of polar lines to a fixed general point (14.8) *y is the pencil on the intersection of the polars of y with respect to C_a and C_b. The locus of poles to a fixed general line ξ is a conic.*

With respect to degenerate conics, all of the points of a double line conic, the intersection point of a two line conic, and the point of a point conic, are sometimes called its singular points. It can be shown, more particularly, that when the pencil has real base points, the exceptions in (14.8) occur for y when it is a singular point of a degenerate conic in the pencil, and for ξ when it is through such a singular point.

Exercises

[10.1] Given the polarity $\xi_1 = 2x_1 - x_3$, $\xi_2 = x_2 + x_3$, $\xi_3 = -x_1 + x_2$, find the self-conjugate locus. Find the pole of the line $\varepsilon = (1,1,1)$, and the involution induced on ε. Is the polarity elliptic or hyperbolic?

[10.2] Prove: A polarity is determined when a self-polar triangle and one additional pole and polar are known.

[11.1] Find the polarity which defines the conic $x_1^2 - x_2^2 - x_3^2 = 0$. Separate the line $\xi = (1,0,2)$ into no-tangent points and two-tangent points. Find the tangents through a two-tangent point. Check that its polar cuts ξ in a no-tangent point.

[11.2] Find the projectivity between the pencils $a : (1,0,1)$ and $b : (1,1,0)$ defined by the conic in problem [11.1]. Check that it is non-perspective and that the tangent at a corresponds to $a \times b$ in the pencil at b. (This exercise properly belongs to Section 12.)

[11.3] Put $\omega(x,y) = \sum a_{ik}x_iy_k$, $a_{ik} = a_{ki}$, and verify that
$$\omega(\lambda x + \mu y, \lambda x + \mu y) = \lambda^2\omega(x,x) + 2\,\lambda\mu\omega(x,y) + \mu^2\omega(y,y).$$

[11.4] Use [11.3] to show: if $\omega(x,x) = 0$ represents a conic C, then $x \times y$ is a secant, tangent, or non-intersector of C according as $\omega(x,x)\omega(y,y) - \omega^2(x,y) < 0$, $= 0$, > 0.

[11.5] Use [11.4] to show: if y is a two-tangent point of the conic $\omega(x,x) = 0$, then the degenerate conic consisting of the two tangents from y to the conic has the equation $\omega(x,x)\omega(y,y) - \omega^2(x,y) = 0$.

[11.6] Use [11.5] to find the tangents from $(4,3,0)$ to $2x_1^2 + 4x_2^2 - x_3^2 = 0$.

[11.7] If a quadruple lies on a conic C, then the three intersections of pairs of opposite sides form a self-polar triangle.

[12.1] Find the conic through $(1,0,1)$, $(0,1,1)$, $(0,-1,1)$ with $x_1 - x_3 = 0$ and $x_2 - x_3 = 0$ as tangent lines.

[12.2] Show that, if four fixed points on a conic are joined to a variable point on C, the cross ratio of the four lines is constant. Dualize this.

[12.3] Prove the converse of Pascal's theorem.

[12.4] Given four points on a conic and the tangent at one of them construct the tangents at the three other points.

[12.5] Find the general form in line and point coordinates of a conic tangent to δ_1, δ_2 and δ_3.

[12.6] Pappus' theorem is Pascal's theorem for a degenerate point conic. Find the analogue to Brianchon's theorem for a degenerate line conic.

[12.7] Using Figure 1 for Desargues' theorem show that if x,y,z,x',y',z' lie on a conic, then the line $a \times b$ is the polar to the point w.

[12.8] If a hexagon H_1 is inscribed in a conic C and H_2 is the hexagon circumscribed about C, and tangent to C at the vertices of the first hexagon, then the "Pascal line" of H_1 is the polar line to the "Brianchon point" of H_2.

[13.1] Take a specific conic C and select a point a and a line α which are pole and polar. Determine the harmonic homology Φ with center a and axis α and show that it carries C into itself. Determine a polarity γ such that $a = \alpha\gamma$ and check that Φ and γ commute.

[13.2] For the conic C of [13.1] select a projectivity π of C into itself and determine the center and axis.

[13.3] The product of two involutions on a conic C is parabolic (that is, has exactly one fixed point) if and only if the line joining the centers of the involutions is tangent to C.

[14.1] Show that for a correlation of a plane on itself the points which lie on their image lines satisfy a quadratic equation. Give conditions under which this equation represents a real non-degenerate conic.

[14.2] Find the degenerate forms of a line conic.

[14.3] Check (14.8) with an example.

CHAPTER III

Affine Geometry

15. The Content of Affine Geometry. The Affine Group

From the Euclidean point of view, the extended plane of Section 1 is a plane in which distances exist, except along the line at infinity, and in which parallels also exist as lines intersecting at infinity. On the other hand, from the projective viewpoint neither parallelism or distances exist in the plane and the special character of the line at infinity is entirely lost. Between these extremes, it is natural to inquire what results may be obtained by specializing a line in the projective plane, analogous to the line at infinity, but without introducing the distance concept. Under this convention, the resulting plane, neither Euclidean nor projective, is called the *affine plane*, and its properties, namely *affine geometry*, form the subject of this chapter.

As just stated, the affine plane P_a may be obtained by distinguishing a line ζ in the projective plane P. Equally well, P_a results when a line ζ is deleted from P. *There is no need to decide in favor of one of these plans and against the other since they are equivalent and both geometric interpretations are helpful.* In the first light, parallel lines in P_a are lines which intersect on ζ, and, in the second, are lines which do not intersect at all.

Corresponding to the two interpretations, an affine transformation, or an *affinity*, is a collineation of P on itself which, from the first point of view, leaves ζ invariant, and, from the second, carries P_a into itself. In either case, it is clear that the inverse of an affinity, or the product of two affinities, is again an affine transformation. Hence, the affinities form a sub-group, Γ_a, of the collineation group Γ. Just as projective geometry consists of the concepts and properties invariant under collineations, *the theorems of affine geometry are those which remain true under affinities.*[1]

Since the line ζ is specialized in P it is natural to distinguish it by the choice of coordinates, and this will be done by always taking it to be δ_3, that is, $x_3 = 0$. Using the first point of view, then, every point of P_a

---

[1] The idea of defining the content of a geometry as the properties invariant under a certain group was first formulated by F. Klein (1849-1925) in the Erlanger program (1872) which has had a profound influence on geometry.

has a non-zero third coordinate and so may be represented as $(x_1,x_2,1)$. The numbers x_1 and x_2 are called affine coordinates and we will refer to the point (x_1,x_2) meaning the point $(x_1,x_2,1)$.[2]

The elements in Γ_a, namely the affinities, consist of the collineations of P,

$$(15.1) \qquad x_i' = \sum_k a_{ik}x_k, \qquad |a_{ik}| \neq 0, \qquad i = 1,2,3,$$

which carry the line $x_3 = 0$ into itself. They are thus the line transformations,

$$\xi_i' = \sum_k A_{ik}\xi_k, \qquad |A_{ik}| \neq 0, \qquad i = 1,2,3,$$

which have $(0,0,1)$ as a fixed element. The result of substitution shows that for this line to be fixed A_{13} and A_{23} must vanish, while A_{33} must not. Hence, necessary conditions for (15.1) to carry δ_3 into itself are

$$\begin{aligned} a_{11}a_{32} - a_{12}a_{31} &= 0 \\ a_{21}a_{32} - a_{22}a_{31} &= 0 \end{aligned} \quad \text{and} \quad \begin{vmatrix} a_{11} & a_{12} \\ a_{21} & a_{22} \end{vmatrix} = A_{33} \neq 0.$$

Regarding these as equations in a_{32} and a_{31} the only solutions possible, since $A_{33} \neq 0$, are $a_{32} = a_{31} = 0$. Thus, the last equation in the collineation (15.1) becomes $x_3' = a_{33}x_3$, with $a_{33} \neq 0$. Conversely, if the last equation is of this form, then δ_3 is obviously invariant. Putting $a_{33} = 1$, and using affine coordinates, where $x_3 = 1$, we then have:

(15.2)   THEOREM: *The affinities of P_a have the form*

$$(15.3) \qquad \begin{aligned} x_1' &= a_{11}x_1 + a_{12}x_2 + a_1 \\ x_2' &= a_{21}x_1 + a_{22}x_2 + a_2 \end{aligned} \qquad \Delta = \begin{vmatrix} a_{11} & a_{12} \\ a_{21} & a_{22} \end{vmatrix} \neq 0.$$

Of particular interest among affinities are those which not only leave ζ invariant but pointwise invariant, and so carry every line into a parallel line. Clearly these affinities form a sub-group of Γ_a, and by (9.12), this sub-group has a representation in affine coordinates of the form

$$(15.4) \qquad \begin{aligned} x_1' &= bx_1 + a_1 \\ x_2' &= bx_2 + a_2 \end{aligned} \qquad b \neq 0.$$

For $b = 1$, and only then, (15.4) is an elation, given by

$$(15.5) \qquad \begin{aligned} x_1' &= x_1 + a_1 \\ x_2' &= x_2 + a_2. \end{aligned}$$

This special type of affine elation is called a *translation*. It carries every line ξ into a parallel line ξ', and has no fixed points in P_a except when it is the identity. The center of the translation is $(a_1,a_2,0)$ on δ_3.

---

[2] If x and y are in affine coordinates, $x \times y$ means $(x_1,x_2,1) \times (y_1,y_2,1)$.

Since (15.4) was obtained as a collineation leaving δ_3 pointwise invariant, it is a homology when not an elation. These homologies, corresponding to $b \neq 1$, are called *dilations*. The center is at $(a_1/(1-b), a_2/(1-'b))$, and when this is chosen to be the origin, the reason for the name becomes apparent from the form of the mapping,

$$(15.6) \qquad x_1' = bx_1, \qquad x_2' = bx_2, \qquad b \neq 0.$$

More particularly still, for $b = -1$ the homology is harmonic (see (9.14)), and is said to be a *central reflection*, or a reflection about a point, in counter distinction to reflection about a line. Again, if the origin is taken as the center, the reason for this terminology is at once suggested by the form which the reflection takes, namely $x_1' = -x_1, x_2' = -x_2$.

In (9.10) it was shown that the product of two harmonic homologies, with the same axis, is an elation. An affine form of this fact is:

(15.7) *The product of two central reflections is a translation, and every translation is the product of two central reflections.*

The theorem is illustrated in the figure: the translation taking a,b into

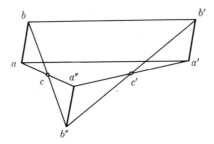

Fig. 27

a',b' is the product of two reflections, that with center c sending a,b into a'',b'', and that with center c' carrying a'',b'' into a',b'.

In the foregoing transformations, we have a hierarchy of groups. The affinities (15.3) form the group Γ_a, in which δ_3 is merely invariant in P. The additional requirement that δ_3 be pointwise invariant yields Γ_s, the sub-group of Γ_a given in (15.4), sometimes called the group of *similitudes*. The elements of Γ_s which are elations then form the sub-group of translations Γ_t. Since translations are elations with a common axis, the group Γ_t is Abelian (see (9.15)).

The translations have another interesting property in the fact that they form an *invariant sub-group* of the group Γ_a. In general, a group Γ_1 is called an invariant, *or normal*, sub-group of the group Γ if, corresponding to any two elements Φ_1 and Φ in Γ_1 and Γ respectively, an element Φ_1'

in Γ_1 exists such that $\Phi_1\Phi = \Phi\Phi_1'$. Of course, if Γ is Abelian, Φ_1 will serve for Φ_1', so every sub-group of an Abelian group is invariant. With respect to Γ_t this invariance means that if two figures A and A' are images under a translation, they are carried by every affinity into figures which are again images under a translation, though in general a different one. For if $A' = A\Phi_1$, where Φ_1 is in Γ_t, the images under a general affinity Φ are $A\Phi$ and $A'\Phi$. Because Γ_t is invariant, it contains a translation Φ_1' such that $A'\Phi = A\Phi_1\Phi = (A\Phi)\,\Phi_1'$, which shows that $A'\Phi$ originates from $A\Phi$ by a translation.

To see that Γ_t is invariant, let an arbitrary translation and affinity, Φ_1 and Φ respectively, be given by

$$\Phi_1 : \begin{array}{l} x_1' = x_1 + a_1 \\ x_2' = x_2 + a_2 \end{array} \qquad \Phi : \begin{array}{l} x_1' = a_{11}x_1 + a_{12}x_2 + a_{13} \\ x_2' = a_{21}x_1 + a_{22}x_2 + a_{23} \end{array} \qquad \begin{vmatrix} a_{11} & a_{12} \\ a_{21} & a_{22} \end{vmatrix} \neq 0.$$

Then $\Phi_1\Phi$ is the transformation,

$$(15.8) \qquad \begin{aligned} x_1' &= a_{11}(x_1 + a_1) + a_{12}(x_2 + a_2) + a_{13} \\ &= a_{11}x_1 + a_{12}x_2 + (a_{11}a_1 + a_{12}a_2 + a_{13}), \\ x_2' &= a_{21}(x_1 + a_1) + a_{22}(x_2 + a_2) + a_{23} \\ &= a_{21}x_1 + a_{22}x_2 + (a_{21}a_1 + a_{22}a_2 + a_{23}). \end{aligned}$$

If Φ_1' denotes the translation,

$$x_1' = x_1 + (a_{11}a_1 + a_{12}a_2), \qquad x_2' = x_2 + (a_{21}a_1 + a_{22}a_2),$$

then $\Phi\Phi_1'$ is the affinity,

$$\begin{aligned} x_1' &= (a_{11}x_1 + a_{12}x_2 + a_{13}) + (a_{11}a_1 + a_{12}a_2) \\ x_2' &= (a_{21}x_1 + a_{22}x_2 + a_{23}) + (a_{21}a_1 + a_{22}a_2) \end{aligned}$$

which coincides with (15.8). Thus:

(15.9)   *The translations form an invariant, Abelian sub-group of the group of affinities.*

If the affine plane is thought of as containing an ideal line ζ, then every ordinary line ξ of the plane contains one ideal point p_ζ, or p_∞, on ζ.[3] Three distinct ordinary points, a,b,c of ξ then determine an *affine ratio* $A(a,b,c)$, defined by

$$(15.10) \qquad\qquad A(a,b,c) = R(p_\infty,a,b,c).$$

This ratio is invariant under all affinities, for if ξ' is the image of ξ under an affinity, and a', b' and c' are the points corresponding to a, b and c, then

$$A(a,b,c) = R(p_\infty,a,b,c) = R(p_\infty',a',b',c') = A(a',b',c').$$

---

[3]The use of p_∞ instead of p_ζ helps in suggesting the Euclidean interpretation. For the same reason ζ is often called the line at infinity.

To evaluate $A(a,b,c)$ we make use of the fact that every ordinary affine point x has a "normalized" projective reprensentation $(x_1,x_2,1)$. Taking a and b as $(a_1,a_2,1)$, $(b_1,b_2,1)$, where $a \curlyvee b$, every point x of $a \times b$, other than p_∞, is given in the normalized form by

(15.11)
$$x = \lambda a + \mu b, \qquad \lambda + \mu = 1$$
or $\quad x = (1 - \mu)a + \mu b.$

In this representation each point of the line corresponds to a unique value of μ. For $x - a = \mu(b - a)$ yields

$$\mu = \frac{x_1 - a_1}{b_1 - a_1} = \frac{x_2 - a_2}{b_2 - a_2}.$$

Affine coordinates of a point are unique, so the terms x_i, a_i, b_i, $i = 1,2$, are fixed. Both denominators cannot vanish, because $a \curlyvee b$, so one of the equalities determines μ. Should one denominator vanish, of course, the corresponding numerator does too. Since $a - b$ is on $a \times b$ and on $x_3 = 0$, it is the point p_∞. Then

$$b = -p_\infty + a, \qquad x = a - \mu a + \mu b = a - \mu(a - b) = a - \mu p_\infty, \qquad \text{and}$$
$$A(a,b,x) = R(p_\infty,a,b,x) = R(p_\infty,a, -p_\infty + a, -\mu p_\infty + a) = -\mu / -1 = \mu.$$

Therefore,

(15.12)   *The affine ratio of three distinct collinear points, a,b,x, where $x = (1 - \mu)a + \mu b$, is*
$$A(a,b,x) = \mu = \frac{x_1 - a_1}{b_1 - a_1} = \frac{x_2 - a_2}{b_2 - a_2}.$$

If x_1 and x_2 are interpreted as rectangular coordinates in a Euclidean plane, then $A(a,b,x)$ is the ratio of signed distances $\overrightarrow{ax}/\overrightarrow{ab}$. In agreement with Euclidean usage, we say that the point $x = (1 - \mu)a + \mu b$, "divides the segment, a,b" in the ratio $\mu/(1 - \mu)$, and call x the *affine center* of a and b when this ratio is 1, that is when $\mu = 1/2$. The relation

$$R(a,b,(a + b)/2,p_\infty) = R(a,b,(a + b)/2,a - b) = (1/2)/(- 1/2) = - 1$$

implies:

(15.13)   *The affine center of a,b is the harmonic conjugate of p_∞ with respect to a,b.*

A harmonic homology with ζ as axis, and with the affine center of a,b as its center, will therefore interchange a and b. Since central reflections have the same axis, and are harmonic, it follows that:

(15.14)   *The affine center of a,b is c if the reflection about c carries a into b.*

In Euclidean geometry it is a familiar fact that parallel lines intercept proportional segments on transversals. The invariance of cross ratio under projection yields the affine analogue :

(15.15)   *If the distinct lines ξ and ξ', intersecting at u, are cut by three parallel lines in the points x,y,z and x',y',z' respectively, then $A(u,x,y) = A(u,x',y')$ and $A(x,y,z) = A(x',y',z')$.*

16. The Affine Theory of Conics

In affine geometry non-degenerate conics may be classified by their relationship to the ideal line ζ. According as this line is a non-intersector,

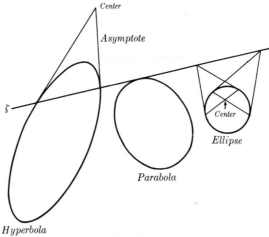

Fig. 28

a tangent, or a secant, the conic is called an *ellipse*, a *parabola*, or a *hyperbola* (Figure 28). A general conic has the form

(16.1) $$a_{11}x_1^2 + 2a_{12}x_1x_2 + a_{22}x_2^2 + 2a_{13}x_1 + 2a_{23}x_2 + a_{33} = 0$$
$$a_{ik} = a_{ki}, \qquad |a_{ik}| \neq 0.$$

To determine its type becomes, analytically, the problem of finding how the line $x_3 = 0$ intersects the conic in the projective form

(16.2) $$\sum_{i,k=1}^{3} a_{ik}x_ix_k = 0, \qquad a_{ik} = a_{ki}, \qquad |a_{ik}| \neq 0.$$

Putting $x_3 = 0$ in (16.2) yields

$$a_{11}x_1^2 + 2a_{12}x_1x_2 + a_{22}x_2^2 = 0,$$

and this equation has two, one, or no solutions, distinct from $(0,0)$, depending on whether the discriminant, $\Delta = a_{11}a_{22} - a_{12}^2$, is negative, zero, or positive. For the first two cases, the conic has points, and is a hyperbola in the first instance and a parabola in the second. When $\Delta > 0$, there are two possibilities to consider. The conic may have points, but simply none on ζ, in which case it is an ellipse, or it may have no real trace at all, in which case we take it to be an *imaginary ellipse*. The imaginary case occurs when the polarity which defines the conic, $\xi_i = \sum_k a_{ik}x_k$, $a_{ki} = a_{ki}$, is elliptic, that is (see (10.23)), when $|a_{ik}|$ and a_{11} have the same sign. Thus, we have:

(16.3)   *The equation (16.1) represents: a hyperbola when $\Delta = a_{11}a_{22} - a_{12}^2 < 0$; a parabola when $\Delta = 0$; an ellipse if $\Delta > 0$ and $|a_{ik}| \cdot a_{11} < 0$; and no trace (or an imaginary ellipse) if $\Delta > 0$ and $|a_{ik}| \cdot a_{11} > 0$.*

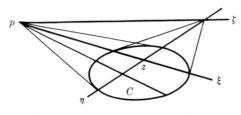

Fig. 29

For the remainder of this section let z be used to denote the pole of ζ with respect to a conic C. If p is any point of ζ, but distinct from z if z is on ζ, consider the pencil through p (Figure 29). This pencil cuts C in a family of parallel chords, and by (15.13), the affine center of each of these chords is the harmonic conjugate of p with respect to the end points of the chord. Since this also implies that the centers are conjugate to p, the centers lie on η, the polar of p, which is a line through z. This shows:

(16.4)   *The parallel chords of a conic, which lie on lines of a pencil p, distinct from z, have their centers on the polar of p, which passes through z.*

When z is on ζ, the conic is tangent to ζ at z and is therefore a parabola. The lines through z in this case are parallel secants.

To obtain a standard form for the equation of a parabola, we choose d_3 as a finite point of C and take δ_1 to be the tangent at d_3. The line ζ, that is δ_3, is tangent to all parabolas, and here we take d_1 as the point of tangency. As shown in (11.11) the projective form of C is then $x_2^2 - kx_1x_3 = 0$, $k \neq 0$.

By choosing $(1,1,1)$ on C this reduces to $x_2^2 - x_1 x_3 = 0$, which, in affine form, is $x_2^2 - x_1 = 0$. Therefore:

(16.5)   *In suitable affine coordinates a given parabola has the equation* $x_2^2 = x_1$.

As in projective geometry, the equations for an affinity and those for a transformation of affine coordinates are formally identical. Thus (16.5) implies that there is always an affinity mapping one parabola on a second, hence different kinds of parabolas cannot be distinguished in the affine plane.

When C is not a parabola, ζ does not contain its pole z. The harmonic homology with center z and axis ζ leaves C invariant. In affine termin-

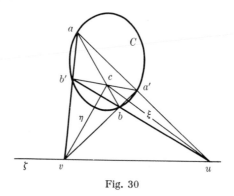

Fig. 30

ology, the conic goes into itself under reflection in z. Because z is the affine center of every chord containing it, it is called the *center of the conic*, and, correspondingly, every line through z is called a *diameter* (Figure 29). Since the pencil of lines conjugate to a diameter ξ contains one line η which is also a diameter, it is natural to call ξ and η *conjugate diameters*. Clearly every pair of conjugate diameters forms with ζ a self-polar triangle. Thus (16.4) may be restated:

(16.6)   *A diameter of an ellipse or a hyperbola is the affine bisector of all chords parallel to the conjugate diameter.*

The construction for the polar of a point shows that if the given diameter cuts C at g and g', then the tangents at these points are parallel to the conjugate diameter and the family of bisected chords.

From the many appealing theorems about conjugate diameters we select the following example (Figures 30 and 33).

If the vertices a, a', b, b' of a parallelogram lie on a conic, then the conic is an ellipse or a hyperbola whose affine center is

$$c = (a \times b) \times (a' \times b'),$$

(16.7)   *the intersection point of the diagonals. The lines ξ and η, through c and parallel respectively to the sides $a \times a'$ and $a \times b'$, are conjugate diameters.*

PROOF: By hypothesis, the points

$$u = (a \times a') \times (b \times b') \quad \text{and} \quad v = (a \times b') \times (a' \times b)$$

lie on ζ. The point c, from its definition, is a finite point, so $\xi = c \times u$ and $\eta = c \times v$ are ordinary lines. The construction given in Section 11 for the polar of a point shows that ξ is the polar of v and that η is the polar of u. Therefore $c = \xi \times \eta$ is the pole of ζ and the center of the conic. Since it has a center, C is an ellipse or a hyperbola. The diameters ξ and η contain each other's poles and so are conjugate.

To obtain standard representations for an ellipse and hyperbola, let the center of a conic C be $d_3 = (0,0,1)$ and take δ_1 and δ_2 to be conjugate diameters. Because the reference triangle $\delta_1, \delta_2, \delta_3 = \zeta$ is self-polar, the equation of C has then the projective form $\sum_i b_i x_i^2 = 0$, $b_i \neq 0$, $i = 1,2,3$.

Setting $b_3 = -1$, the affine equation of C is

(16.8) $b_1 x_1^2 + b_2 x_2^2 = 1$, $b_1 \neq 0$, $b_2 \neq 0$.[4]

Because the conic is real, b_1 and b_2 cannot both be negative. If both are positive, the projective locus of C, $b_1 x_1^2 + b_2 x_2^2 - x_3^2 = 0$ does not intersect $x_3 = 0$, and C is an ellipse. Applying the affinity $x_1' = \sqrt{b_1} x_1$, $x_2' = \sqrt{b_2} x_2$ yields:

(16.9)   *In suitable affine coordinates a given ellipse has the equation* $x_1^2 + x_2^2 = 1$.[4]

When b_1 and b_2 are of opposite sign, the affinity $x_1' = x_2$ and $x_2' = x_1$ interchanges them, so there is no loss of generality in supposing $b_1 > 0$ and $b_2 < 0$. In the projective plane C cuts δ_3 in the points $(\sqrt{-b_2}, \pm\sqrt{b_1}, 0)$, so the conic is a hyperbola. Using the affinity $x_1' = \sqrt{b_1} x_1$, $x_2' = \sqrt{-b_2} x_2$ yields:

(16.10)   *Every hyperbola has a representation* $x_1^2 - x_2^2 = 1$.[4]

As was the case with parabolas, special ellipses such as circles and particular hyperbolas, like those which are equilateral, cannot be distinguished in the affine plane.

---

[4]It is important to remember that the coordinate axes are an arbitrary pair of conjugate diameters.

We note some further affine characteristics. If C is an ellipse, ζ is a non-intersector. The pole of ζ is then a no-tangent point, so every line through the center of C is a secant. When C is a hyperbola, the center of C is a two-tangent point, and the tangents through the center are called *asymptotes* of the conic. If γ is the polarity defining C, then $\gamma(z)$, the induced involution on the pencil at the center, has the asymptotes for fixed elements and hence is hyperbolic. By definition, $\gamma(z)$ assiociates conjugate diameters and so is an involution. This, with (8.12), implies:

(16.11)   *The asymptotes of a hyperbola, together with any two conjugate diameters, form a harmonic set.*

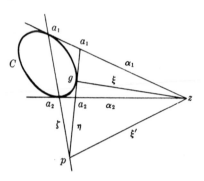

Fig. 31

Again, we can present only one example to indicate the many affine theorems concerning hyperbolas (Figure 31).

(16.12)   *A tangent to a hyperbola cuts the asymptotes in a pair of points whose affine center is the contact point of the tangent.*

Let the tangent η, touching C at g, cut the asymptotes α_1, α_2 at a_1 and a_2 respectively, and take $p = \eta \times \zeta$. The diameter $\xi = g \times z$ has for its conjugate the diameter ξ', which is parallel to the tangents at the intersection points of ξ and C (see the comments following (16.6)). Hence ξ' is parallel to η and so contains p. By the previous theorem, $\alpha_1, \alpha_2, \xi, \xi'$ form a harmonic set, and hence intersect η in the harmonic quadruple p, g, a_1, a_2. Because p is on ζ, g is then the affine center of a_1, a_2 (see (15.13)).

In (11.11) a form for a conic was derived, analogous to that of an equilateral hyperbola $xy = k$. If for a hyperbola the coordinate triangle is taken so that δ_1 and δ_2 are asymptotes, the conic is tangent to these lines at d_2 and d_1 on $\delta_3 = \zeta$. As shown in (11.11), then, the projective form of the conic becomes $x_3^2 - k x_1 x_2 = 0$, $k \neq 0$. Choosing $(1,1,1)$ as a point of

C gives k the value 1, and the hyperbola, referred to its asymptotes, has the affine representation

$$(16.13) \qquad\qquad\qquad x_1 x_2 = 1.$$

The analytic expression for a tangent to a conic in P_a is easily obtained. If C is expressed in projective coordinates, it has the form $\sum a_{ik} x_i x_k = 0$, $a_{ik} = a_{ki}$, and the polar of any point y is the line $\sum a_{ik} x_i y_k = 0$. When y is on C this line is the tangent at y. Putting $x_3 = 1$, the *tangent at the point* (y_1, y_2) *on C is therefore given by*

$$(16.14) \qquad \begin{aligned} (a_{11}y_1 + a_{12}y_2 + a_{13})x_1 + (a_{22}y_1 + a_{22}y_2 + a_{23})x_2 \\ + (a_{31}y_1 + a_{32}y_2 + a_{33}) = 0. \end{aligned}$$

17. Convex Sets

If a and b are two points in the affine plane, the set of points $(1 - \theta)a + \theta b$, where $0 < \theta < 1$, is called the *open segment* $S^*(a,b)$. With the same definition but with $0 \leqslant \theta \leqslant 1$, the points form the *closed segment* $S(a,b)$. These sets correspond to the Euclidean idea of the points between a and b, with the end points belonging to the closed interval, but not the open one. We also speak, in a familiar way, of the segment $S(a,b)$ from a to b.

A set μ in the affine plane is said to be *convex* if for each pair of points a,b belonging to μ, the entire segment $S(a,b)$ also belongs to μ. Simple examples of such sets are: the whole plane, a segment, an open segment, a point.

(17.1) *The no-tangent points of an ellipse or a parabola form a convex set.*

For if a and b are no-tangent points, the line $a \times b$ is a secant (see (11.5)). Because of (11.12), if $S(a, b)$ did not consist entirely of no-tangent points, then on $a \times b$, as a projective line, the no-tangent points would include $\zeta \times (a \times b)$. The line ζ would then be a secant, and the conic a hyperbola. Because of (17.1), the no-tangent points of an ellipse or parabola are called interior points of the conic.

(17.2) *The no-tangent points of an ellipse or parabola C, together with the points of C, form a convex set.*

Because of (17.1) it is only necessary to establish the property that if a is on C and b is on or interior to C, then $S^*(a,b)$ consists of no-tangent points. Since a tangent contains only one point of C, and fails to contain no-tangent points, the line $a \times b$ is a secant when b is on or interior to C, and $b \ne a$. If the desired property did not hold, then the argument of (17.1) shows that the conic would have to be a hyperbola.

If $L(x)$ is defined by $L(x) = a_1x_1 + a_2x_2 + a_3$, then the line, given by $L(x) = 0$, determines four convex sets: its *two sides*, $L(x) > 0$ and $L(x) < 0$, and the *two half-planes*, $L(x) \geqslant 0$ and $L(x) \leqslant 0$. The formal proof of this obvious fact is the same in all four cases and is based on the following relation, which holds for all θ:

$$
\begin{aligned}
& L[(1 - \theta)y + \theta z] \\
(17.3) \quad & = a_1[(1 - \theta)y_1 + \theta z_1] + a_2[(1 - \theta)y_2 + \theta z_2] + a_3 \\
& = (1 - \theta)[a_1y_1 + a_2y_2 + a_3] + \theta[a_1z_1 + a_2z_2 + a_3] \\
& = (1 - \theta)L(y) + \theta L(z).
\end{aligned}
$$

If, for instance, y and z belong to the set $L(x) < 0$, so $L(y) < 0$ and $L(z) < 0$, then for $0 < \theta < 1$ it is clear that $(1 - \theta)L(y) + \theta L(z) < 0$. From (17.3) it follows that every point $p = (1 - \theta)y + \theta z$, $0 < \theta < 1$, satisfies $L(p) < 0$, so the set $L(x) < 0$ is convex. The other cases are similar.

Two points y and z on different sides of ξ are said to be *separated by ξ*.

(17.4)   *If the line ξ separates y and z, then $S(y,z)$ intersects ξ.*

For suppose $L(y) < 0$ and $L(z) > 0$. The line joining y and z, namely $u = (1 - \theta)y + \theta z$, cuts ξ in the point for which $L(u) = 0$. From (17.3), this corresponds to $\dfrac{1 - \theta}{\theta} = \dfrac{L(z)}{-L(y)}$. Hence $\dfrac{1}{\theta} = 1 - \dfrac{L(z)}{L(y)} > 1$, because $L(z)/L(y) < 0$. Therefore $0 < \theta < 1$, so the intersection belongs to $S(y,z)$.

Given two or more sets of points, convex or not, the set formed by those points which are common to all the sets is called the *intersection* of the sets (which may be empty).

(17.5)   *The intersection, D, of any aggregate of convex sets is either convex or empty.*

If a and b are points of D they belong to each set of the aggregate. Each set, being convex, contains $S(a,b)$, which, being in each set, is then in D.

As an application of (17.5), consider a non-collinear triple of points, a_1, a_2, a_3. Let ξ_i denote the side of the triangle a_1, a_2, a_3 opposite a_i, and take L_i^* and L_i respectively as the side and the half-plane of ξ_i which contain a_i, $i = 1, 2, 3$. Then the intersection of L_1^*, L_2^*, L_3^* is a convex set $T^*(a_1, a_2, a_3)$, and the intersection of L_1, L_2, L_3 is a convex set $T(a_1, a_2, a_3)$. The "*triangular*" set $T(a_1, a_2, a_3)$ consists of its "interior", $T^*(a_1, a_2, a_3)$, and the three segments $S(a_1, a_2)$, $S(a_2, a_3)$ and $S(a_3, a_1)$. If a_1, a_2, a_3 belong to a convex set K, then $T(a_1, a_2, a_3)$ is contained in K.

Triangular sets may be used to define a more general notion. A point of a convex set μ is called an *interior point* of the set if it belongs to the interior of a triangular set whose vertices belong to μ. The non-interior points of μ are called its *boundary points*. In these terms, the whole plane, which is convex, consists only of interior points. The no-tangent points of parabolas and ellipses are also interior in this sense, the curves themselves consisting of boundary points. To see that p is interior to a convex set it suffices to show that it belongs to two non-collinear open segments $S^*(a,b)$ and $S^*(c,d)$ which lie in the set.

The following simple fact, too, is often useful.

(17.6)   *If a is an interior point of the convex set μ and b is any point of μ, then every point of the open segment $S^*(a,b)$ is interior to μ.*

For, by definition, there exists a triangular set $T(c,d,e)$ with a belonging to $T^*(c,d,e)$. Choose one of the vertices, say d, not collinear with a and b, and let $a \times d$ cut $S^*(e,c)$ in d'. Since b,d,d' are non-collinear, and a is a point of $S^*(d,d')$, $S^*(a,b)$ belongs to $T^*(b,d,d')$ and hence is interior to μ. As a corollary of this:

(17.7)   *If a convex set has interior points, these points form a convex set.*

Another basic notion associated with convex sets is that of a *supporting line*. The line η is a supporting line of the convex set μ if the line and the set have at least one point in common and μ lies entirely in one of the half planes defined by η. This last requirement is equivalent to saying that η does not separate any pair of points in μ. From its definition, a supporting line can contain no interior point of μ and must contain at least one boundary point. If μ contains no boundary points it has no supporting lines.

The no-tangent points of an ellipse C, for example, form a set with no supporting lines. The points of C and μ together, however, have every tangent of C as a supporting line since, excepting its contact point, a tangent consists of two-tangent points. In this example there is a unique supporting line at each boundary point, but this need not always be the case. The set $T(a_1,a_2,a_3)$ has not only the two sides through a vertex a_i as supporting lines, but also all lines of the pencil at a_i which do not cut the opposite segment.

Though there may be many supporting lines at a boundary point, it is an important fact that there must always be at least one.

(17.8)   *If b is a boundary point of the convex set μ, there exists a supporting line of μ through b.*

PROOF: When μ has no interior points, the entire set lies on a line, which is then a supporting line of the set. We suppose, therefore, that μ contains an interior point a (Figure 32). Then μ contains interior points, p_1 and p_2, not collinear with b, such that a is on $S^*(p_1,p_2)$. Let c be any point of $a \times b$ such that b lies on $S^*(a, c)$, and define $q_1 = (p_1 \times c) \times (p_2 \times b)$, and $q_2 = (p_2 \times c) \times (p_1 \times b)$. For any two distinct points, y, z, we now introduce the notation $R^*(y,z)$ to denote the open ray emanating from y and containing z but not y.

(1) $\qquad\qquad R^*(b,c), R^*(b,q_1), R^*(b,q_2)$ *contain no points of* μ.

If z, on $R^*(b,c)$, belonged to μ, then, by (17.6), all points of $S^*(a,z)$, and hence b, would be interior to μ. A similar argument holds for $R^*(b,q_1)$ and $R^*(b,q_2)$.

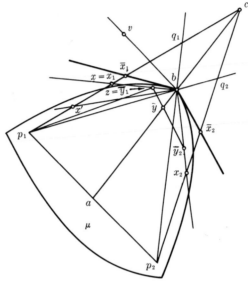

Fig. 32

Let the points x of $S(p_1,c)$, for which $R^*(b,x)$ intersects μ, form the set L, and those for which $R^*(b,x)$ contains no points of μ form the set R. Then p_1 belongs to L, while q_1 and c are in R.

(2) $\qquad\qquad$ *If x belongs to L, every point of $S(p_1,x)$ belongs to L.*

For let x' be any point of $S(p_1,x)$, and denote by z a point of μ on $R^*(b,x)$. Since $S(b,z)$ belongs to μ, there exists a point z' of μ which belongs to both the sets $S^*(b,z)$ and $S^*(b,x)$. Then all points of $S(p_1,z')$ belong to μ, and, by construction, $R^*(b,x')$ must intersect $S(p_1,z')$. Hence x' is in L, and (2) follows from the arbitrariness of x' on $S(p_1,x)$. Formal logic now yields:

(3) $\qquad\qquad$ *If x belongs to R, all points of $S(x,c)$ belong to R.*

Since all points of $S(p_1,c)$ belong to L or R, which are mutually exclusive, non-empty sets, it follows from (2) and (3) that there exists a unique point $\bar{x}_1$ on $S(p_1,c)$ which divides L from R. That is, all x on $S^*(p_1,\bar{x}_1)$ belong to L and all x on $S^*(\bar{x}_1,c)$ belong to R.

(4) $\bar{x}_1$ is on $S(p_1,q_1)$ and is not p_1.

The first part follows from (3) and the fact that q_1 belongs to R. The second part is evident from the fact that p_1 is interior to μ. By a completely symmetrical argument, there exists $\bar{x}_2$ on $S(p_2,q_2)$, $\bar{x}_2 \not\sim p_2$, such that, for all x on $S^*(p_2,\bar{x}_2)$, $R^*(b,x)$ intersects μ, while, for all x on $S^*(\bar{x}_2,c)$, $R^*(b,x)$ contains no point of μ.

Let x_i be any point of $S^*(p_i,\bar{x}_i)$, $i = 1,2$. By the same argument as that which established (2), $R^*(b,x_i)$ contains an interior point y_i of μ. Hence $S^*(b,y_i)$ consists entirely of points interior to μ and, in particular, contains a point $\bar{y}_i$ of μ which lies in $T(a,c,p_i)$, $i = 1,2$. The segment $S(\bar{y}_1,\bar{y}_2)$ also consists of interior points, so $\bar{y}$, the intersection of $S(\bar{y}_1,\bar{y}_2)$ with $S(a,c)$, is an interior point. Thus the set comprised of b and the rays $R^*(b,x_i)$, $i = 1,2$, bounds a convex set which contains a. Since this is true for x_i arbitrarily close to $\bar{x}_i$, $i = 1,2$, it is also true for b and the rays $R^*(b,\bar{x}_i)$, $i = 1,2$.[5] This convex set H, bounded by b and $R^*(b,\bar{x}_i)$, contains μ. For if v is any point not in H, then, by construction, $R^*(b,v)$ passes through c or else intersects one of the sets $S^*(c,\bar{x}_i)$, $i = 1,2$. Hence $R^*(b,v)$ contains no point of μ and v is not in μ. The two lines $\eta_i = b \times \bar{x}_i$, $i = 1,2$, are therefore supporting lines of μ.

In this construction there are two possibilities for the relationship of $\bar{x}_1$ and $\bar{x}_2$. If they are collinear with b, $\eta_1 \sim \eta_2$ and there is only one supporting line of μ through b. When $\bar{x}_1,\bar{x}_2,b$ are not collinear, $\eta_1 \not\sim \eta_2$, and all the lines $b \times x$, for x on $S(\bar{x}_1,t)$, $t = (p_1 \times q_1) \times (b \times \bar{x}_2)$, are supporting lines of μ, with η_1,η_2 distinguished as "extreme" supporting lines.

The present definition admits rather artificial convex sets, for example, the interior of an ellipse plus one boundary point. To avoid such irregularity, and also trivial cases, we introduce the notion of a domain.

DEFINITION: *A convex domain is a convex set with interior points, which is not the entire plane, and which contains $S(a,b)$ if it contains $S^*(a,b)$.*

When a convex domain K is not a strip bounded by two parallel lines, its boundary points are said to form a *convex curve* C. The terminology is purely associative for in general C is not a convex set. The supporting lines of K are also called *supporting lines of* C. In these terms (17.8) states that at each of its points a convex curve C has at least one supporting line. When there is only one supporting line at a point p of C, it is called

---

[5]Since $x_i = \theta p_i + (1 - \theta)\bar{x}_i$, $0 < \theta < 1$, "x_i arbitrarily close to $\bar{x}_i$" means "θ sufficiently near zero."

the *tangent of C* at *p*. At a point of *C* where no tangent exists, the extreme supporting lines are often referred to as the one-sided, or right and left, tangents at the point.

It is also convenient to have various classifications for convex curves. Thus, *C* is called *strictly convex* if it contains no segment or, what is equivalent, if each of its supporting lines intersects it in only one point. *C* is said to be *differentiable* if it has a unique supporting line, that is, a tangent, at each of its points. Either an ellipse or a parabola affords an instance of a convex curve which is both strictly convex and differentiable, while the convex curve bounding a triangular set is neither strictly convex nor differentiable.

If, for a given convex set μ, an ellipse exists which contains μ in its interior, then μ is said to be bounded; otherwise it is said to be unbounded. A convex curve is defined to be *closed* or *open* according as its associated convex domain is bounded or unbounded. Thus an ellipse and a triangle are closed convex curves, while a line and a parabola are open.

18. The Equiaffine Group. Area

In elementary geometry area is defined in terms of distance. It is remarkable, however, that area may be definable in situations which are too general to permit a definition of distance (see next chapter).

Such a situation arises in connection with affine geometry. A subgroup of Γ_a, which has only one parameter less than Γ_a and is therefore too general to leave a reasonable distance invariant, does leave a reasonable area invariant. We are led to it by the following considerations. Let x_1, x_2 be an arbitrary, but fixed, system of affine coordinates, and let

$$\Phi : x_i' = a_{i1}x_1 + a_{i2}x_2 + a_{i3}, \qquad i = 1,2, \qquad \Delta = \begin{vmatrix} a_{11} & a_{12} \\ a_{21} & a_{22} \end{vmatrix} \neq 0$$

be an affinity of P_a on itself. By definition Φ is a one-to-one mapping, and the jacobian of the transformation does not vanish since it is Δ. Suppose then that D is a region in P_a and D' is the region into which D is carried by Φ. If the area of D, namely $A(D)$, is defined by

$$(18.1) \qquad A(D) = \left| \iint_D dx_1\, dx_2 \right|,$$

the customary rules for transforming a double integral gives the area of D' in the form

$$(18.2) \quad A(D') = \left| \iint_{D'} dx_1'\, dx_2' \right| = \left| \iint_D \frac{\partial(x_1', x_2')}{\partial(x_1, x_2)}\, dx_1\, dx_2 \right| = \left| \iint_D \Delta\, dx_1\, dx_2 \right|.$$

Though (18.2) shows that area, as defined in (18.1), is not preserved by a general affinity, it also shows that it is preserved for $\Delta = \pm 1$. If the determinants of the affinities Φ_1 and Φ_2 are Δ_1 and Δ_2, then the determinant of $\Phi_1^{-1} = 1/\Delta_1$ while that $\Phi_1\Phi_2$ is $\Delta_1\Delta_2$. Thus the affinities with determinants ± 1 form a sub-group Γ_{a_1}, of the group Γ_a. The elements of Γ_{a_1} are called *equiaffinities* (also unimodular affinities) and the theorems which remain valid under such transformations form the content of *equiaffine geometry*. *Area, as defined in (18.1), is thus an equiaffine invariant.*[6]

It should be observed that the exact definition of $\int\int F(x,y)\,dx\,dy$ is purely analytical and does not depend on x and y being rectangular coordinates. If it should seem that we are begging the question of area by using the integral, this is only due to the fact that in elementary calculus the integral is frequently introduced as area. The same remarks will apply later when we discuss angles. It is true that functions like $\sin x$ were originally derived from the concept of an angular measure x. However, $\sin x$ can be defined, independently of the angle concept, for instance as the solution of the differential equation $y'' + y = 0$ which satisfies the conditions $y(0) = 0$ and $y'(0) = 1$. This approach has the advantage that it also yields $\cos x$ as the solution for which $y(0) = 1$ and $y'(0) = 0$. The relation $\sin^2 x + \cos^2 x = 1$ and the general addition theorems follow immediately.

The area defined in (18.1) is not invariant under general affinities. If area is to retain certain natural properties, so that a parallelogram, for example, has positive area and a proper sub-parallelogram has smaller area, then evidently no area can exist which is invariant under all affinities. For $x_1' = x/2$, $x_2' = x_2$ will transform the parallelogram $0 \leqslant x_1 \leqslant 1$, $0 \leqslant x_2 \leqslant 1$ into the sub-parallelogram $0 \leqslant x_1' \leqslant 1/2$, $0 \leqslant x_2 \leqslant 1$.

Since $A(D)$ is defined by the same integral used in ordinary geometry, the results are the same. For example, the area of a triangle with vertices (x_1,x_2), (y_1,y_2), (z_1,z_2) is given by

$$(18.3) \qquad A[\,T(x,y,z)] = \pm\, 1/2 \begin{vmatrix} x_1 & x_2 & 1 \\ y_1 & y_2 & 1 \\ z_1 & z_2 & 1 \end{vmatrix}.$$

Care must be exercised, however, in determining *when the geometric interpretation has an affine sense*. For instance, there is no affine meaning in the statement that two triangles with equal base and altitude have the same area, since length and perpendicularity are undefined in the affine plane. A correct statement for equiaffine geometry is that two triangles $T(a_1,b,c)$ and $T(a_2,b,c)$ have the same area when $a_1 \times a_2$ is parallel to

---

[6]The choice of a fixed system of affine coordinates amounts to selecting a unit of area.

$b \times c$. As another illustration of a theorem in this geometry, if z is any point of the open segment $S^*(b,c)$, and a is not collinear with b and c, the affine ratio $|A(b,c,z)|$, defined in (15.10), is equal to the ratio of areas $A[T(a,b,z)]/A[T(a,b,c)]$.

In equiaffine geometry it is possible to distinguish different ellipses by the area they bound. For an ellipse in a general affine representation, (16.1), can always be transformed to the form $x_1^2 + x_2^2 = 1$, (16.9), by the affinity Φ whose determinant $\Delta \neq 0$. The affinity γ, given by $x_i' = |\Delta|^{-\frac{1}{2}} x_i$, $i = 1,2$, has determinant $|\Delta|^{-1}$, so $\Phi\gamma$ is in Γ_{al}. Under $\Phi\gamma$ the ellipse then takes the form

(18.4) $$x_1^2 + x_2^2 = a^2,$$

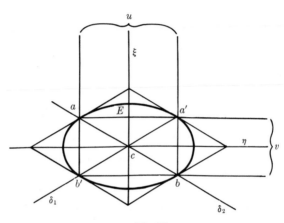

Fig. 33

where $a = |\Delta|^{-\frac{1}{2}}$. It is clear that integration yields πa^2 for the area bounded by this ellipse E, but an affine interpretation of a^2 must be given if this statement is to have an affine sense.

The form (18.4) refers E to δ_1 and δ_2 as conjugate diameters (Figure 33). Since the tangents at the intersection points of a diameter are parallel to the conjugate diameter, (16.6), the four tangents at the intersection points of δ_1 and δ_2 form a parallelogram circumscribed about E. With δ_1, δ_2 interpreted as a rectangular system this is a square of area $4a^2$, as are all the parallelograms of E defined this way. But tangency and parallelism are preserved by every affinity, while area is preserved by all equiaffinities. Hence we can state as a theorem of equiaffine geometry:

> (18.5) *The parallelograms circumscribed about a given ellipse, such that the affine centers of the sides are points of contact, all have the same area B, and the area of the ellipse itself is $\pi B/4$.*

A familiar property of hyperbolas also holds in equiaffine geometry.

(18.6) *The triangle formed by the asymptotes and any tangent to a hyperbola has constant area.*

If the hyperbola is referred to δ_1 and δ_2 as asymptotes, its representation takes the form $x_1 x_2 = k/2$, $k > 0$ (see (16.3)). At the point $g = (g_1, g_2)$ on C the equation of the tangent η is $x_1 g_2 + x_2 g_1 = k$, so $a_1 = (k/g_2, 0)$ and $a_2 = (0, k/g_1)$ are the intersection points of η with $x_2 = 0$ and $x_1 = 0$ respectively. Because of (18.3) the area of the triangle $T(d_3, a_1, a_2)$ is $(1/2)(k/g_2)(k/g_1)$, which has the constant value k because $g_1 g_2 = k/2$.

A more interesting fact concerning area in equiaffine geometry, and one which will be of great importance later, is the following theorem of C. Loewner (born 1893) and F. Behrend (born 1911):

(18.7) *If K is a closed, convex curve, then among the ellipses which have a given center z and contain K there is exactly one ellipse E with minimum area.*

PROOF: We have to show that any two ellipses, E and E', with the given property coincide. The involutions of conjugate diameters of E and of E' are both elliptic and so, by (13.17), have a common pair of elements. If such a pair be taken for the x_1 and x_2 axes, then they, together with the ideal line ζ, form a triangle which is self-polar with respect to both E and E'. The equations of the conics then become:

$$E : a_1 x_1^2 + a_2 x_2^2 = 1, \qquad E' : a_1' x_1^2 + a_2' x_2^2 = 1,$$

where $a_i > 0$ and $a_i' > 0$, $i = 1, 2$. The fact that K is contained in both E and E' means, algebraically, that

(18.8) $\quad a_1 x_1^2 + a_2 x_2^2 \leqslant 1$ *and* $a_1' x_1^2 + a_2' x_2^2 \leqslant 1$, *whenever x lies in K*.

Now $\dfrac{(a_1 + a_1') x_1^2}{2} + \dfrac{(a_2 + a_2') x_2^2}{2} = 1$ is also an ellipse, E_0, with center z, and (18.8) implies that $\dfrac{(a_1 + a_1') x_1^2}{2} + \dfrac{(a_2 + a_2') x_2^2}{2} \leqslant 1$, for x in K, so E_0 contains K. The areas of E, E' and E_0 are, respectively,

$$\frac{\pi}{(a_1 a_2)^{1/2}}, \ \frac{\pi}{(a_1' a_2')^{1/2}}, \quad \text{and} \quad \frac{2\pi}{[(a_1 + a_1')(a_2 + a_2')]^{1/2}}.$$

The first two areas must be equal, since E and E' were assumed to be minimal. For the same reason the area of E_0 must be at least as large as that of E and E'. Algebraically then,

(18.9) $\quad (a_1 + a_1')(a_2 + a_2') \leqslant 4 a_1 a_2 = 4 a_1' a_2'$, *so*

$\qquad\qquad (a_1 + a_1')(a_2 + a_2') \leqslant 4 \sqrt{a_1 a_2 a_1' a_2'}$.

On the other hand, from elementary algebra we have the fact that $(a_i + a_i') \geqslant 2\sqrt{a_i a_i'}$, $i = 1,2$, where the equality holds only if $a_i = a_i'$, $i = 1,2$. This, with (18.9), forces $a_i = a_i'$, $i = 1,2$, and $E = E'$.

Any affinity which carries K and z into itself will also take E into itself. For if K goes into itself, the area it bounds is preserved hence the affinity automatically has determinant ± 1. The image E' of E, which is also an ellipse, has the same area as E and contains K. Hence, from (18.7), $E = E'$.

Exercises

[15.1]   If a,b,c and a',b',c' are non-collinear triples of points in the affine plane, there is exactly one affinity carrying a, b and c into a', b' and c' respectively.

[15.2]   If, in two projectively related pencils, three pairs of corresponding lines are parallel, then all corresponding pairs are parallel.

[15.3]   If corresponding sides of two triangles are parallel, then the lines joining corresponding vertices are concurrent or parallel. Construct examples for both cases.

[15.4]   The diagonals of a parallelogram intersect at the affine center of each diagonal.

[15.5]   State and prove the affine form of the theorem that the medians of a triangle are concurrent at a point which divides each median in the ratio 1 : 2.

[15.6]   Prove the theorem of Menelaus (1st century): If a,b,c are not collinear and no one of them is on ξ, and $d = \xi \times (b \times c)$, $e = \xi \times (c \times a)$, $f = \xi \times (a \times b)$, then

$$A(f,a,b) \cdot A(d,b,c) \cdot A(e,c,a) = 1.$$

Hint: draw the parallel to ξ through c and use (15.15).

[16.1]   No two tangents to a parabola are parallel.

[16.2]   If a parallelogram is circumscribed about a central conic, the diagonals are conjugate diameters of the conic.

[16.3]   If a hexagon inscribed in a conic has two pairs of opposite sides parallel, then the third pair of opposite sides are parallel.

[16.4]   If c is the center of the ellipse C, defined by the polarity Ψ, and Φ is the reflection in c, show that $x \to x\Psi\Phi$ is an elliptic polarity. Formulate this projectively.

[16.5]   Use theorem (27.3) to show that if a tangent to a hyperbola cuts the asymptotes at a and a', then any pair of parallel lines, through a and a' respectively, are conjugate.

[16.6]   If the chord $S(a,b)$ of a hyperbola meets the asymptotes in p and q, then $A(p,q,b) = A(q,p,a)$.

[16.7]   Show that $x_1^2 + 3x_1x_2 - 4x_2^2 + 2x_1 - 10x_2 = 0$ is a hyperbola. Find an affinity which transforms it to the form $x_1^2 - x_2^2 = 1$.

[16.8]   Let C be a parabola and a and b be points such that $a \times b$ does not pass through the point at infinity on C. Show that there is exactly one translation (possibly the identity) which carries C into a parabola through a and b.

[17.1]   A parallelogram bounds a convex set.

[17.2]   Let K_1 and K_2 be convex sets, and let the point x_i range over K_i, $i = 1,2$. Prove that the set of all points K_θ consisting of $(1 - \theta)x_1 + \theta x_2$, where $0 \leqslant \theta \leqslant 1$, θ fixed, is a convex set.

[17.3]   In [17.2], if K_1 and K_2 are parallel segments, then K_θ is a segment parallel to K_1. If K_1 and K_2 are non-parallel segments, then K_θ is a parallelogram.

[17.4]   Each branch of a hyperbola is a strictly convex, open, differentiable curve.

[17.5]   If the intersection of an aggregate of convex domains contains interior points, then it is a convex domain.

[17.6]   If p lies on $S^*(a,b)$ and $S^*(c,d)$, two non-collinear segments in the convex set K, then p is an interior point of K.

[17.7]   Construct an example of a closed, differentiable, convex curve which is not strictly convex.

[17.8]   A convex domain which contains a straight line is either a half plane or a strip bounded by two parallel lines.

[18.1]   In a fixed ellipse, a triangle, whose vertices consist of the center of the ellipse and the end points of two conjugate diameters, has constant area.

[18.2]* If $\eta_{|1}$, $\eta_{|2}$ and $\eta_{|3}$ are the tangents at the three consecutive points a_1, a_2 and a_3 on a parabola, and b_1, b_2 and b_3 are the intersection points of the tangents in pairs, with b_i not on $\eta_{|i}$, $i = 1,2,3$, then

$$A[T(a_1,a_2,b_3)]^{1/3} + A[T(a_2,a_3,b_1)]^{1/3} = A[T(a_1,a_3,b_2)]^{1/3}$$

(where A indicates area, and T a triangular set).

[18.3]   Theorem of Ceva: Let a, b, c and p be a quadrangular set and put

$$d = (a \times p) \times (b \times c), \quad e = (b \times p) \times (c \times a)$$

and $f = (c \times p) \times (a \times b)$.

Then $A(f,a,b)A(d,b,c)A(e,c,a) = -1$.

(Hint: Apply the relation $| A(b,c,z) | = A[T(a,z,b)]/A[T(a,b,c)]$ of the text to the triangles $T(p,a,b)$, $T(p,b,c)$ and $T(p,c,a)$.) Show that Ceva's theorem is an affine and not only an equiaffine theorem.

[18.4]   For the parallelogram with vertices ($\pm$ 1, $\pm$ 1) find the Loewner ellipse (see (18.7)) with the center of the parallelogram as center. (Hint: Show first that because of the uniqueness of the ellipse, it must have the form $\dfrac{x_1^2}{a^2} + \dfrac{x_2^2}{b^2} = 1$.)

[18.5]   Among the ellipses circumscribed about a given parallelogram, the ellipse, for which the diagonals are conjugate diameters, has the smallest area. (Hint: use [18.4].)

[18.6]   Derive from [18.5]: If the ellipse E with center z contains the convex curve C, also having center z, and if E touches C in the endpoints of two conjugate diameters, then E is the Loewner ellipse for C.

[18.7]   Can different types of parabolas be distinguished in equiaffine geometry?

Chapter IV

Projective Metrics

19. Metric Spaces

The specialization of a line in the projective plane produced the affine plane and reintroduced the notion of parallelism. It is logical to consider next by what steps distance can be recovered, that is, how a distance, or metric, can be obtained from the affine plane. However, we wish to study this problem in a much more general framework. To that end, some general concepts associated with "*distance*" will be developed.

By whatever rule the number ab is to be chosen for the distance corresponding to the points a and b, the usual intuition of distance demands the following properties.

I.   *The distance ab is a real, non-negative number, and $ab = 0$ if and only if a and b coincide.*

II. *The distance from a to b is the same as from b to a, that is, $ab = ba$.*

III. *Distance means shortest distance. All triples, a,b,c satisfy the "triangle inequality" $ab + bc \geqslant ac$.*

These postulates are still much too general to provide a basis for an interesting geometry. For instance, if R_1 is any collection of objects $a,b,c,\cdots$, called points, and if we define $ab = 1$ for $a \neq b$ and $ab = 0$ for $a = b$ then I, II and III are satisfied, though in a trivial way. Despite the possibility of a case such as R_1, we introduce some notions solely in terms of the basic postulates.

A set R of points, in which distance is defined for each pair of points so that I, II and III are satisfied, is called a metric space.

In a metric space R the sequence of points $x_1, x_2, \cdots, x_n, \cdots$ is said to *converge to the point x, or to tend to x,* if the $\lim\limits_{n \to \infty} x_n x = 0$.

In R_1 the only sequences which are convergent are of the type $a,a,a,\cdots$ For a less trivial space, let R_2 consist of all possible n-tuples, $(x_1,x_2,\cdots,x_n)$, of real numbers, with the distance between the n-tuples, or points, x and y defined by

$$d_2(x,y) = \max_{i = 1, 2, \cdots, n} \{\, |x_i - y_i| \,\}.$$

With the same set of n-tuples another metric space, R_3, is obtained with distance given by

$$d_3(x,y) = \sum_{i=1}^{n} |x_i - y_i|.$$

That properties I and II hold for d_2 and d_3 is clear, also that d_3 satisfies III. To see that III holds for d_2, let x,y,z be any three points in R_2, and choose i_0 such that

$$d_2(x,z) = \max_{i=1,2,\cdots,n} \{|x_i - z_i|\} = |x_{i_0} - z_{i_0}|.$$

Then $d_2(x,z) = |x_{i_0} - z_{i_0}| \leqslant |x_{i_0} - y_{i_0}| + |y_{i_0} - z_{i_0}|$

$$\leqslant \max_{i=1,\cdots,n} \{|x_i - y_i|\} + \max_{i=1,\cdots,n} \{|y_i - z_i|\} = d_2(x,y) + d_2(y,z).$$

The same space of points, but with distance given by

$$d_4(x,y) = e(x,y) = \left[\sum_{i=1}^{n} (x_i - y_i)^2 \right]^{1/2}$$

defines R_4, *the so-called n-dimensional Euclidean space*, often denoted by E^n. Properties I and II are easily verified for d_4. The triangle inequality, however, is more recondite. To obtain it we make use of the so-called *Cauchy-Schwarz inequality*:

If $a_1, a_2, \cdots, a_n$ and $b_1, b_2, \cdots, b_n$ are real numbers then

(19.1)
$$\left(\sum_{1}^{n} a_i b_i \right)^2 \leqslant \left(\sum_{1}^{n} a_i^2 \right) \left(\sum_{1}^{n} b_i^2 \right),$$

the equality holding when, and only when, all a_i are zero, or all b_i are zero, or $b_i = \mu a_i$ for all i.

PROOF: Since the equality holds when all a_i are zero, or when all b_i are zero, suppose the a_i are not all zero and consider the quadratic expression,

$$\sum_{1}^{n} (\lambda a_i + b_i)^2 = \lambda^2 \sum_{1}^{n} a_i^2 + 2\lambda \sum_{1}^{n} a_i b_i + \sum_{1}^{n} b_i^2.$$

Because of its form, the expression is never negative hence its discriminant is not positive,

or
$$4 \left(\sum_{1}^{n} a_i b_i \right)^2 - 4 \left(\sum_{1}^{n} a_i^2 \right) \left(\sum_{1}^{n} b_i^2 \right) \leqslant 0.$$

that is
$$\left(\sum_{1}^{n} a_i b_i \right)^2 \leqslant \left(\sum_{1}^{n} a_i^2 \right) \left(\sum_{1}^{n} b_i^2 \right).$$

The discriminant is actually negative unless the quadratic in λ has a real root, which occurs only when $\lambda a_i + b_i = 0$, for all i, or $b_i = -\lambda a_i = \mu a_i$. All b_i are zero when μ is zero and conversely.

Returning to the triangle inequality for E^n, it must be shown that any three points, x,y,z, satisfy the relation,

$$\left[\sum_1^n (x_i - z_i)^2\right]^{1/2} \leq \left[\sum_1^n (x_i - y_i)^2\right]^{1/2} + \left[\sum_1^n (y_i - z_i)^2\right]^{1/2}.$$

With $(x_i - y_i)$ and $(y_i - z_i)$ playing the roles of a_i and b_i, (19.1) implies

$$\left[\sum_1^n (x_i - y_i)(y_i - z_i)\right]^2 \leq \left(\sum_1^n (x_i - y_i)^2\right)\left(\sum_1^n (y_i - z_i)^2\right),$$

or $$\sum_1^n (x_i - y_i)(y_i - z_i) \leq e(x,y)e(y,z).$$

The equality holds only when, for all i, $x_i = y_i$, or $y_i = z_i$, or $x_i - y_i = \mu(y_i - z_i)$, with $\mu \geq 0$. Then

$$e^2(x,y) + 2\sum (x_i - y_i)(y_i - z_i) + e^2(y,z) \leq e^2(x,y) + 2e(x,y)e(y,z) + e^2(y,z).$$

The left side of this inequality is just $\sum_1^n (x_i - z_i)^2$, as can be seen by writing

$$\sum_1^n (x_i - z_i)^2 = \sum_1^n [(x_i - y_i) + (y_i - z_i)]^2,$$

therefore

$$\sum (x_i - z_i)^2 \leq [e(x,y) + e(y,z)]^2, \text{ or } e(x,z) \leq e(x,y) + e(y,z).$$

The equality sign in this last expression holds when, for all i,

$$y_i = \frac{x_i + \mu z_i}{1 + \mu} = (1 - t)x_i + tz_i, \quad t = \frac{\mu}{1 + \mu} \text{ and } 0 \leq t \leq 1.$$

The situations where $x_i = y_i$ or $y_i = z_i$ correspond to $t = 0$ and $t = 1$ respectively. It follows from this that in E^n the equality

$$e(x,y) + e(y,z) = e(x,z)$$

holds only when y lies on the "*segment $S(x,z)$.*"

For $n = 1$ the distances d_2, d_3 and d_4 coincide. For $n > 1$, however, they differ and so present three different metrizations, or distances, imposed on the same set of points. *Two distances d' and d'', for the same set R, are* *called equivalent* when convergence of a sequence $\{x^i\}$ to a point x under d' implies convergence of the sequence to the point x under d'', and conversely. That is, the $\lim_{i \to \infty} d'(x,x^i) = 0$ if and only if the $\lim_{i \to \infty} d''(x,x^i) = 0$. That d_2, d_3 and d_4 are equivalent metrics can be deduced from the fact that $d_j(x,x^i) \to 0$ if and only if $x_k^i \to x_k$ for $k = 1,2, \cdots, n$. The equivalence also follows directly from the following inequalities which hold for any two n-tuples, a and b, and for any fixed, positive, integral n :

$$(1/n) \max \{|a_i - b_i|\} \leqslant (1/n)\Big[\sum (a_i - b_i)^2\Big]^{1/2} \leqslant (1/n)\sum|a_i - b_i|$$
$$\leqslant \max \{|a_i - b_i|\} \leqslant \Big[\sum (a_i - b_i)^2\Big]^{1/2} \leqslant \sum|a_i - b_i|.$$

In general, if xy has the properties I, II and III, then $\sqrt{xy}$ does also. By inspection, I and II can be seen to hold for $\sqrt{xy}$, and III follows from the fact that $\big(\sqrt{xy} + \sqrt{yz}\big)^2 = xy + yz + 2\sqrt{xy}\sqrt{yz} \geqslant \big(\sqrt{xz}\big)^2$. Here the equality sign occurs only when $x = y$ or $z = y$. Moreover, the distances xy and $\sqrt{xy}$ are always equivalent. In particular, d_i and $\sqrt{d_i}$ are equivalent metrizations of R_i, $i = 2,3,4$.

If R' is a subset of the metric space R, then in terms of the distance already assigned to its points R' is also a metric space. For instance, suppose R' to be the set $K(a,\delta)$ consisting of the points in R at a distance δ from the point a. In any space, $K(a,0) = a$. In a trivial space the locus $K(a,\delta)$ may be uninteresting. In R_1, for example, the sphere contains no points when δ is greater than zero and different from 1, while for $\delta = 1$ it contains all points but a.

In the n-dimensional Euclidean space, $K(a,r)$ is the locus

$$\sum_{i=1}^{n} (x_i - a_i)^2 = r^2, \quad r > 0,$$

which for $n = 2$ and $n = 3$ yields the ordinary circle and sphere respectively. For points x,y on a circle a second distance may be defined as the shorter of the two circular arcs joining x and y. If a is the origin then $x_1^2 + x_2^2 = r^2$ is the equation of the circle. The points x and y determine radii r_x and r_y which make angles θ_x and θ_y with the initial radius to $(r,0)$. Since x_1/r and x_2/r are the cosine and sine of θ_x, the formula for the cosine of the difference of two angles gives $[(x_1y_1 + x_2y_2)/r^2]$ as the cosine of θ the angle between r_x and r_y. Hence

$$r\theta = r \text{ Arc cos } [(x_1y_1 + x_2y_2)/r^2]$$

is the length of the shorter arc from x to y, where the Arc cos denotes (here, and in the future) the principal value, that is $0 \leqslant \text{Arc cos} \leqslant \pi$.

Similarly the points of the ordinary sphere may be remetrized by taking the distance from x to y as the shorter of the two great circle arcs joining them. If a is again the origin, the direction cosines of r_x and r_y are respectively x_i/r and y_i/r, $i = 1,2,3$. The cosine of the angle θ is $\sum x_i y_i / r^2$ and the length of the shorter great circle arc is

$$r\theta = r \text{ Arc cos} \left[\sum x_i y_i / r^2 \right].$$

More generally, *the $(n-1)$-dimensional, spherical space $S_r^{(n-1)}$,* centered at the origin in E^n, is the locus satisfying the equation

$$\sum_{i=1}^{n} x_i^2 = r^2.$$

On $S_r^{(n-1)}$ the spherical distance is given by

$$s(x,y) = r \text{ Arc cos} \left[\sum_1^n x_i y_i / r^2 \right].$$

The fact that this distance is always defined, even in the general case, namely that $|\sum x_i y_i| \leqslant r^2$, follows from (19.1) which gives

$$\left(\sum x_i y_i \right)^2 \leqslant \left(\sum x_i^2 \right) \left(\sum y_i^2 \right) = r^4.$$

However, the proof that $s(x,y)$ satisfies the triangle inequality is complicated and will be omitted here.

In many branches of modern mathematics the distance concept enters in the following way. A certain type of convergence is defined for the elements of a space. The question then arises: *does a suitable metric exist such that convergence in the metric sense is equivalent to convergence in the previous sense?* For instance, let R be the space whose elements are the real functions $f(t)$, which are defined and continuous on the interval $a \leqslant t \leqslant b$, and let convergence in this space mean uniform convergence. If $f(t)$ and $g(t)$ are elements, or "points", of R, the function $|f(t) - g(t)|$ is also a continuous function and so assumes a maximum value on the closed interval from a to b. This value may be taken to define the distance from f to g, that is,

$$d(f,g) = \max_t |f(t) - g(t)|.$$

Clearly $d(f,g)$ has properties I and II, and the triangle inequality is established in exactly the same way as for d_2. Convergence in the sense of d and uniform convergence are now equivalent. For if $\{f_i(t)\}$ converges uniformly to $f(t)$, then $d(f_i,f) \to 0$ and conversely. In the language of

modern topology this result would be expressed by saying that with convergence defined as uniform convergence the space R is metrizable.

This same problem arises in connection with the projective plane, or,
more generally, in connection with the *n-dimensional projective space P^n*.
The definition of the latter is analogous to that for P^2: the points of P^n
are the classes $[x]$ of $(n + 1)$–tuples, $(x_1, x_2, \cdots, x_{n+1})$, of real numbers; the
zero class is excluded as before; and x and y belong to the same class, or
$x \sim y$, if and only if $\lambda \neq 0$ exists such that $x_i = \lambda y_i$, $i = 1, 2, \cdots, n + 1$.
In P^n there is a natural concept of convergence, namely that the points
$p^i = (p_1^i, p_2^i, \cdots, p_{n+1}^i)$, $i = 1, 2, \cdots$, tend to p if representations p^{i*} and p^*
exist such that the $\lim_{i \to \infty} p_k^{i*} = p_k^*$, $k = 1, 2, \cdots, n + 1$. *The question is
whether or not this convergence is expressible in terms of a suitable
metric.*

That such a metric does exist may be seen in the following way. For

any point x, let $|x|$ denote $\left[\sum_1^{n+1} x_i^2 \right]^{1/2}$. Because the zero class is excluded,

$|x|$ is never zero. For any two representations, x and λx, of the same

point it follows that $\dfrac{x}{|x|} = \pm \dfrac{\lambda x}{|\lambda x|}$, where the sign depends on whether λ

is positive or negative. Now if a sequence p^i converges to p in the sense
defined above, so that for certain representations $p^{i*} \to p^*$, then clearly
the sequence of numbers $|p_i^*| \to |p^*|$. Therefore the sequence of points

in the representation $\dfrac{p^{i*}}{|p^{i*}|}$ converges to $\dfrac{p^*}{|p^*|}$. But for general represen-

tations, $\dfrac{p^i}{|p^i|}$ and $\dfrac{p^{i*}}{|p^{i*}|}$ differ at most in sign and the same is true of $\dfrac{p}{|p|}$

and $\dfrac{p^*}{|p^*|}$. Therefore, with the proper choice of sign,

(19.2) $\pm \dfrac{p_k^i}{|p^i|} \to \dfrac{p_k}{|p|}$, $k = 1, 2, \cdots, n + 1$,

where the sign depends on i but not on k. Another way of saying this is
that if a_i is the smaller of the two numbers,

$$\sum_{k=1}^{n+1} \left| \frac{p_k^i}{|p^i|} - \frac{p_k}{|p|} \right| \text{ and } \sum_{k=1}^{n+1} \left| \frac{p_k^i}{|p^i|} + \frac{p_k}{|p|} \right|,$$

then $p^i \to p$ implies $a_i \to 0$. Conversely, let $a_i \to 0$. Because of (19.2), if p^*
is taken as $\dfrac{p}{|p|}$ and p^{i*} as $\pm \dfrac{p^i}{|p^i|}$, then for these representations $p^{i*} \to p^*$,

which is the definition of $p^i \to p$. It follows that if the distance between x and y is defined by

$$\pi(x,y) = \min \left\{ \sum_{i=1}^{n+1} \left| \frac{x_i}{|x|} - \frac{y_i}{|y|} \right|, \quad \sum_{i=1}^{n+1} \left| \frac{x_i}{|x|} + \frac{y_i}{|y|} \right| \right\},$$

then convergence in the metric sense and in the sense first defined are equivalent. That $\pi(x,y)$ is actually a metric is easily verified (exercise [19.5]).

The conventions for P^2 carry over in a natural way to P^n. Thus if a and b are $(n+1)$-tuples, then $a \cdot b$ means $\sum_{i=1}^{n+1} a_i b_i$. The locus of points x satisfying $x \cdot \xi = 0$, where $\xi \neq 0$, is called a *hyperplane*, and it is point and hyperplane which play dual roles in P^n.

Analogous to the affine plane, P_a or A^2, the n-dimensional affine space A^n is derived from P^n by specializing one hyperplane in the projective space. If this is chosen as $x_{n+1} = 0$, and we regard it as deleted from P^n, then every point of A^n has a non-zero last coordinate. As before, this coordinate may be taken to be 1. Then x, in the normalized form $\bar{x} = (\bar{x}_1, \bar{x}_2, \cdots, \bar{x}_n, 1)$, has a unique representation, and $\bar{x}_1, \cdots, \bar{x}_n$ are called affine coordinates. That x^i is a sequence in A^n converging to x in A^n means that x^i and x have representations, x^{i*} and x^*, in P^n for which $x_k^{i*} \to x_k^*$, $k = 1, 2, \cdots, n+1$. In particular, $x_{n+1}^{i*} \to x_{n+1}^*$. Since $x_{n+1}^{i*} \neq 0$ and $x_{n+1}^* \neq 0$ it follows that

$$\frac{x_k^{i*}}{x_{n+1}^{i*}} \to \frac{x_k^*}{x_{n+1}^*}, \quad k = 1, 2, \cdots, n,$$

hence that $\bar{x}_k^i \to \bar{x}_k$, $k = 1, 2, \cdots, n$. Obviously $\bar{x}_k^i \to \bar{x}_k$, $k = 1, 2, \cdots, n$ implies that x^i as a sequence in P^n converges to x in P^n, since $\bar{x}^i$ and $\bar{x}$ are simply particular representations. It follows from the preceding discussion that $\bar{x}^i \to \bar{x}$ in A^n is equivalent to $\pi(x^i, x) \to 0$.

Though $\pi(x,y)$ shows that P^n is metrizable, the metric geometry to which it leads is not fruitful. From this point of view, the most interesting distance for P^2 (or P^n) is that of elliptic geometry, which is discussed in the next chapter.

20. Segments, Straight Lines, Great Circles.
Projective–Metric Spaces

In a metric space, if the point $x(t)$, $\alpha \leqslant t \leqslant \beta$, depends continuously on the parameter t, that is if $t_i \to t_0$ implies $x(t_i) \to x(t_0)$, then the set $x(t)$ is called a *curve* from $x(\alpha)$ to $x(\beta)$. In the space R_1 no two distinct points can be connected by a curve. In the spaces R_2, R_3, R_4, and in the space R_5 of n-tuples, with $d_5(x,y) = \left[\sum_{i=1}^{n} (x_i - y_i)^2 \right]^{1/4} = \sqrt{e(x,y)}$, any two points y and z can be connected by a continuous curve. This may be done, for instance, by

$$x(t) = (1 - t)y + tz$$
$$= ((1-t)y_1 + tz_1, \; (1-t)y_2 + tz_2, \; (1-t)y_3 + tz_3, \; \cdots, \; (1-t)y_n + tz_n)$$

where $0 \leqslant t \leqslant 1$.

The *length of a curve* $x(t)$, $\alpha \leqslant t \leqslant \beta$, in a metric space is defined in the ordinary way : Let $\Delta : t_0 = \alpha < t_1 < t_2 < \cdots < t_n = \beta$ be a finite decomposition of the interval $\alpha \leqslant t \leqslant \beta$, and put

$$L(\Delta) = \sum_{i=1}^{n} x(t_i)x(t_{i-1}).$$

Then the length $L\{x(t)\}$ of $x(t)$ is defined as the least upper bound (which may be ∞) of the set of numbers $L(\Delta)$ corresponding to all possible divisions Δ. It can be deduced from the triangle inequality that if Δ_i is any sequence of decompositions, $\Delta_i : t_0^i = \alpha < t_1^i < t_2^i < \cdots < t_n^i = \beta$, in which $\max_k (t_{k+i}^i - t_k^i)$ tends to zero as $i \to \infty$, then

$$\lim_{i \to \infty} L(\Delta_i) = L\{x(t)\}.$$

We will not prove this here since it will not be used. For any Δ, repeated use of the triangle inequality justifies the relation:

(20.1) $$L\{x(t)\} \geqslant L(\Delta) = \sum_{i=1}^{n} x(t_i)x(t_{i-1}) \geqslant x(\alpha)x(\beta).$$

When the equality holds, that is when $L\{x(t)\} = x(\alpha)x(\beta)$, then $x(t)$ is called a *segment from* $x(\alpha)$ *to* $x(\beta)$. From (20.1) it follows that *a segment is always a shortest connection of its endpoints.*

In the trivial case where $x(t) = x(\alpha)$ for $\alpha \leqslant t \leqslant \beta$ the condition $L\{x(t)\} = x(\alpha)x(\beta) = 0$ is always satisfied. As the examples R_1 and R_5

show, there may be no other segments in a space than these trivial ones. This is obvious for R_1. In R_5 if $x(\alpha) \neq x(\beta)$ and t_1 is such that $x(t_1)$ is not $x(\alpha)$ or $x(\beta)$, then

$$L\{x(t)\} \geqslant x(\alpha)x(t_1) + x(t_1)x(\beta) > x(\alpha)x(\beta).$$

In the spaces R_2, R_3 and R_4, the curve $x(t) = (1-t)y + tz$, $0 \leqslant t \leqslant 1$, is a segment joining y and z, and in the case of $R_4 = E^n$ *it was shown to be the only segment connecting these points. In R_2 and R_3 there will, in general, be infinitely many segments connecting y and z.* In R_3, for example, any curve $x(t) = (x_1(t), x_2(t), \cdots, x_n(t))$ from y to z will be a segment if the function $x_i(t)$ varies monotonely from y_i to z_i, that is if:

$$y_i \leqslant z_i \text{ and } t_1 \leqslant t_2 \text{ imply } x_i(t_1) \leqslant x_i(t_2),$$
$$y_i \geqslant z_i \text{ and } t_1 \leqslant t_2 \text{ imply } x_i(t_1) \geqslant x_i(t_2).$$

(See exercise [20.2].)

If $x(t)$, $\alpha \leqslant t \leqslant \beta$, is a segment and $\alpha \leqslant t_1 < t_2 < t_3 \leqslant \beta$, then (20.1) implies :

(20.2) $$x(t_1)x(t_2) + x(t_2)x(t_3) = x(t_1)x(t_3).$$

Therefore if $\bar{x}(t)$, $\bar{\alpha} \leqslant t \leqslant \bar{\beta}$ and $\alpha \leqslant \bar{\alpha} < \bar{\beta} \leqslant \beta$, denotes a subarc of $x(t)$, and $\bar{\Delta} : \bar{t}_0 = \bar{\alpha} < \bar{t}_1 < \bar{t}_2 < \cdots < \bar{t}_n = \bar{\beta}$ is a decomposition of $\bar{\alpha} \leqslant t \leqslant \bar{\beta}$, then repeated application of (20.2) yields :

(20.3) $$L(\bar{\Delta}) = \sum_{i=1}^{n} \bar{x}(\bar{t}_i)\bar{x}(\bar{t}_{i-1}) = \bar{x}(\bar{\alpha})\bar{x}(\bar{\beta}) = x(\bar{\alpha})x(\bar{\beta}).$$

Hence $\bar{x}(t)$ is also a segment.

(20.4)   *A subarc of a segment is a segment.*

The relation (20.3) shows that the length s of the arc $x(t)$ between $x(\alpha)$ and $x(t)$ equals $x(\alpha)x(t)$. Introducing this variable in place of t as a parameter changes the representation from $x(t)$ to $y(s)$ defined by the condition that $y(s) = x(t)$ if $s = x(\alpha)x(t)$. In the new representation,

$$y(s_1)y(s_2) = |s_1 - s_2|, \qquad 0 \leqslant s \leqslant x(\alpha)x(\beta).$$

Thus the association $s \leftrightarrows y(s)$ sets up a one-to-one correspondence between the segment and the interval $0 \leqslant s \leqslant x(\alpha)x(\beta)$ on the real s-axis, and this association preserves distance. Such a distance preserving association is called an *isometry*, and *the segment is said to be isometric to the interval.* This is a basic concept in metric spaces and is more fully developed in a later section on motions.

The converse of the above situation is also true, namely that in a metric space *a set isometric to an interval,* $\alpha \leqslant s \leqslant \beta$, *of the real axis is a segment.* For if $x(s)$ denotes the image of s under the isometry, then

$$x(s_1)x(s_2) = |s_1 - s_2|,$$

and it follows immediately that $L\{x(s)\} = \beta - \alpha = x(\alpha)x(\beta)$.

These results suggest the *definition of a (metric) straight line in a metric space as a set isometric to the real s-axis.* If $x(s)$ denotes the image of s under such an isometry, then

(20.5) $x(s_1)x(s_2) = |s_1 - s_2|$ *for any pair* s_1, s_2.

Any subarc, $\alpha \leqslant s \leqslant \beta$, of a straight line $x(s)$ is then a segment. In R_2, R_3 and R_4 the locus

$$x(t) = (1 - t)y + tz, \qquad y \neq z, \qquad -\infty < t < \infty$$

is a straight line. The mapping $t \to x(t)$ is not an isometry of the t-axis on the set $x(t)$, but the parameter transformation

$$s = t d_i(y,z)$$

yields a representation $x(s)$ of the locus such that $s \to x(s)$ is an isometry of the set and the real s-axis. This can be checked, using

$$x(s) = \frac{[d_i(y,z) - s]y + sz}{d_i(y,z)}.$$

Again, for $R_4 = E^n$, these loci are the only straight lines, whereas in R_2 and R_3 there are many others.

Besides the straight lines, the *sets isometric to a circle* S_r^1 for some $r > 0$ are important. With the sphere in mind, such sets *will be called great circles.* If $x(s)$, $0 \leqslant s < 2\pi r$ is a representation of S_r^1 in terms of arc length, then

(20.6) $x(s_1)x(s_2) = \min \{|s_1 - s_2|, 2\pi r - |s_1 - s_2|\}$.

The existence of a representation in which distance has this form is typical of great circles. For any pair of distinct points on it, a great circle contains two arcs joining them and at least one of these is a segment. They are both segments if and only if the points have the distance πr (corresponding to diametrically opposite points).

We are now in a position to connect the topics of the first chapters with metric notions and to define the type of metric space in which we will henceforth be interested. As already noted, a hyperplane in P^n is a locus of the form $x \cdot \xi = 0$, $\xi \neq 0$. The *line* $x \times y$, $x \not\vdash y$, in P^n is now defined to be the locus of points $\lambda x + \mu y$, where $(\lambda, \mu) \neq (0,0)$. The spaces R, with which we will be concerned, belong to a class defined by the following properties:

1).   *R is a subset of P^n which does not lie in a hyperplane.*
2).   *R is a metric space (with distance xy).*
3).   *The strict triangle inequality, $xy + yz > xz$, holds whenever x, y and z are distinct and non-collinear (z is not on $x \times y$).*
4).   *If x and y are distinct points of R the intersection, $\lambda(x,y)$, of $x \times y$ with R is a metric straight line or a great circle.*

Because of 1) and 2) such spaces will be called *n-dimensional projective-metric spaces.* The distance xy is also called a *projective metric.*

Merely requiring that R be a metric space seems inadequate, and that to be interesting the extra condition should be imposed that xy is equivalent to $\pi(x,y)$, that is, that $x^i x \to 0$ if and only if $\pi(x^i,x) \to 0$. Actually, this condition need not be assumed, since it is a consequence of 3) and 4). Due to the peculiar structure of the projective space, the proof is simple only when R is a subset of the affine plane. For that case a proof, for $n = 2$, is given in the next section, and for $n = 3$ in Section 47.

The condition 1) above is necessary to justify the term *n-dimensional.* If it is omitted R may lie in a hyperplane of P^n, in which case it would be at most $(n - 1)$-dimensional. The second and third conditions are natural ones that need no justification. Since both a great circle and a straight line contain a segment joining a given pair of their points, condition 4) implies *that two points y,z, of R can be connected by at least one segment, and at most two, lying on $y \times z$. Property* 3) *insures the fact that no other segments from y to z exist.*

When at least one point is removed from a projective line there can be at most one arc on the line joining two of the remaining points. Therefore $\lambda(x,y)$ must be a straight line when it is a proper subset of $x \times y$. When $\lambda(x,y)$ coincides with $x \times y$ it contains two arcs from x to y and is a great circle.

(20.7) $\lambda(x,y)$ *coincides with $x \times y$ if and only if it is a great circle.*

When $\lambda(x,y)$ is a proper subset of $x \times y$ it is a subarc of $x \times y$ without endpoints. That it is an arc follows from the fact that $\lambda(x,y)$ is a straight line, $x(s)$, of the form (20.5) and therefore contains, with any two points, a segment connecting them which is an arc on $x \times y$. It cannot have endpoints since $x(0)x(n) = x(0)x(-n) = n$ tends to ∞ with n which means that the sequences $\{x(n)\}$ and $\{x(-n)\}$ have no limit points.

(20.8)   *If $\lambda(x,y)$ is a proper subset of $x \times y$ it originates from $x \times y$ by the removal of a closed interval which may be a point.*

Among the metric spaces discussed in the preceding section, only E^n is a projective-metric space (R_2 and R_3 are eliminated by condition 3)). To see that the class is still very general, we give another example. The space R_6 which is the strip $-\pi/2 < x_2 < \pi/2$ in E^2, with the ordinary metric $e(x,y)$, is not a projective-metric space. For if x and y are points of R_6 for which $x_2 \neq y_2$, then they are not contained in a metric line. However, under the metrization $d(x,y) = e(x,y) + \rho(x,y)$, where $\rho(x,y) = |\tan x_2 - \tan y_2|$, R_6 is a projective-metric space. The function $\tan x_2$ increases monotonically with x_2 so that if x, y and z lie on $x \times y$, with $x_2 < y_2 < z_2$, then $\rho(x,y) + \rho(y,z) = \rho(x,z)$. Since $e(x,y)$ is also additive on such a line, $d(x,y)$ is. For horizontal lines $\rho(x,y) = 0$ and $d(x,y)$ is addititive since $e(x,y)$ is. On these lines $e(x,y)$ becomes infinite with $|x_1 - y_1|$. On others $e(x,y)$ is bounded, but $\rho(x,y) \to +\infty$ for fixed x and $y \to \pm \pi/2$.

In these examples all $\lambda(x,y)$ are straight lines. The original definition of a projective-metric space admits a priori the possibility that the space contains both straight lines and great circles. According to a theorem of Hamel (born 1877), however, *both types cannot coexist in the same projective-metric space*: either all metric lines are straight lines or they are all great circles of the same length.[1] In the second case R is the whole projective space. In the first case there is a hyperplane in P^n which has no common point with R. In the case of P^2, where a hyperplane is a straight line, if a non-intersector of R is chosen for the line at infinity, R may be regarded as a subset of the affine plane.

Hamel's theorem will not be proved here since it will not be used and because its true character becomes clear only when non-projective metrics are also considered.[2] It explains, however, why R will appear in all examples as either the whole projective space or else as a subset of the affine space. To have short terms, spaces of the former type will be called *closed* and those of the latter type will be called *open*.

For a better understanding of the *meaning of Hamel's theorem*, consider the projective plane P^2 from which a closed interval of one projective line has been removed and let $\overline{R}$ denote the resulting subset of P^2. If x and y are any two distinct points of $\overline{R}$, and $\overline{\lambda}(x,y)$ is the intersection of $\overline{R}$ with $x \times y$, then $\overline{\lambda}(x,y)$ either coincides with $x \times y$ or has the property (20.8). Hamel's theorem implies that in spite of this there is no metric $d(x,y)$ in terms of which $\overline{R}$ is a projective-metric space.

The problem of determining all projective metrics is one of a famous set

---

[1] G. Hamel, "Über die Geometrien in denen die Geraden die Kürzesten sind". *Math. Ann.*, vol. 57 (1903), pp. 231-264. The above result is in § 4 of this paper.

[2] H. Busemann, "On Spaces in which Two Points Determine a Geodesic". *Trans. Am. Math. Soc.*, vol. 54, No. 2, pp. 171-184, 1943.

of problems which Hilbert (1862-1943) proposed in 1900. A first solution, in the sense that it provided a general, analytic procedure for constructing all projective metrics, was given by Hamel. Later, other answers were found which were geometrically more satisfying. We cannot treat the problem here since it requires the methods of the calculus of variations and advanced differential geometry. We will, however, discuss some particularly interesting examples of projective metrics.

21. Perpendiculars in Open Two–Spaces

In the remainder of this chapter, and in the next, only projective metric spaces of two dimensions will be considered. Three-dimensional spaces will be discussed in the last chapter of the book.

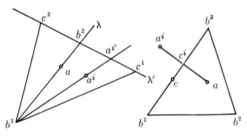

Fig. 34

Let R be an open, two-dimensional, projective-metric space. Then R may be considered as a subset of the affine plane A^2. Our previous remarks show that any two points, x and y in R, can be connected by exactly one segment and that this coincides with $S(x,y)$, the segment in the sense of affine geometry. *Hence R is a (proper or improper) convex subset of the affine plane.*

In addition, R consists entirely of interior points. For if z is any point of R, points x and y exist in R such that x, y and z are not collinear (condition 1)), hence $\lambda(x,z)$ and $\lambda(y,z)$ contain segments $S(x,x')$ and $S(y,y')$ respectively, where z is neither x' nor y'. Since $S(x,x')$ and $S(y,y')$ belong to R this is sufficient, as was noted prior to (17.6), to imply that z is interior to R. Therefore R is either the entire affine plane, or a strip between parallel lines (excluding the lines themselves), or else its boundary is a convex curve C_R. The space E^2 and the example given in the last section show that the first two types of domains can be given a projective metrization. It will appear that in the last case, also, a projective metric always exists.

We now show that convergence in the sense of R is equivalent to convergence in the sense of P^2.

(21.1)   *If R is a projective-metric space in a convex subset of the affine plane, and if a and the sequence $\{a^i\}$ lie in R, then $aa^i \to 0$ if and only if $\pi(a,a^i) \to 0$.*

a) Let $\pi(a,a^i) \to 0$ (Figure 34). It must be shown that $aa^i < \varepsilon$ ıor $i > N(\varepsilon)$. Choose any line λ through a, and on λ take b^1 and b^2 so that a is between them and $ab^i < \varepsilon/5$, $i = 1,2$. Let λ' be any line through b^2 which is distinct from λ and take c^1 and c^2 on λ' with b^2 between them and with $b^2c^i < \varepsilon/5$. Because $\pi(a^i,a) \to 0$, for all i greater than a suitable N the points a^i will be in $T(b^1,c^1,c^2)$. Let the line determined by b^1 and any such a^i intersect $S(c^1,c^2)$ in $a^{i'}$. Then for $i > N$ it follows that

$$a^ia < a^ia^{i'} + a^{i'}b^2 + b^2a \leqslant b^1a^{i'} + a^{i'}b^2 + b^2a$$
$$\leqslant a^{i'}b^2 + b^1b^2 + a^{i'}b^2 + b^2a \leqslant 2a^{i'}b^2 + 2ab^2 + ab^1 < 5\varepsilon/5 = \varepsilon.$$

b) Let $\bar{a}^ia \to 0$. Assume $\pi(\bar{a}^i,a)$ does not converge to zero. Then there is $a\delta > 0$ and a subsequence of $\{\bar{a}^i\}$, say $\{a^i\}$, such that for all i, $\pi(a^i,a) > \delta$. There is then a triangle $T(b^1,b^2,b^3)$ which has a as an interior point but is so small that all a^i lie outside it. Let c^i denote the point in which $S(a,a^i)$ intersects the boundary of $T(b^1,b^2,b^3)$. Since $\bar{a}^ia \to 0$, it follows that $a^ia \to 0$. This, with $aa^i > ac^i$, implies that $ac^i \to 0$. On the other hand, there is a subsequence $\{c^{i'}\}$ of $\{c^i\}$ and a point c on the boundary of T such that $\pi(c^{i'},c) \to 0$. Part a) of this proof shows then that $c^{i'}c \to 0$. Since also $c^{i'}a \to 0$ and $ac \leqslant ac^{i'} + cc^{i'}$ it follows that $ac = 0$. Because $a \neq c$ this contradicts axiom I for a metric. Hence $\pi(\bar{a}^i,a) \to 0$.

A two-dimensional, projective-metric space R is part of the projective plane, and theorems on R can and will be derived by using theorems of projective geometry or other properties of the larger space. Thus no attempt is made to think in terms of R alone. *For a deeper geometric understanding, however, it is important that the ability to think intrinsically in an unfamiliar space be developed since not all spaces can be conveniently imbedded in some familiar space.* One aim in this, and in Section 23, is to convey an idea of how such general investigations can be carried out. Although a projective metric may be kept in mind, most theorems of these sections are of such a nature that they hold for any subset R of the affine plane, whose metric is equivalent to $\pi(x,y)$ and is such that any two of its points are contained in exactly one straight line in the metric sense, that is, a line representable in the form (20.5). It is not necessary that this line be part of a projective line. The theorems which hold under these general conditions are marked with an asterisk.

Let $p(t)$ represent a metric line L in R, and let g be any point of R.[3] Then

$$(21.2)^* \qquad\qquad gp(t) \to \infty \qquad \text{as} \qquad t \to \pm\infty.$$

The triangle inequality gives

$$gp(t) + gp(0) \geqslant p(t)p(0) = |t|,$$

implying (21.2). The function $gp(t)$ is also continuous, since

$$gp(t_1) + p(t_1)p(t_2) \geqslant gp(t_2) \quad \text{and} \quad gp(t_2) + p(t_1)p(t_2) \geqslant gp(t_1)$$

imply $|gp(t_1) - gp(t_2)| \leqslant p(t_1)p(t_2) = |t_1 - t_2|$. This continuity of the (real) function $gp(t)$, together with (21.2), implies that $gp(t)$ reaches a minimum at some point $f = p(t_0)$. In any metric space R the point f is called a *foot* of the point x on the set μ, if f is in μ and $xy \geqslant xf$ for all y in μ. The result just obtained states that a point g has a foot on a given line L. A point g, not on L, may have several feet on L. However:

$(21.3)^*$ *If f is a foot of g on the set μ, and g is not in μ, then f is the unique foot in μ of any point z, distinct from g, on $S(g,f)$.*

For $z = f$ the theorem is obvious. For $z \neq f$ let f_1 be any point of μ distinct from f. Then $zf_1 > gf_1 - gz \geqslant gf - gz = zf$ shows that f is unique.

$(21.4)^*$ *If the points g and g', on different sides of the line L, have a common foot f on L, then g, f and g' are collinear. (Clearly f is unique.)*

For otherwise $S(g,g')$ would intersect L in a point y such that

$$gy + yg' = gg' < gf + fg',$$

which is impossible in view of $gf \leqslant gy$ and $fg' \leqslant yg'$ (Figure 35).

$(21.5)^*$ *If f, the foot of g on L, is unique when g belongs to the set μ, then f depends continuously on g (in μ).*

For any point and sequence, g and $\{g_i\}$, in μ it is to be shown that $gg_i \to 0$ implies $ff_i \to 0$, where f is the foot on L of g and f_i is the foot of g_i. Since $g_i \to g$, the distances gg_i are bounded (Figure 35). Then $gg_i + gf \geqslant g_if \geqslant g_if_i$ shows that the distances g_if_i are bounded. This, with $gg_i + g_if_i \geqslant gf_i$, implies a bound for $\{gf_i\}$, hence, from (21.2), the sequence $\{f_i\}$ must lie on a bounded interval of L. There exists, then, a convergent subsequence of $\{f_i\}$, say $f_i' \to f'$, and it suffices to show that for all such sequences $f' = f$. For an indirect proof, suppose $f' \neq f$. Let $\{g_i'\}$ indicate the subsequence of $\{g_i\}$, correspond-

---

[3]Capital Roman letters will be used to distinguish metric lines from the projective lines $x \times y$.

ing to $\}f_i'\{$, and define δ by means of $0 < gf' - gf = 4\delta$. From the triangle law,

$$g_i'f_i' \geqq gf' - g_i'g - f_i'f' = gf + 4\delta - g_i'g - f_i'f' \geqq g_i'f + 4\delta - 2g_i'g - f_i'f'.$$

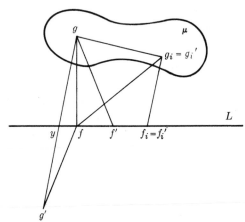

Fig. 35

Since $g_i' \to g$ and $f_i' \to f'$, it follows that $g_i'g < \delta$ and $f_i'f' < \delta$ for sufficiently large i, hence, for such i, $g_i'f_i' \geqq g_i'f + \delta$ and f_i' is not the foot on L of g_i'.

(21.6)*   *If the points of R have unique feet on L, and the point g, not on L, has f as foot on L, then the locus of points having f as foot on L is the line L′ through f and g.*

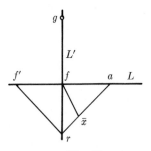

Fig. 36

PROOF: Let r be a point of L' on the side of L opposite to that containing g (Figure 36). Then f', the foot of r on L, is also f. For suppose $f' \neq f$. Take point a on L so that f lies between a and f'. As a variable point x traverses the segment $S(r,a)$ from r to a, then, by (21.5), its foot varies continuously from f' to a. Hence for some position $\bar{x}$ on $S(r,a)$, other than r, the foot of $\bar{x}$ is f, and since $\bar{x}, f$ and g are not collinear this contradicts (21.4). It follows that $f' = f$ and that every point of L', on the side of L opposite to g, has f as foot. By the same token, every point of L', on the side of L opposite to r, has f as foot so all points of L' have this property. A point off L' cannot have f as foot in virtue of (21.4).

If the line L' intersects L at f and every point of L' has f as foot on L, then L' is said to be perpendicular to L, and this is indicated by $L' \perp L$. It will be seen from examples that $L' \perp L$ need not imply $L \perp L'$. In these terms (21.6) can be reformulated:

(21.7)*   *Through every point g of R there exists a line perpendicular to a given line L if and only if every point of R, not on L, has a unique foot on L.*

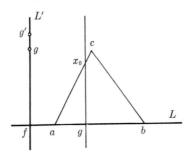

Fig. 37

For if every point of R has a unique foot on L, and g is any point not on L, then g and its foot f on L determine a line which, by (21.6), is perpendicular to L (Figure 37). If g is on L, choose a and b on L so that g is between a and b and take a point c not on L. As a variable point x traverses $S(a,c)$ from a to c and then traverses $S(c,b)$ from c to b, its foot moves continuously from a to b and hence through g. There is thus on $S(a,c)$ or $S(c,b)$ a point x_0 whose foot is g, and the line through x_0 and g is perpendicular to L.

Conversely, if through every point g, not on L, there passes a line L' perpendicular to L, then f, the intersection of L and L', is the unique foot of g on L. For take g' on L' so that g is between f and g'. By the definition of a perpendicular, f is a foot on L of g', hence, by (21.3), is the unique foot on L of g.

We show next that *the existence of perpendiculars in R is equivalent to the convexity of its circles*. The latter are defined in the usual way, the circle $K(p,\delta)$ being the locus of points x for which $px = \delta > 0$. If L is any line through p, then, because of (20.3), $K(p,\delta)$ intersects L in exactly two points so $K(p,\delta)$ is always a simple, closed curve. The interior, $U(p,\delta)$, of $K(p,\delta)$ is defined to be the set of points x for which $px < \delta$. The union of the interior and the circle itself, that is, the points x for which $px \leqslant \delta$, is called the disc $\overline{U}(p,\delta)$. Convex curve was defined in Section 17, and it follows from that definition that $K(p,\delta)$ is convex if, and only if, $U(p,\delta)$ and $\overline{U}(p,\delta)$ are convex sets.[4]

(21.8)*   *If all circles in R are convex, then they are all strictly convex.*

For assume $K(p,\delta)$ is convex, but not strictly convex (Figure 38). Then it contains a proper segment $S(a,b)$. Let c be a point of this segment, distinct from a and b, and choose g on $L(p,c)$ so that p lies between g and c. Then, $ga < gp + pa = gp + pc = gc$ and similarly $gb < gc$. If $\sigma = \max \{ ga, gb \}$, then $\overline{U}(g,\sigma)$ contains a and b but not c and hence not $S(a,b)$. Because

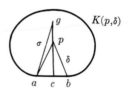

Fig. 38

$\overline{U}(g,\sigma)$ is not convex it follows that $K(g,\sigma)$ is not convex, contradicting the given conditions.

(21.9)*   *The circles of R are convex if and only if any given point has a unique foot on any given line.*

PROOF: Suppose the circles are convex. If a point g has two feet, f_1 and f_2, on a line L, then for all x on $S(f_1,f_2)$ this implies $gx > gf_1 = gf_2 = \sigma$. The convexity of $\overline{U}(g,\sigma)$, however, implies that $S(f_1,f_2)$ belongs to $\overline{U}(g,\sigma)$, hence that $gx \leqslant \sigma$. Combined with the last inequality this gives $gx = \sigma$ so $S(f_1,f_2)$ belongs to $K(g,\sigma)$, which is then weakly convex in contradiction to (21.8).

Conversely, suppose every point has a unique foot on every line. If a non-convex circle $K(g,\delta)$ exists, then $\overline{U}(g,\delta)$ contains a pair of points, $\bar{a}$ and $\bar{c}$, such that some point b on $S(\bar{a},\bar{c})$ does not belong to $\overline{U}(g,\delta)$ (Figure 39). The distance gx varies continuously with x on the line L through $\bar{a}$ and $\bar{c}$. Since $g\bar{a} \leqslant \delta$ and $gb > \delta$ it follows that a point a exists on $S(\bar{a},b)$

---

[4]For the more general theory, treated in the theorems with asterisks, convexity of $K(p,\delta)$ is defined to mean that $U(p,\delta)$ contains, with any pair a,b, the metric segment $S(a,b)$.

such that $ga = \delta$. Similarly a point c exists on $S(b,\bar{c})$ such that $gc = \delta$. By assumption g has a unique foot f on L, hence $f \neq a$ and $f \neq c$, while $gb > \delta$ shows $f \neq b$. It may be supposed that f lies on the same side of b as a. As a point x traverses $S(g,c)$ from g to c, its foot on L moves continuously from f to c and hence through b. Let x_0 be a point whose foot is b. Then $x_0 b < x_0 c$, but also $x_0 b \geqslant gb - gx_0 \geqslant gc - gx_0 = x_0 c$. From this contradiction it follows that $K(g,\circ)$ is convex.

If perpendiculars do exist, and a is any point on a line L, then a line L' through a is easily constructed such that L is perpendicular to L'. For if b is any point of L distinct from a, then L is perpendicular to any supporting line of $K(b,ab)$ at a. If L' is such a supporting line then for all x on L' distinct from a the relation $bx > ba$ shows that a is the foot of b on L', hence that $L \perp L'$. This corresponds to the fact in ordinary geometry that the radius of a circle, drawn to the contact point of a tangent, is perpendicular to the tangent.

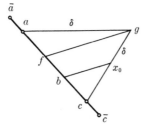

Fig. 39

We conclude this discussion with an example which shows that *circles, in general, are not convex*. Let R_7 be the entire affine plane and define distance, in terms of affine coordinates, by

$$d(x,y) = [(x_1 - y_1)^2 + (x_2 - y_2)^2]^{\frac{1}{2}} + |f(x_1) - f(y_1)| + |f(x_2) - f(y_2)|,$$

where f is the following function :

$$f(t) = 0 \;\; \text{for } t < 0$$
$$f(t) = 3t \;\; \text{for } 0 \leqslant t \leqslant 1$$
$$f(t) = 3 \;\; \text{for } t > 1.$$

This is easily shown to be a projective metric (exercise [21.9]). If g, a, b and c denote respectively the points $(0,0)$, $(0,2)$, $(1,1)$ and $(2,0)$, then $K(g,5)$ passes through a and c, since

$$d(g,a) = d(g,c) = 2 + |f(2) - f(0)| = 5.$$

But b, which lies on $S(a,c)$, is not in $\overline{U}(g,5)$, since

$$d(g,b) = \sqrt{2} + 2 |f(1) - f(0)| = \sqrt{2} + 6 > 5.$$

22. **Motions**

As mentioned briefly in Section 20, the notion of an isometry is one of the most important and basic concepts in distance geometries. *The mapping Φ of the subset μ of the metric space R on the subset μ' of the metric space R' is isometric, or is an isometry, if it preserves distance,* that is if $xy = x\Phi y\Phi$. An isometry is always a one-to-one mapping since $x \neq y$ implies $x\Phi y\Phi = xy \neq 0$, hence $x\Phi \neq y\Phi$. Therefore the inverse of an isometry is always defined and is clearly also an isometry. Moreover, if Φ is an isometry of μ on μ' and Φ' is an isometry of μ' on μ'', then $\Phi\Phi'$ is an isometry of μ on μ''. It follows at once:

(22.1)   *The isometric mappings of a metric set μ on itself form a group.*

An isometry of a metric space R on (all of)[5] itself is called a motion of R. Hence (22.1) implies :

(22.2)   *The motions of a metric space form a group.*

For the trivial space R_1, every one-to-one mapping of R_1 on itself is a motion. For the spaces R_2, R_3 and R_4, the translations, $x_i' = x_i + a_i$, are motions since $x_i' - y_i' = x_i - y_i$, and the reflection in $p, x_i' = 2p_i - x_i$ is also a motion because $y_i' - x_i' = x_i - y_i$. *Not every metric space, however, has a motion (other than the identity).* For instance, in E^2 with the metric $e(x,y)$ the hyperbola $x_1^2 - x_2^2 = 1$ is a metric space which has the reflection in either axis, and in the origin, as a motion. If the point $(2,0)$ is added to the space then the only motion (not the identity) is the reflection in the x_1-axis. Adding $(2,1)$ instead of $(2,0)$ yields a space with only the identity motion.

Obvious, but important, is the following remark.

(22.3)   *The motions of a metric space R which leave a set μ in R invariant, form a subgroup of the group of all motions of R. The same is true for the motions of R which leave μ pointwise invariant.*

Two sets μ and μ', in R and R' respectively, are called isometric or congruent if an isometry of μ on μ' exists. If R and R' are themselves isometric then all their geometric properties are the same. A familiar example, expressed in the present terms, will show the meaning of this statement. Consider the set of all pairs (r,θ), $r \geqslant 0$, $0 \leqslant \theta < 2\pi$, and $(0,\theta_1) = (0,\theta_2)$ for any two values θ_1 and θ_2. Let a metric be defined by

(22.4) $d[(r_1,\theta_1), (r_2,\theta_2)] = [r_1^2 + r_2^2 - 2r_1r_2 \cos(\theta_1 - \theta_2)]^{1/2}$.

---

[5]The mapping $t' = t + 1$ of the positive t axis with $|t_1 - t_2|$ as distance is isometric, but is not a motion since it maps $t > 0$ on the proper subset $t > 1$.

Let E^2 have its usual meaning, with distance given by

$$(22.5) \qquad d[(x_1,x_2), (x_1',x_2')] = [(x_1 - x_1')^2 + (x_2 - x_2')^2]^{1/2}.$$

Then the mapping of the (r,θ)-plane on the (x_1,x_2)-plane defined by

$$x_1 = r \cos \theta, \qquad x_2 = r \sin \theta,$$

is an isometry. In fact, if r and θ are considered as polar coordinates in the (x_1,x_2)-plane, then (22.4) is the distance (22.5) expressed in the new coordinates. Therefore it is trivial that the geometric properties of the two planes are the same.

Since they have the same geometry, two isometric spaces should, in some sense, have the same group of motions. The concept, from algebra, of isomorphism of groups makes this sense explicit. *A one-to-one mapping, $a \to a'$, of the elements of the group Γ on the group Γ' is called an isomorphism if it maps the product ab on the product $a'b'$.* If e and $\bar{e}$ are the unit elements of the groups Γ and Γ', then, by definition, $ae \to a'e'$. But also $ae = a \to a'$, hence $a'e' = a'$, so $e' = \bar{e}$, and the unit elements are corresponding. Similarly $aa^{-1} = e$, so $aa^{-1} \to a'(a^{-1})' = e'$, and the isomorphism maps the inverse of a onto the inverse of the image of a. From the point of view of algebraic structure, two isomorphic[6] groups are indistinguishable just as two congruent spaces are metrically identical.

As one might expect, if two metric spaces R and R' are isometric then their groups of motions are isomorphic. For suppose Ψ is the correspondence of the congruence, that is $x \to x' = x\Psi$ with $d(x,y) = d'(x',y')$. If $x \to x\Phi$ is a motion of R on itself, then Φ' defined by $x' = x\Psi \to (x\Phi)\Psi$ is a motion of R' on itself. First, since Ψ is one-to-one, Φ' is defined at every point of R', and is a motion of R' because

$$d'(x_1',x_2') = d(x_1,x_2) = d(x_1\Phi, x_2\Phi) = d'(x_1\Phi\Psi, x_2\Phi\Psi) = d'(x_1\Phi',x_2\Phi').$$

In the same way, every motion of R' induces a motion in R so $\Phi \to \Phi'$ is a one-to-one correspondence of the groups. It will be an isomorphism if $(\Phi_1\Phi_2) \to \Phi_1'\Phi_2'$, that is, if $[x(\Phi_1\Phi_2)]\Psi = x'(\Phi_1'\Phi_2')$ for all x in R. This is so, since $x'(\Phi_1'\Phi_2') = (x'\Phi_1')\Phi_2' = (x\Phi_1\Psi)\Phi_2' = [(x\Phi_1)\Phi_2]\Psi = [x(\Phi_1\Phi_2)]\Psi$. We have, thus, established:

(22.6) *If $\Gamma = \{ \Phi \}$ and $\Gamma' = \{ \Phi' \}$ denote respectively the group of motions in the metric spaces R and R', then a congruence of the spaces, $x \to x\Psi = x'$, induces the isomorphism $x\Phi \to x\Phi\Psi$ of Γ on Γ'.*

The structures of the groups of motions in congruent spaces are therefore the same.

Another connection between isometry and isomorphism is given by:

---

[6]That is, groups for which an isomorphism of one group on the other exists.

(22.7)   *If a motion Ψ of R exists which carries the set μ into the set μ' and if Γ and Γ' denote the subgroups of the motions of R which respectively leave μ and μ' invariant (or which leave them pointwise invariant), then Γ and Γ' are isomorphic.*

PROOF: Let Φ be any motion of R which leaves fixed each point of μ. Then $\Psi^{-1}\Phi\Psi$ leaves every point x' in μ' fixed. For Ψ^{-1} carries x' into a point x in μ which is left fixed by Φ and carried back to x' by Ψ. Therefore $\Psi^{-1}\Phi\Psi$ is an element of Γ'. Moreover, every element Φ' of Γ' may be expressed in this form. For the argument just given implies that $\Phi_1 = \Psi\Phi'\Psi^{-1}$ is an element of Γ and

$$\Psi^{-1}\Phi_1\Psi = \Psi^{-1}(\Psi\Phi'\Psi^{-1})\Psi = (\Psi^{-1}\Psi)\Phi'(\Psi^{-1}\Psi) = \Phi'.$$

The mapping $\Phi \to \Psi^{-1}\Phi\Psi = \Phi'$ of Γ on Γ' is one-to-one, since

$$\Psi^{-1}\Phi_1\Psi = \Psi^{-1}\Phi_2\Psi$$

implies $\Phi_1 = \Phi_2$. Finally, the product $\Phi_1\Phi_2$ goes into

$$\Psi^{-1}(\Phi_1\Phi_2)\Psi = (\Psi^{-1}\Phi_1\Psi)(\Psi^{-1}\Phi_2\Psi) = \Phi_1'\Phi_2'.$$

The non-parenthetical assertion in (22.7) is established in the same way. If Φ carries μ into itself, then $\Psi^{-1}\Phi\Psi$ carries μ' into itself. The proof that $\Phi \to \Psi^{-1}\Phi\Psi$ is an isomorphism of Γ on Γ' goes through as before.

Since a motion leaves distance invariant, it preserves any concept expressible solely in terms of distance. As an instance of this:

(22.8)   *A motion Φ of a metric space carries a segment $S(x,y)$ into a segment $S(x\Phi,y\Phi)$; a metric line goes into a metric line, and a great circle into a great circle.*

For if $p(t)$, $\alpha \leqslant t \leqslant \beta$, represents the segment $S(x,y)$, then $p'(t) = p(t)\Phi$ is again a segment, since $p'(t_1)p'(t_2) = p(t_1)\Phi p(t_2)\Phi = p(t_1)p(t_2) = |t_1 - t_2|$, $\alpha \leqslant t \leqslant \beta$, etc. Similarly Φ transforms a foot of g on the line L into the foot of $g\Phi$ on $L\Phi$.

Now let R be a two-dimensional projective-metric space. Since the metric line L through x and y in R lies on $x \times y$, a motion Φ of R will transform L into L' through $x\Phi$ and $y\Phi$. Hence L' is a subset of $(x\Phi) \times (y\Phi)$. If the space R is closed, then L coincides with $x \times y$ and (9.1) yields:

(22.9)   *A motion of a closed, two-dimensional, projective-metric space is a collineation.*

If R is open, a motion of R cannot, of course, be a collineation. However:

(22.10)   *A motion of an open, two-dimensional, projective-metric space R is induced by one and only one collineation.*

Since Φ preserves collinearity it carries a quadrangular set a,b,c,d in R into a quadrangular set a',b',c',d' in R. There is one, and only one, collineation Ψ of the projective plane into itself determined also by these quadrangular sets. That $x\Phi = x\Psi$ for all x in R follows from the same kind of argument as in (9.1) making use of the fact that metric lines are contained in projective lines, and that both Φ and Ψ are one-to-one transformations. We note, incidentally, that *if R is not the entire affine plane then Ψ need not be an affinity.*

23. Motions of Open Two-Spaces

Although the developments corresponding to open and closed spaces present many analogies, they are different enough to make a separate treatment preferable. In the present section, therefore, the space is restricted to being open and two-dimensional. Again, most of the results do not depend on the assumption that metric lines lie on projective lines. Those, with this greater generality, are marked, as before, with an asterisk.

An obvious fact that is constantly useful is:

(23.1)*   *A motion which leaves two points of a straight line fixed leaves every point of the line fixed.*

As a consequence:

(23.2)*   *A motion which leaves three, non-collinear points a, b and c fixed is the identity.*

For if d is any point of $S(a,c)$, distinct from a and c, then every point z of R lies on a line L through d and x where x belongs to $S(a,b)$ or $S(b,c)$. Since all points on the sides of $T(a,b,c)$ are fixed, both d and x are fixed, hence all points of L, including z, are invariant.

(23.3)*   *Three non-collinear points and their images determine a motion uniquely.*

For if the motions Φ and Ψ both carry the non-collinear triple a,b,c into a',b',c', the mapping $\Phi\Psi^{-1}$, which leaves a, b and c fixed is the identity, hence $\Phi = \Psi$.

A consequence of (23.1) and (23.2) is that a motion Φ has either no fixed points, one fixed point, a line of fixed points, or is the identity. When it is involutory Φ always has fixed points. For if $a' = a\Phi$ differs from a, then $a'\Phi = a\Phi^2 = a$ implies that the center of $S(a,a')$ is fixed. It follows that there is just one fixed point x or else a line of fixed points, L. In the first case Φ is called *the reflection in x* and in the second *the reflec-*

tion in L. The use of the definite article is justified by the fact that *a reflection in a point or in a line is unique.* For reflection in a point x this is obvious since the image a' of any point a is uniquely determined by the condition that x be the center of a and a'. That there are not two reflections in a line follows from:

(23.4)* *If a reflection Φ in L exists then every point a not on L has exactly one foot f on L and f is the center of a and $a' = a\Phi$.*

For the center of a and a' remains fixed under Φ and hence is on L. If this center is denoted by f, and x is any point of L distinct from f, then

$$2ax = ax + x(a\Phi) > aa' = 2af$$

implies $ax > af$, hence f is the foot of a.

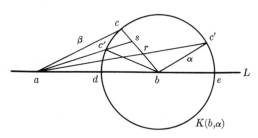

Fig. 40

In a general metric space let $B(a,a')$, where $a \neq a'$, denote the locus of points x such that $ax = a'x$. Clearly (from (23.4)):

(23.5)* *If the reflection Φ in a line L exists, then $L = B(a,a\Phi)$ for every point a not on L.*

From (21.7) and (23.4) we obtain:

(23.6)* *If the reflection, Φ, in L exists, then perpendiculars to L exist and Φ maps each perpendicular to L on itself.*

In the further discussion we need the following lemma, which is interesting in itself.

(23.7)* *Given two distinct points a,b and two numbers α,β satisfying $\alpha + \beta > ab$ and $|\alpha - \beta| < ab$, then there exists on each side of the line L through a and b one and only one point c such that $ac = \beta$ and $bc = \alpha$.*

PROOF: Let $ab = \gamma$. The circle $K(b,\alpha)$ intersects L in two points d and e so named that $ad = |\gamma - \alpha| < \beta$ and $ae = \alpha + \gamma > \beta$ (Figure 40). As a point x travels from d to e, along either of the semi-circles on opposite sides of L, the distance ax changes continuously from $ad < \beta$ to $ae > \beta$ and so for some point c assumes the value β, that is, $ac = \beta$. There cannot be two points c and c' on the same side of L such that $ac = ac' = \beta$ and $bc = bc' = \alpha$. For with this property, c' clearly cannot lie on either the line through a and c or that through b and c. If it lies in the interior of the triangle $T(a,b,c)$, let s denote the intersection of the line through a and c' with that through b and c. Then

$$\beta + \alpha = ac' + c'b < ac' + c's + sb = as + sb < ac + cs + sb = \alpha + \beta.$$

By a symmetrical argument, c cannot be on or interior to the triangle $T(a,b,c')$.

The only other possibility is that two sides of the triangles, for instance $S(b,c)$ and $S(a,c')$, intersect in a point r. As the figure shows, this implies $\alpha - cr + rc' = br + rc' > bc' = \alpha$, or $rc' > rc$. But, in contradiction, $\beta - rc' + rc = ar + rc > ac = \beta$, or $rc > rc'$. Hence c, on a given side of L, is unique.

A first consequence of this lemma is:

(23.8)*   *A motion $\Phi \neq 1$ which leaves every point of a line L fixed is the reflection in L.*

For if a and b are distinct points of L, and c is any point not on L, then, from (23.2), $c' = c\Phi$ is not c since Φ is not the identity. Because $c'a = ca$ and $c'b = cb$ and, by (23.7), there is only one such point distinct from c', it follows that $c'\Phi = c^2\Phi = c$, hence Φ^2 is the identity.

(23.9)*   *If the circles are convex and the reflection in L exists, then $L' \perp L$ implies $L \perp L'$.*

For if f is the intersection of L and L', let x be any point of L distinct from f and let g be its foot on L'. By (23.6), Φ carries g into g' on L' with $xg = xg'$. The uniqueness of g then implies $g' = g$, hence $g = f$.

Combining (9.7), (23.8) and (23.6) yields:

(23.10)   *If R is an open, two-dimensional, projective-metric space, the reflection of R in the point x is induced by a harmonic homology whose axis ξ does not intersect R. The reflection of R in the line L is induced by a harmonic homology whose axis carries L and whose center x is not in R. The lines through x carry the perpendiculars to L.*

If a motion of R carries a, b and c into a', b' and c', these triples must, of course, be congruent, that is, $ab = a'b'$, $bc = b'c'$ and $ac = a'c'$. When a,b,c are non-collinear, the motion, if it exists, is unique. The space R will therefore have its greatest possible degree of mobility when for any pair of non-collinear, congruent triples a motion exists carrying one into the other.

(23.11)*   *There will be a motion, for every pair of congruent triples, which carries one into the other if, and only if, the reflections in all lines exist. If all reflections do exist, every motion is expressible as the product of three, or fewer, reflections.*[7]

PROOF: Suppose motions exist for pairs of congruent triples, and let L be any line of R. Take a and b as distinct points of L and c any point not on L. If $\alpha = cb$, $\beta = ca$, and $\gamma = ab$, then $\alpha + \beta > \gamma$ and $|\alpha - \beta| < \gamma$. From (23.7) there is then a unique point c', on the side of L opposite c,

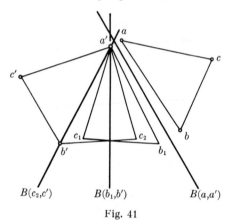

Fig. 41

such that $ac' = \beta$ and $bc' = \alpha$. Since the triples a,b,c and a,b,c' are congruent there exists, by assumption, a motion Φ for which $a\Phi = a, b\Phi = b$ and $c\Phi = c'$. Because $c' \neq c$, Φ is not the identity. But it leaves all points of L fixed because it leaves a and b fixed. By (23.8)*, then, Φ is the reflection about L.

To prove the converse, we first observe that for a distinct pair, a,a', lemma (23.7) permits the construction of infinitely many points of the equidistant locus $B(a,a')$. If p and q are any two points of $B(a,a')$ and if the reflection in L', through p and q, exists, then, from (23.5), $L' = B(a,a')$. Therefore if all line-reflections exist, any point a can be moved to any point a' by the reflection in $B(a,a')$. If, now, a,b,c and a',b',c' are given congruent triples, define Φ_1 to be the identity if $a = a'$, and the reflection in $B(a,a')$ if $a \neq a'$ (Figure 41). Since Φ_1 is a motion, the images

---

[7]As the asterisk indicates, the proof of this theorem does not use the fact that the metric is projective. On the other hand, the assumptions are so strong, that any space satisfying them is projective-metric, namely is either Euclidean or hyperbolic (see Section 29). The proof of this fact, however, does not fall within the framework of this book.

of a, b and c form a triple a_1, b_1 and c_1 also congruent to a',b',c', and $a_1 = a'$. If $b_1 = b'$, let Φ_2 be the identity, otherwise take it as the reflection in $B(b_1,b')$. Under Φ_2, the point b_1 goes to $b_2 = b'$. Because $a_1 b_1 = a' b_1 = a' b'$, the point $a_1 = a'$ lies on $B(b_1,b')$, hence a_1 goes into itself, and $a_2 = a_1 = a'$. The triples a_1,b_1,c_1 and a_2,b_2,c_2 are congruent because Φ_2 is a motion. Finally, if $c_2 = c'$ let Φ_3 be the identity, otherwise the reflection in $B(c_2,c')$. Repeating the argument shows the image triple a_3,b_3,c_3 is the triple a',b',c'.

This theorem shows that any motion can be generated from reflections in lines if these exist. The section on Minkowskian geometry will demonstrate that they need not exist, even if the reflection in every point exists.

As a consequence of the last theorem:

(23.12)*   *If reflections in every line of R exist, and if Ψ is an isometric mapping of the set μ in R on the set μ' in R, then a motion Φ of R exists which coincides with Ψ on μ, that is $x\Psi = x\Phi$ for x in μ.*

For if a,b,c is a non-collinear triple in μ, then $a\Psi$, $b\Psi$ and $c\Psi$ is a congruent, non-collinear triple in μ'. From (23.11) there is a motion Φ carrying the first triple into the second, and, from (23.3), Φ is unique. For x in μ, and on any side of the triangle $T(a,b,c)$, it is clear that $x\Psi = x\Phi$. When x is not on a side of the triangle, then, from lemma (23.7), if $x\Phi$ were not $x\Psi$, it would be the image of $x\Psi$ under reflection in the line through $a\Psi$ and $b\Psi$, and by the same token it would also be the image of $x\Psi$ under reflection about the line through $a\Psi$ and $c\Psi$, which is impossible since these images differ. Hence $x\Phi = x\Psi$ for all x in μ. Should the points of μ all be collinear, a similar argument shows the existence of Φ, but not as a unique motion.

A motion Ψ which has only one fixed point p is called a *rotation about p*. This includes the reflection in p.

(23.13)*   *A rotation Ψ is determined by its center p and by one distinct pair a and aΨ.*

We distinguish two cases, according as p is, or is not, collinear with a and $a\Psi = a'$. When the three are collinear, $p\Psi = p$ and $a\Psi = a'$ imply $a = a'\Psi$. Thus Ψ is involutory on the line L through a and p, and all points of L are fixed under Ψ^2. For any point g, not on L, the distance relations $ag = a(g\Psi^2)$ and $a'g = a'(g\Psi^2)$ imply, by (23.7), that $g\Psi^2$ is either g or its image under reflection about L. But Ψ leaves L invariant hence either maintains the sides of L or else interchanges them. In either case, Ψ^2 maintains the sides, hence $g\Psi^2 = g$, and Ψ^2 is the identity. Thus Ψ is involutory and, hence, is the reflection in p. For the case where p, a and a'

are not collinear, let b indicate the center of $S(a,a')$ (Figure 42). From (23.7) there is only one point c, distinct from b, such that $pc = pb$ and $a'c = a'b$. But since Ψ is a motion, with $p' = p$ and $a\Psi = a'$, it follows, with $b' = b\Psi$, that $pb' = pb$ and $a'b' = ab = a'b$. Therefore b', which cannot be b since only p is fixed, must be c. Thus p, a and a' determine b and b' uniquely. The non-collinear triples, p,a,b and p,a',b' then determine Ψ uniquely.

The inverse of a rotation is clearly also a rotation with the same center. Moreover:

(23.14)*   *The product of two rotations, Ψ_1, Ψ_2, with the same center p is either the identity or else a rotation about p.*

The point p is fixed under $\Psi = \Psi_1\Psi_2$, so Ψ is either a rotation about p or else there is a second point g, fixed under Ψ. In the latter event,

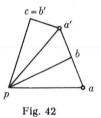

Fig. 42

$g\Psi_1 = g\Psi_2^{-1}$. By (23.13), then, $\Psi_1 = \Psi_2^{-1}$, since they are determined by the same pair, g and $g\Psi_1$. Hence $\Psi = \Psi_1\Psi_2 = \Psi_1\Psi_1^{-1} = 1$.

(23.15)*   *If Ψ_1 and Ψ_2 are respectively the reflections in the lines L_1 and L_2, which are distinct and which intersect at the point p, then $\Psi_1\Psi_2$ is a rotation about p.*

The mapping $\Psi = \Psi_1\Psi_2$ is a motion and leaves p fixed. For Ψ to leave a second point g fixed, would imply $g\Psi_1 = g\Psi_2^{-1} = g\Psi_2$. The center of $S(g,g\Psi_1)$ would then be on both L_1 and L_2, hence would be p, which is impossible. Hence Ψ leaves only p fixed and is a rotation.

(23.16)*   *If the reflections in all lines of a pencil through p exist, and if a and a' are a pair of points distinct from p such that $pa = pa'$, then a rotation about p exists which carries a into a'. The rotation is expressible as the product of the reflection in two lines of the pencil where one of the two lines is arbitrary.*

For if Ψ_1 is the reflection in the arbitrary line L_1 of the pencil, let b be the center of $S(a\Psi_1,a')$. In the general case, then, the reflection Ψ_2 in the line through p and b takes $a\Psi_1$ into a', so $a\Psi_1\Psi_2 = a'$. That $\Psi_1\Psi_2$ is a rotation is a consequence of (23.15).

24. Minkowskian Geometry

Having discussed some general aspects of projective metrics, we will now consider some specific cases. In Section 15 it was shown that the affine ratio $A(x,y,z)$ of three collinear points with affine coordinates satisfies the relation

$$|A(x,y,z)| = d_4(x,z)/d_4(x,y).$$

It is easily seen that d_4 could be replaced in this equality by d_2 or d_3. We now pose the problem of finding all projective metrics, $m(x,y)$, defined in the whole affine plane, and such that

(24.1) $$|A(x,y,z)| = m(x,z)/m(x,y).$$

These metrics are called *Minkowskian,* after their discoverer, H. Minkowski (1864-1909). The Euclidean metric, d_4, is Minkowskian, but, as was previously observed, neither d_2 nor d_3 is a projective metric.

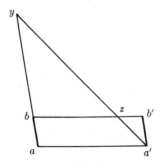

Fig. 43

Unless otherwise stated, all lengths referred to in this section will be understood to be Minkowskian. We show first that :

(24.2)   *Opposite sides of a parallelogram have equal lengtns.*

PROOF: As in Figure 43, let a,b and a',b' define the opposite sides of a parallelogram. Choose y on the extension of $S(a,b)$ through b, and let z indicate the intersection of $S(b,b')$ and $S(y,a')$. Then (24.1) and (15.15) imply that

$$m(a,b)/m(a,y) = |A(a,y,b)| = |A(a',y,z)| = m(a',z)/m(a',y),$$

or,

(24.3) $$m(a,y)/m(a',y) = m(a,b)/m(a',z).$$

Applied to a,a',y the triangle law justifies

$$m(a,y) - m(a,a') \leqslant m(a',y) \leqslant m(a,y) + m(a,a').$$

Dividing through by $m(a,y)$, and taking the limit as $m(a,y)$ increases, yields:

$$(24.4) \qquad m(a',y)/m(a,y) \to 1 \text{ as } m(a,y) \to \infty .$$

A similar application to the triangle a',z,b' shows that

$$m(a',b') - m(z,b') \leqslant m(a',z) \leqslant m(a',b') + m(z,b').$$

If this is divided through by $m(a',b')$, then since (21.1) implies that $m(z,b') \to 0$ as $m(a,'y) \to \infty$, it follows that $m(a',z)/m(a',b') \to 1$ as $m(a,y) \to \infty$. This relation together with (24.3) and (24.4) gives $m(a,b) = m(a',b')$ and establishes (24.2).

The translation $x_i' = x_i + c_i$, $i = 1,2$, carries two points a,b into points a',b' such that a,b and a',b' define the opposite sides of a, possibly degenerate, parallelogram.

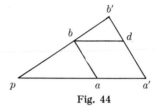

Fig. 44

This is also true of $x_i' = - x_i + 2y_i$, $i = 1,2$, the reflection in the point (y_1,y_2). In virtue of (24.2), this shows:

(24.5) *The translations and central reflections are motions of any Minkowski space.*

As another consequence of (24.2):

(24.6) *If a lies on $S(p,a')$, b on $S(p,b')$, and if $a \times b$ is parallel to $a' \times b'$ then $m(p,a)/m(p,a') = m(p,b)/m(p,b') = m(a,b)/m(a',b')$.*

For let the line through b parallel to $p \times a'$ cut $a' \times b'$ at d (Figure 44). Because of (15.15), $A(p,a',a) = A(p,b',b) = A(a',b',d)$. Substituting in these equalities from (24.1), and replacing $m(a',d)$ by $m(a,b)$ gives the stated result.

With a,a',b,b' and p situated as in (24.6), the similitude, with p as center and defined by

$$(24.7) \quad x_i' = \lambda x_i + p_i(1 - \lambda), \qquad i = 1,2, \qquad \lambda = m(p,a')/m(p,a),$$

carries a into a' and b into b'. For if a'' indicates the image of a, then $a'' = (1 - \lambda)p + \lambda a$. Hence, from (15.12), $A(p,a,a'') = \lambda = m(p,a')/m(p,a)$, and a'' coincides with a'. Similarly the image of b coincides with b'. For any pair of points a,b with images a',b', (24.6) shows then that

$m(a',b')/m(a,b) = \lambda$. Admitting negative values for λ, we may state this in a more general form.

(24.8)   *Under a similitude, $x'_i = \lambda x_i + a_i$, $i = 1,2$, $\lambda \neq 0$, all Min-kowskian distances are multiplied by $|\lambda|$.*

We consider Minkowskian circles next, using as before the notation $K(p,\delta)$ for the circle with center p and radius δ, with $\delta > 0$ understood. We prove first:

(24.9)   *The circles are strictly convex.*

For a direct proof,[8] let $K(g,\delta)$ be an arbitrary circle and let a and b be any two distinct points such that $m(a,g) \leqslant \delta$ and $m(b,g) \leqslant \delta$. The strict convexity of $K(g,\delta)$ will follow if every point c on $S^*(a,b)$ satisfies $m(g,c) < \delta$. This is obviously true when g, a and b are collinear and $a \neq b$. Assume,

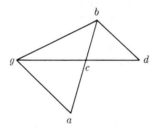

Fig. 45

therefore, that they are not collinear, and suppose that $m(g,b) \geqslant m(g,a)$ (Figure 45). Let d be the intersection point of $g \times c$ and the line through b parallel to $g \times a$. Then (24.2) and (24.8) imply that

$$m(b,d)/m(g,a) = m(c,d)/m(g,c),$$

and this with $m(g,b) \geqslant m(g,a)$ shows that

$$m(b,d)/m(g,b) \leqslant m(c,d)/m(g,c).$$

Adding 1 to both sides of this last relation gives

$$m(g,d)/m(g,b) < [m(g,b) + m(b,d)]/m(g,b) \leqslant [m(g,c) + m(c,d)]/m(g,c)$$
$$= m(g,d)/m(g,c), \quad \text{or} \quad m(g,c) < m(g,b) \leqslant \delta, \qquad \text{q.e.d.}$$

From (24.5) it follows that a translation or a central reflection which carries g into g' takes $K(g,\delta)$ into $K(g',\delta)$. Similarly, the similitude of (24.8) transforms $K(g,\delta)$ to $K(g',|\lambda|\cdot\delta)$. Two figures, such that one is

---

[8]Because of (21.8) it would suffice to show that all circles are convex.

the image of the other in a similitude, are called *homothetic*. We may therefore say:

(24.10)   *Any two Minkowskian circles are homothetic, so that if the "unit" circle $K(z,1)$, where $z = (0,0)$, is known, then all circles are known and $m(a,b)$ is determined for any pair a,b.*

An actual method for calculating $m(a,b)$ is the following. The known translation $x_i' = x_i + a_i$, $i = 1,2$, carries $K(z,1)$ into a curve cutting the ray from a through b in a point c such that $m(a,b) = A(a,c,b)$.

The previous construction can also be used to define a metric. That is, suppose K is a closed, strictly convex curve in the affine plane, and that K has an affine center z. This means that K goes into itself under a reflection in z. For any pair of points, a,b, we locate c by the given construction, with K replacing $K(z,1)$, and define a function $d(a,b)$ by $d(a,b) = |A(a,c,b)|$. Denote by $K(g,\delta)$ the locus of points x satisfying $d(g,x) = \delta$, where $\delta > 0$. Then the definition of $d(a,b)$ implies that the similitude $x_i' = \delta x_i + g_i$, $i = 1,2$, carries $K(z,1)$, into $K(g,\delta)$, hence that any two loci $K(g,\delta_1)$ and $K(g,\delta_2)$ are homothetic. The fact that K has z as an affine center shows also that $d(a,b)$ is invariant under reflections in a point. In particular, it is invariant under the reflection in $(a + b)/2$, so that $d(a,b) = d(b,a)$. These remarks indicate that, under the similitude of (24.8), $d(a,b)$ is multiplied by $|\lambda|$ and so is invariant for $|\lambda| = 1$. Two given points a and b lie on the metric straight line

$$p(t) = [(d(a,b) - t)a + tb]/d(a,b)$$

which coincides with $a \times b$. The triangle inequality holds trivially for any three points on $a \times b$. In order to show that $d(x,y)$ is actually a projective metric it suffices to verify that $d(g,b) + d(b,d) > d(g,d)$ for any point b not on $g \times d$. This may be done by the same method as that in the proof of (24.9). On the ray from g (Figure 45) which is parallel to the ray from b through d, take a so that $\delta = d(g,b) = d(g,a)$, and put $c = (g \times d) \times (a \times b)$. Then, because of the behavior of $d(x,y)$ under similitudes and reflections in a point,

(24.11) $d(b,d)/d(g,b) = d(b,d)/d(g,a) = d(c,d)/d(g,c)$.

Since $K(g,\delta)$ is homothetic to K it is strictly convex, hence c lies inside $K(g,\delta)$, (the points a and b lie on $K(g,\delta)$), so $d(g,c) < \delta = d(g,b)$. From (24.11) it then follows that $d(c,d) < d(b,d)$. Therefore

$$d(g,d) = d(g,c) + d(c,d) < d(g,b) + d(b,d).$$

The function $d(x,y)$ is thus a Minkowskian distance. Eliminating the special role of the origin, we can formulate the following important result.

(24.12)

In the affine plane A^2 let K be a closed, strictly convex curve with affine center c. For any two points, a,b, in A^2 define $m(a,b) = A(a,y,b)$, where y is the intersection point of the ray from a through b with the image of K under the translation $x_i' = x_i + (a_i - c_i)$, $i = 1,2$. Then $m(a,b)$ is a Minkowskian metric for which $K = K(c,1)$.[9]

Because the Minkowskian circles are strictly convex, the results of Section 21 on perpendiculars can be applied. For instance, perpendiculars to the same line L cannot intersect, hence they are parallel. If L' through g and f is perpendicular to L at f, then $K(g,gf) = K(g,\delta)$ has L as a supporting line at f (Figure 46). Because circles are homothetic, L' is also perpendicular to any line parallel to L. If $K(g,\delta)$ has a tangent at f, then L must be the tangent and there is no other line through f to which L' is

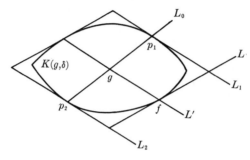

Fig. 46

perpendicular. When a second supporting line L also exists at f, at most one of the lines L and L can be perpendicular to L', showing that *in general perpendicularity is not symmetric*. Actually the symmetry will not hold, in general, even when the circles are differentiable. To construct the line perpendicular to L' at g, draw L_1 and L_2, the two supporting lines of $K(g,\delta)$ which are parallel to L'. The line L_0, which connects p_1 and p_2, the respective contact points of L_1 and L_2 with $K(g,\delta)$, passes through g and is perpendicular to L_1 and to L_2 and hence to L'. It is clear that, in general, L_0 will not be parallel to L, so L will not be perpendicular to L'. For obviously the shape of $K(g,\delta)$ can be altered in the neighborhood of f so as to change the direction of L without affecting L_1 and L_2 at all. The inference, then, is that:

---

[9]The preceding discussion also contains the result: If the whole affine plane is so metrized that (24.1) holds and the locus $m(c,x) = 1$ is strictly convex for some c, then the metric is Minkowskian.

(24.13)
In Minkowskian geometry perpendicularity is always symmetric if and only if any parallelogram circumscribed about a circle is such that when the midpoints of two opposite sides are contact points, then the midpoints of the other two sides are also contact points.

Because symmetry of perpendicularity assures unique supporting lines to circles, it implies that circles are differentiable. Theorem (16.9) shows that (24.13) holds if the circles are ellipses. However, these are not the only curves which yield symmetric perpendicularity.[10]

We turn next to the *definition of area in Minkowskian geometry*. Because perpendicularity is not as simple as in Euclidean geometry, it is natural to use the unit circle (instead of some analogue to the square) for the normalization of area. We choose affine coordinates, x_1 and x_2, so that

$$\iint_{\overline{U}(z,1)} dx_1 dx_2 = \pi,$$ and define the area $A(D)$ of a general domain D by

$$A(D) = \iint_D dx_1 dx_2.$$

A reasonable area in any metric space must be invariant under all motions of the space. The area just defined has this property. For a motion Φ of the Minkowski plane, since it maps A^2 on itself and carries lines into lines, is an affinity, and so has the form:

$$\Phi : x_i' = a_{i1}x_1 + a_{i2}x_2 + a_i, \quad i = 1,2, \quad \Delta = \begin{vmatrix} a_{11} & a_{12} \\ a_{21} & a_{22} \end{vmatrix} \neq 0.$$

Therefore Φ can be expressed as the product of

$$\Phi_1 : x_i' = a_{i1}x_1 + a_{i2}x_2, \quad i = 1,2,$$
and
$$\Phi_2 : x_i'' = x_i' + a_i, \quad i = 1,2.$$

Since Φ_2 is a translation, it is a motion, so $\Phi_1 = \Phi\Phi_2^{-1}$ is also a motion. Because Φ_1 leaves z fixed it carries $\overline{U}(z,1)$ into itself and is therefore an equiaffinity. Because Φ_2 is an equiaffinity, Φ is too, which shows:

(24.14)
The motions of a Minkowski plane form a (proper) subgroup of the equiaffine group. Hence Minkowski area is invariant under motions.

That the subgroup is proper, that is that not every equiaffinity is a motion, is obvious from the example $x_1' = 2x_i$, $x_2' = (1/2)x_2$, which sends the segment $S(z,(0,1))$ into $S(z,(0,1/2))$ and therefore does not preserve distance.

The groups of motions belonging to two different Minkowski spaces

---

[10]See J. Radon, "Über eine besondere Art ebener konvexer Kurven". *Ber. Verh. Sächs. Akad.*, Leipzig, vol. 68 (1916), pp. 123-128.

are in general not isomorphic. This follows from the discussion in the
next section where it is shown that reflections in lines exist under some
Minkowski metrizations and not under others.

We mention the following form of Loewner's theorem (18.7):

(24.15)   *Among the ellipses with center z, and containing $\bar{U}(z,1)$, there
is one and only one ellipse E of minimal area. All Minkowski
motions which leave z fixed carry E into itself.*

25. Reflections in Minkowskian Geometry.
Euclidean Geometry.

A Minkowskian geometry, in general, admits only the translations and
central reflections as motions. In Section 23 we saw that the existence of
reflections about lines was the decisive factor for the mobility properties
of the space. For a Minkowskian geometry, we ask first what the existence
of a reflection Φ in one line L implies.

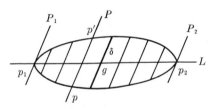

Fig. 47

Let L' be any line parallel to L. If a and a' are arbitrary points on L
and L' respectively, then the translation Υ which carries a into a', takes
L into L'. Then the motion $\Phi' = \Upsilon^{-1}\Phi\Upsilon$ is a reflection about L'. First, it is
clearly not the identity, and secondly it leaves the points of L' fixed
since Υ^{-1} takes L' into L, which is then fixed under Φ and carried back to
L' by Υ. Hence,

(25.1)   *In a Minkowskian geometry, if the reflection in a line L exists, then
the reflections in all lines parallel to L exist.*

But, in general, the reflection in a line not parallel to L will not exist. If
g is any point of L, then since Φ leaves g fixed it carries any circle $K(g,\delta)$
into itself. It also, by (23.6), maps any perpendicular to L onto itself.
Hence, if a line P (Figure 47), perpendicular to L, cuts $K(g,\delta)$ at p on one
side of L, it must have a second intersection at $p' = \Phi p$. The midpoint
of $S(p,p')$ therefore lies on L. In other words, if a reflection about L exists,

then every circle with center on L has a family of parallel chords, the locus of whose affine (or Minkowskian) centers is L. (Note: Two of the family of perpendiculars, P_1 and P_2, are supporting lines of $K(g,\delta)$. Clearly, L must pass through their contact points p_1, p_2, hence $L \perp P_i$, $i = 1,2$. Compare (23.9)*.)

Conversely, suppose that a circle $K(g,\delta)$ has a family F of parallel chords whose centers lie on a line L. Because circles are homothetic, in any concentric circle $K(g,\sigma)$ the chords parallel to those in F will also be bisected by L. Choose L as the x_1 – axis, and the element of F through g as the x_2-axis. Then the affinity $\Phi : x_1' = x_1$, $x_2' = -x_2$ maps any circle with center g onto itself. Since Φ leaves every point of L fixed, and is not the identity, it will be the reflection in L if it is a motion. This will clearly be so if Φ carries every circle into another with equal radius. To obtain this property of Φ let Υ denote the translation which carries x into a point g of L, where x is the center of any circle $K(x,\sigma)$. Because Υ is a motion $K(x,\sigma)\Upsilon = K(g,\sigma)$. This with the invariance under Φ of circles centered on L gives $K(x,\sigma)\Upsilon\Phi = K(g,\sigma)\Phi = K(g,\sigma)$. Since the translations form an invariant subgroup of the group of all affinities (see (15.9)), there is a translation Υ_1 such that $\Upsilon\Phi = \Phi\Upsilon_1$, or $\Phi = \Upsilon\Phi\Upsilon_1^{-1}$. Hence

$$K(x,\sigma)\Phi = K(x,\sigma)\Upsilon\Phi\Upsilon_1^{-1} = K(g,\sigma)\Upsilon_1^{-1}.$$

But Υ_1^{-1} is a motion, being a translation, so $K(g,\sigma)\Upsilon^{-1} = K(x,\sigma)\Phi$ is again a circle of radius σ. This establishes :

(25.2) *In Minkowskian geometry the reflection in a line L exists if and only if in every circle with center on L there is a family of parallel chords bisected by L. (If one circle with center on L has this property then all do.)*

The last theorem, together with (25.1) and (16.6), shows that a reflection exists in every line if $K(g,\delta)$ is an ellipse. This condition is also necessary, that is:

(25.3) *A Minkowskian geometry admits a reflection in every line if and only if the circles are ellipses.*

First, if one circle is an ellipse, then all are, since circles are homothetic. It will suffice, therefore, to prove that if all reflections in lines exist, then the circle $K(z,1)$ is the ellipse E with center z (established in (24.15)), which contains $\overline{U}(z,1)$ and has minimal area. Common to both E and $K(z,1)$ there is at least one point a, otherwise E could be shrunk and still contain $K(z,1)$ (Figure 48). Because E and $K(z,1)$ both have z as center, the point a' of $K(z,1)$ diametrically opposite to a also lies on E. Now let b be any point of $K(z,1)$ distinct from a and a', and let c be the midpoint of $S(a,b)$. Since $m(c,a) = m(c,b)$ and $m(z,a) = m(z,b)$, the lemma (23.7) implies that the reflection Φ in the line L, through z and c, carries a into b. But this reflection is a motion which leaves z fixed, and so, by (24.15), carries E into itself. Therefore $b = a\Phi$ lies on E, hence all points of $K(z,1)$ do, and the circle is an ellipse.

The argument just given also shows that $K(z,1)$ is an ellipse if, and only if, for any pair a,a' on $K(z,1)$ there exists a rotation about z carrying a into a'.

A *Minkowskian geometry in which reflections about every line exist is called Euclidean.* We may put this last result, then, in the form:

(25.4) *The Minkowskian geometries whose circles are ellipses, and only these, are Euclidean.*

This seems to indicate that there are many Euclidean geometries varying with the choice of the ellipse which serves as a unit circle. However, any two Euclidean geometries are isometric (which, according to definition,

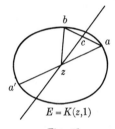

$E = K(z,1)$

Fig. 48

means that a distance preserving mapping of one on the other exists). This is a consequence of the sufficiency part of the theorem:

(25.5) *Two Minkowskian geometries m and m' in the affine plane A^2, with respective unit circles $K(z,1)$ and $K'(z',1)$, are congruent if and only if an affinity of A^2 exists which carries $K(z,1)$ into $K'(z',1)$.*

PROOF : If m and m' are congruent then, by assumption, there exists a one-to-one mapping $\Phi : x \to \bar{x}$, of A^2 on itself such that $m(x,y) = m'(\bar{x},\bar{y})$. Since Φ carries straight lines into straight lines, it is an affinity. Distance is preserved, hence $K(z,1) \to K'(z\Phi,1)$. If $z\Phi \neq z'$, then a translation Υ exists such that $K'(z\Phi,1)\Upsilon = K'(z',1)$ and $\Phi\Upsilon$ is the affinity of the theorem.

Conversely, suppose an affinity $\Phi : x \to \bar{x}$ exists which carries $K(z,1)$ into $K'(z',1)$. Because Φ preserves affine ratio, and z and z' are the respective affine centers of $K(z,1)$ and $K'(z',1)$, it follows that $\bar{z} = z\Phi = z'$. Both the metrics m and m' are invariant under translation, hence to show that Φ preserves distance it will suffice to show that for any x, $m(z,x) = m'(\bar{z},\bar{x})$, where $\bar{x} = x\Phi$. If e is the intersection of $K(z,1)$ with the ray from z through x, and e' the intersection of $K'(\bar{z},1)$ with the ray from $\bar{z}$ through $\bar{x}$, then because Φ carries the first ray into the second, $e' = e\Phi = \bar{e}$. From the invariance of affine ratio, we then have

$$m(z,x) = m(z,x)/m(z,e) = A(z,e,x) = A(\bar{z},\bar{e},\bar{x}) = m'(\bar{z},\bar{x})/m'(\bar{z},\bar{e}) = m'(\bar{z},\bar{x}).$$

It is now easily shown that:

(25.6)   *All Euclidean geometries are congruent.*

For if $K(z,1)$ and $K'(z',1)$ are ellipses with centers at z and z', then, by (16.9), an affinity exists which carries $K(z,1)$ into $K'(z',1)$. From the previous theorem it follows that the Euclidean geometries, corresponding to the ellipses, are congruent.

Henceforth we will speak, therefore, of *the* Euclidean geometry and *the* Euclidean metric.[11]

We next consider how to obtain an analytic expression for the Euclidean metric or, more generally, for a Minkowskian metric, when the unit circle $K(z,1)$, $z = (0,0)$, is given in a definite affine coordinate system x_1,x_2. The Minkowskian distance $m(z,x)$ is a function $F(x_1,x_2)$ such that

(25.7) $F(x_1,x_2) > 0$ for $z \neq x$, and $F(\lambda x_1, \lambda x_2) = |\lambda| F(x_1,x_2)$ for all λ.

The first part comes from the definition of a metric, and the second from $\lambda x = (1 - \lambda)z + \lambda x$ which gives $m(z,\lambda x)/m(z,x) = |A(z,x,\lambda x)| = |\lambda|$. Clearly $m(z,x) = F(x_1,x_2) = 1$ is the unit circle, hence is a strictly convex curve. Since $m(a,b)$ is invariant under translations, in particular under $x = x_i - a_i$, $i = 1,2$, then $m(a,b) = m(z, b - a) = F(b_1 - a_1, b_2 - a_2)$.

Conversely, if a function $F(x_1,x_2)$ is given which has the property (25.7), and if $F(x_1 x_2) = 1$ represents a strictly convex curve, then

(25.8) $m(x,y) = F(y_1 - x_1, y_2 - x_2)$ *is a Minkowski metric.*

Using $F(y - x)$ for $F(y_1 - x_1, y_2 - x_2)$, if $x = (1 - \lambda)a + \lambda b$, then

$$m(a,x)/m(a,b) = F(x - a)/F(b - a) = F(\lambda(b - a))/F(b - a) = |\lambda| = |A(a,b,x)|.$$

Thus (24.1) is satisfied by $m(a,b)$. The unit circle

$$m(z,x) = F(x - z) = F(x_1,x_2) = 1$$

is given as a strictly convex curve, and that z is its affine center follows from (25.7) when λ is taken as -1. With the unit circle chosen for K, then,

---

[11]Actually the Euclidean metric was already defined in Section 19. It will be seen presently that the two definitions agree.

the conditions of (24.12) are satisfied by $m(a,b)$ which is thus a Minkowski metric. To sum up:

> *If* $F(x_1,x_2) > 0$ *when* $(x_1,x_2) \neq (0,0)$, *and* $F(\lambda x_1, \lambda x_2) = |\lambda| \cdot F(x_1,x_2)$, *and if* $F(x_1,x_2) = 1$ *is a strictly convex curve, then*

(25.9)
$$m(x,y) = F(y_1 - x_1, \ y_2 - x_2)$$

> *is a Minkowski metric. Every Minkowski metric* $m(x,y)$ *can be written in this form by defining* $F(x_1,x_2) = m(z,x)$, $z = (0,0)$.

This theorem permits us to write down the general expression for a Euclidean metric. The general form for the equation of an ellipse in the affine plane is

(25.10)
$$a_{11}x_1^2 + 2a_{12}x_1x_2 + a_{22}x_2^2 + 2a_{13}x_1 + 2a_{23}x_2 + a_{33} = 0,$$
$$a_{11} \cdot a_{22} - a_{12}^2 > 0 \text{ and } |a_{ik}| \cdot a_{11} < 0.$$

If the origin is the center, that is the pole of the line at infinity, then, as previously observed, $a_{13} = a_{23} = 0$. Hence $a_{33} \neq 0$ and (25.10) can be put in the form

(25.11) $\quad Ex_1^2 + 2Fx_1x_2 + Gx_2^2 = 1, \quad E > 0, \quad G > 0, \quad EG - F^2 > 0.$

The conditions on E, F and G are more simply expressed by saying that the quadratic form (25.11) is positive definite, that is, it is positive except for $x_1 = x_2 = 0$. The letters E, F and G have been chosen in agreement with their usage in differential geometry where the form (25.11) is of great importance. The left side of (25.11) does not satisfy the condition (25.7) for a function $F(x_1,x_2)$, but its positive square root does. Hence if $\Phi(x)$ is defined by

(25.12)
$$\Phi(x) = Ex_1^2 + 2Fx_1x_2 + Gx_2^2,$$

then, because of (25.9), the Euclidean distance corresponding to (25.10), as the unit circle is expressed by

(25.13)
$$e(x,y) = [\Phi(x - y)]^{1/2}.$$

If a Minkowski metric is given by a function $F(x_1,x_2)$, then x_1, x_2, in general, will not be affine coordinates in terms of which $\sigma = \iint\limits_{\overline{U}(z,1)} dx_1 \, dx_2$

is π, so $\iint\limits_{D} dx_1 \, dx_2$ will not, in general, give the area of a domain D. However, $x_1' = x_1$, $x_2' = \dfrac{\pi}{\sigma} x_2$ are coordinates in terms of which the integral over $\overline{U}(z,1)$, that is the region $F(x_1,x_2) \leqslant 1$, is equal to π. Therefore *the area of a domain D is given by*

$$(25.14) \qquad \iint\limits_{D} dx_1' \, dx_2' = \frac{\pi}{\sigma} \iint\limits_{D} dx_1 \, dx_2.$$

When $F(x_1,x_2)$ is given explicitly, σ can be evaluated as follows. Introduce polar coordinates r,ω by setting $x_1 = r \cos \omega$ and $x_2 = r \sin \omega$.[12] The unit circle then becomes $F(r \cos \omega, r \sin \omega) = rF(\cos \omega, \sin \omega) = 1$, or, alternately $r = [F(\cos \omega, \sin \omega)]^{-1}$, so,

$$(25.15) \quad \sigma = (1/2) \int_0^{2\pi} [F(\cos \omega, \sin \omega)]^{-2} d\omega = \int_{-\pi/2}^{\pi/2} [F(\cos \omega, \sin \omega)]^{-2} d\omega.$$

In the case of the Euclidean distance (25.13),

$$[F(\cos \omega, \sin \omega)]^2 = E \cos^2 \omega + 2F \cos \omega \sin \omega + G \sin^2 \omega.$$

Then

$$\sigma = \int_{-\pi/2}^{\pi/2} [F(\cos \omega, \sin \omega)]^{-2} d\omega$$

$$= \int_{-\pi/2}^{\pi/2} [E \cos^2 \omega + 2F \sin \omega \cos \omega + G \sin^2 \omega]^{-1} d\omega$$

$$= \int_{-\pi/2}^{\pi/2} [E + 2F \tan \omega + G \tan^2 \omega]^{-1} d(\tan \omega)$$

$$= \frac{1}{\sqrt{EG - F^2}} \left[\tan^{-1} \left(\frac{G \tan \omega + F}{\sqrt{EG - F^2}} \right) \right]_{-\pi/2}^{\pi/2} = \frac{\pi}{\sqrt{EG - F^2}}.$$

Using this in (25.14) gives *the formula for the Euclidean area of the domain D* as:

$$(25.16) \qquad \sqrt{EG - F^2} \iint\limits_{D} dx_1 \, dx_2.$$

Because of the frequent occurence of $\sqrt{EG - F^2}$, we introduce the symbol $W = + \sqrt{EG - F^2}$.

26. Angles and Motions in Euclidean Geometry

All the known Euclidean formulas must be expressible in terms of E, F and G since these determine the metric. In obtaining such forms it will prove useful to change the notation of (25.12) by introducing

$$(26.1) \quad \Phi(x,y) = \Phi(x_1,x_2; y_1,y_2) = Ex_1y_1 + F \cdot (x_1y_2 + x_2y_1) + Gx_2y_2,$$

---

[12]See comments on angles preceding (18.3).

so that $\Phi(x,x) = \Phi(x_1,x_2; x_1,x_2)$ has the same meaning as the previous $\Phi(x)$. Then Φ has the property:

$$(26.2) \qquad \begin{aligned} &\Phi(\lambda x + \mu y, \ \lambda'x + \mu'y) \\ &= \lambda\lambda'\Phi(x,x) + \lambda\mu'\Phi(x,y) + \mu\lambda'\Phi(y,x) + \mu\mu'\Phi(y,y). \end{aligned}$$

If a, b and c are the sides of a triangle and α is the angle opposite a, then $a^2 = b^2 + c^2 - 2bc \cos \alpha$ is the ordinary cosine law giving a in terms of b, c and α. Since the metric is invariant under translations, it may be assumed that the vertex opposite a is the origin, and the other two vertices may be denoted by x and y. A special case of (26.2) is

$$(26.3) \qquad \Phi(x - y, \ x - y) = \Phi(x,x) + \Phi(y,y) - 2\Phi(x,y).$$

Since $\Phi(x - y, x - y)$, $\Phi(x,x)$, and $\Phi(y,y)$ are the squares of the lengths of the sides in the triangle, (26.3) must essentially be the law of cosines. The exact analogue is obtained by re-expressing (26.3) in the form

$$(26.4) \qquad \begin{aligned} e^2(x,y) &= \Phi(x - y, \ x - y) \\ &= \Phi(x,x) + \Phi(y,y) - 2\sqrt{\Phi(x,x)}\,\sqrt{\Phi(y,y)}\,\frac{\Phi(x,y)}{\sqrt{\Phi(x,x)},\ \sqrt{\Phi(y,y)}}. \end{aligned}$$

It follows that the rays from the origin to x and y define an angle whose cosine must be

$$(26.5) \qquad \cos(x,y) = \frac{\Phi(x,y)}{\sqrt{\Phi(x,x)}\,\sqrt{\Phi(y,y)}}.$$

The sin (x,y) is obtained in a similar way. From (25.16), the area of the triangle is $(W/2)\,|\,x_1y_2 - x_2y_1\,|$. On the other hand, it is also given by $(1/2)bc\,|\sin(x,y)\,|$, or $(1/2)\sqrt{\Phi(x,x)}\,\sqrt{\Phi(y,y)}\cdot|\sin(x,y)\,|$. With the usual agreement regarding the sign of the sine function, the two results imply that

$$(26.6) \qquad \sin(x,y) = \frac{W \cdot (x_1y_2 - x_2y_1)}{\sqrt{\Phi(x,x)}\,\sqrt{\Phi(y,y)}}.$$

This may also be derived from the relation $\sin^2(x,y) + \cos^2(x,y) = 1$, which is equivalent to the identity

$$(26.7) \qquad W^2 \cdot (x_1y_2 - x_2y_1)^2 + \Phi^2(x,y) = \Phi(x,x) \cdot \Phi(y,y).$$

These conclusions amount to a reformulation, in terms of the general Euclidean distance $\sqrt{\Phi(x - y, \ x - y)}$, of certain known facts from elementary geometry. As such they do not throw any new light into the nature of angle itself. From a more basic beginning, *the angle between two rays S_1, S_2, with the same origin, is a measure of the deviation of the rays.* Like every metric concept it must be invariant under motions. Another natural requirement is that it must be additive in the sense that if the rays S_1, S_2

and S have the same origin and S is between S_1 and S_2, then the angle (S_1,S_2) is the sum of the angles (S_1,S) and (S,S_2). Finally, a normalization of angle is needed. For instance, to obtain radian measure it is necessary that a straight angle have measure π.

If a space fails to possess sufficiently many rotations, the requirements stated may not be adequate to determine the angle measure. But if for a fixed point p and any distinct pair c,c' such that $pc = pc'$ the rotation about p exists which carries c into c', then it can be shown that *there is one and only one angle measure satisfying the above conditions.*

In Euclidean geometry this measure is easily determined. If S_1 and S_2 are rays issuing from p, they cut out a sector $G(S_1,S_2)$ from the disc $\overline{U}(p,1)$. A ray S from p, between S_1 and S_2, subdivides this sector, and because area is additive, $A[G(S_1,S)] + A[G(S,S_2)] = A[G(S_1,S_2)]$, (where $A[G(S_1,S_2)]$ means the area of $G(S_1,S_2)$). Moreover, when S_1 and S_2 are opposite, that is distinct but on a common line, then the sector is semicircular and $A[G(S,S_2)] = \pi/2$. Since area is invariant under motions, which include rotations, the number $2A[G(S_1,S_2)]$ fits all the above requirements for a measure of (S_1,S_2). In view of the previous remarks, it is unique and hence is *the* angle measure.

Taking p as the origin z, and (r_1,ω_1) and (r_2,ω_2) as the polar coordinates of x and y, the intersection points of S_1,S_2 with $K(z,1)$, then, as in the derivation of (25.16), the area of $G(S_1,S_2)$ can be obtained in the form

$$(1/2)W \int_{\omega_1}^{\omega_2} [E + 2F \tan \omega + G \tan^2 \omega]^{-1} d(\tan \omega)$$

$$= (1/2) \left[\arctan \left(\frac{G \tan \omega_2 + F}{W} \right) - \arctan \left(\frac{G \tan \omega_1 + F}{W} \right) \right].$$

Twice this area is the measure of the angle (S_1,S_2). By the usual formula for the tangent of a difference, then,

$$\tan (x,y) = \frac{GW^{-1}(\tan \omega_2 - \tan \omega_1)}{1 + W^{-2}(G \tan \omega_1 + F)(G \tan \omega_2 + F)}.$$

Putting $\tan \omega_1 = x_2/x_1$, $\tan \omega_2 = y_2/y_1$, this becomes

$$\tan (x,y) = \frac{W \cdot (x_1y_2 - x_2y_1)}{Ex_1y_1 + F \cdot (x_1y_2 + x_2y_1) + Gx_2y_2} = \frac{W \cdot (x_1y_2 - x_2y_1)}{\Phi(x,y)},$$

which leads back to (26.5) and (26.6).

If a line L, through the origin z and a point a, is perpendicular to a line L' through z and b, then the properties of perpendiculars, together with (16.6), imply that L and L' are conjugate diameters of the ellipse $Ex_1^2 + 2Fx_1x_2 + Gx_2^2 = 1$. The condition that $(a_1,a_2,0)$ and $(b_1,b_2,0)$ be conjugate points is that $Ea_1b_1 + F(a_1b_2 + a_2b_1) + Ga_2b_2 = 0$, and hence that

(26.8) $\Phi(a,b) = 0,$ *where* $a \neq z,$ $b \neq z.$

The expression (26.5) for the cosine of the angle (a,b) shows that this amounts to taking $\pi/2$, (or $3\pi/2$), as the measure of the angle.

Introducing rectangular coordinates in the usual sense means, from our present point of view, applying an affinity which transforms the ellipse $\Phi(x,x) = 1$, to

(26.9) $\Phi_0(x,x) = x_1^2 + x_2^2 = 1,$ *or* $E_0 = G_0 = 1,$ $F_0 = 0.$

Distance then takes the ordinary form

$$\sqrt{\Phi_0(x - y, \, x - y)} = \sqrt{(x_1 - y_1)^2 + (x_2 - y_2)^2},$$

and (25.6) proves that the space with this distance is congruent to the space with $\sqrt{\Phi(x-y, \, x-y)}$ as metric. When $E_0 = F_0 = 1$ and $G_0 = 0$ then $W_0 = 1$ and previous formulas reduce to standard form.

The motions of the Euclidean plane, with distance

$$d(x,y) = \sqrt{\Phi(x - y, \, x - y)},$$

are the affinities

(26.10) $x_i' = a_{i1}x_1 + a_{i2}x_2 + a_{i3},$ $i = 1,2,$ $\Delta = \begin{vmatrix} a_{11} & a_{12} \\ a_{21} & a_{22} \end{vmatrix} \neq 0,$

for which

(26.11) $d(x,y) = d(x',y').$

Since translations preserve distance, and since the general affinity (26.10) can be expressed as the product of

(26.12) $x_i' = a_{i1}x_1 + a_{i2}x_2,$ $i = 1,2,$ $\Delta \neq 0,$

and a translation, it suffices to investigate when (26.12) is a motion. By the same reasoning, we can further simplify the problem by taking y in (26.11) to be the origin. Hence (26.12) will be a motion when $\Phi(x,x) = \Phi(x',x')$, that is, when

$$Ex_1^2 + 2Fx_1x_2 + Gx_2^2 \equiv Ex_1'^2 + 2Fx_1'x_2' + Gx_2'^2$$
$$= E(a_{11}x_1 + a_{12}x_2)^2 + 2F(a_{11}x_1 + a_{12}x_2)(a_{21}x_1 + a_{22}x_2) + G(a_{21}x_1 + a_{22}x_2)^2$$
$$= x_1^2(Ea_{11}^2 + 2Fa_{11}a_{21} + Ga_{21}^2) + 2x_1x_2[Ea_{11}a_{12} + F(a_{11}a_{22} + a_{12}a_{21}) + Ga_{21}a_{22}]$$
$$+ \, x_2^2(Ea_{12}^2 + 2Fa_{12}a_{22} + Ga_{22}^2).$$

Comparing coefficients, the condition for (26.12), and hence for (26.10), to be a motion is that

(26.13) $\Phi(a_{11},a_{21}; \, a_{11},a_{21}) = E,$ $\Phi(a_{11},a_{21}; \, a_{12},a_{22}) = F,$
            and $\Phi(a_{12},a_{22}; \, a_{12},a_{22}) = G.$

For general E, F and G it would, of course, be highly unpleasant to solve the equations (26.13) for the coefficients a_{ij}. However, the groups of motions corresponding to Φ and Φ_0 are isomorphic. Hence the same information about the motions can be obtained by using Φ_0. For this case, where $E_0 = G_0 = 1$ and $F_0 = 0$, the system (26.13) reduces to

$$(26.14) \quad a_{11}^2 + a_{21}^2 = 1, \qquad a_{11}a_{12} + a_{21}a_{22} = 0, \qquad a_{12}^2 + a_{22}^2 = 1.$$

From the first and third of these equations, there exist numbers α and β such that

$$a_{11} = \cos \alpha, \qquad a_{21} = \sin \alpha, \qquad a_{12} = \cos \beta, \quad \text{and} \quad a_{22} = \sin \beta.$$

The second equation of (26.14) is then

$$\cos \alpha \cos \beta + \sin \alpha \sin \beta = \cos (\alpha - \beta) = 0,$$

hence $\alpha - \beta = \dfrac{\pi}{2} + 2n\pi$, or else $\alpha - \beta = \dfrac{3\pi}{2} + 2n\pi$, $n = 0, \pm 1, \pm 2, \cdots$ For the second of these two cases, $\sin \alpha = -\cos \beta$ and $\cos \alpha = \sin \beta$, while for the first, $\sin \alpha = \cos \beta$ and $\cos \alpha = -\sin \beta$. *A motion of the Euclidean plane, therefore, has one of the following forms:*

$$(26.15) \qquad \begin{aligned} x_1' &= x_1 \cos \alpha - x_2 \sin \alpha + a, \\ x_2' &= x_1 \sin \alpha + x_2 \cos \alpha + b, \end{aligned}$$

or,

$$(26.16) \qquad \begin{aligned} x_1' &= x_1 \cos \alpha + x_2 \sin \alpha + a, \\ x_2' &= x_1 \sin \alpha - x_2 \cos \alpha + b. \end{aligned}$$

The determinant $\Delta = \begin{vmatrix} a_{11} & a_{12} \\ a_{21} & a_{22} \end{vmatrix}$ is 1 for (26.15) and is -1 for (26.16).

Setting $x_i' = x_i$, $i = 1,2$, the fixed points of (26.15) are given by the system of equations:

$$\begin{aligned} x_1(\cos \alpha - 1) - x_2 \sin \alpha &= -a, \\ x_1 \sin \alpha + x_2(\cos \alpha - 1) &= -b. \end{aligned}$$

The determinant of this system has the value $2(1 - \cos \alpha)$. When this is not zero, there is a unique solution to the system and the motion is a rotation. When $2(1 - \cos \alpha) = 0$, then $\cos \alpha = 1$, $\sin \alpha = 0$ and the motion is a translation, which is the identity when a and b are both zero.

The fixed points of (26.16) are determined by the system:

$$\begin{aligned} x_1(\cos \alpha - 1) + x_2 \sin \alpha &= -a, \\ x_1 \sin \alpha - x_2(1 + \cos \alpha) &= -b. \end{aligned}$$

Since the determinant of the system has the value $-\cos^2 \alpha + 1 - \sin^2 \alpha = 0$, the equations represent the same line L, or else they are incompatible. In the former case, all points of L are fixed. Because $\Delta = -1$ the motion is not the identity, hence it is the reflection in L. For the incompatible system, we may write (26.16) as the product of two motions. The first,

$$x_1' = x_1 \cos \alpha + x_2 \sin \alpha, \qquad x_2' = x_1 \sin \alpha - x_2 \cos \alpha,$$

is the reflection in the line $x_2 = \dfrac{1 - \cos \alpha}{\sin \alpha} x_1,$ or $x_2 = x_1 \tan \alpha/2,$ and the second is the translation $x_1'' = x_1' + a,\ x_2'' = x_2' + b.$

This last case can be more simply described by observing that the reflection in a line L followed by a translation can also be represented as the reflection in a line parallel to L followed by a translation in the direction of L. For let $x \to x'$ represent the reflection in L and $x' \to x''$ represent the translation. Take any point x_1 not on L (Figure 49) and draw L_1

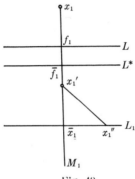

Fig. 49

through x_1'' parallel to L. Then L_1 cuts the line M_1 through x_1 and x_1' in a point $\bar{x}_1$. Designate the intersection of L and M_1 by f_1, and at $\bar{f}_1$, the midpoint of $S(x_1,\bar{x}_1)$, construct L^* perpendicular to M_1. Then it can easily be shown that the distance $e(x_1',\bar{x}_1)$ is the same for all choices of x_1. The length $e(f_1,\bar{f}_1) = e(x_1',\bar{x}_1)/2$ is therefore constant. It follows that L^*, parallel to L, is independent of the initial choice x_1. The original product motion $x \to x''$ is therefore the same as the product of the reflection in L^* with the translation taking $\bar{x}_1$ into x_1''.

Thus the discussion of (26.15) and (26.16) yields:

(26.17)   *A motion of the Euclidean plane, which is not the identity, is a rotation, a translation, a reflection in a line, or the product of a reflection in a line with a translation in the direction of the line.*

The *congruence theorems of elementary plane geometry* can be obtained, in the present approach, from the fact of (23.12) that two sets μ and μ' are congruent if and only if a motion exists which maps μ on μ'. Theorem

(23.11) also shows that the triple a,b,c can be moved into the congruent triple a',b',c' by a motion. Since angle measure is invariant under this mapping, corresponding angles have the same measure, and we express this in the usual way as

$$\measuredangle\, abc = \measuredangle\, a'b'c', \qquad \measuredangle\, acb = \measuredangle\, a'c'b', \qquad \measuredangle\, bac = \measuredangle\, b'a'c'.$$

That the corresponding equalities of "two sides and the included angle" implies congruence also follows. For suppose $e(c,a) = e(c',a')$, $e(b,a) = e(b',a')$, and $\measuredangle\, cab = \measuredangle\, c'a'b'$. The angle equality guarantees the existence of a motion Ψ which carries β and γ, the rays from a through c and b respectively, into β' and γ', the rays from a' through c' and b' respectively. Hence b on γ must go into b' on γ' because $e(a,b) = e(a',b')$. Similarly c on γ must go into c' on γ' because $e(a,c) = e(a',c')$. Since Ψ is

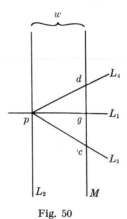

Fig. 50

a motion the remaining parts are equal. Euclid's proof of this theorem, or at least the proof attributed to him, is meaningless since it employs superposition without ever defining motions. The present approach corrects this in a natural way.

Since we have the triangle-congruence theorems, we may, and will, use all the standard theorems and concepts of Euclidean geometry. Also, some of the results of projective geometry, properly interpreted, yield new Euclidean theorems. The following is an example that will have later application.

(26.18)   *If the lines L_1, L_2, L_3 and L_4 are concurrent, and if two of them, say L_1 and L_2, are perpendicular, then the four lines form a harmonic set if and only if L_1 and L_2 bisect the two angles formed by L_3 and L_4.*

PROOF: (Figure 50). Let p be the center of the pencil and suppose that L_1 bisects one of the angles between L_3 and L_4. At a point g, distinct from p, on L_1 construct M perpendicular to L_1 and let c and d denote the respective intersections of M with L_3 and L_4. It is clear that g is the midpoint of $S(c,d)$, hence its harmonic conjugate with respect to c and d is w, the point at infinity on M. But since L_2 is parallel to M it passes through w. Thus the transversal M cuts the set of lines in a harmonic set of points which implies the lines are harmonic. The argument is reversible. If the lines are harmonic, then g, the harmonic conjugate of w with respect to c and d, is the midpoint of $S(c,d)$. Because the triangles c,p,g and d,p,g are then congruent, L_1 is a bisector of one angle formed by L_3 and L_4. The other bisector must be perpendicular to L_1, hence is L_2.

27. The Euclidean Theory of Conics[13]

Since the reader is familiar with the ordinary theory of conics, our interest will not lie in the development of that theory, but rather in how it appears from our present point of view.

An *axis* of a conic C is a line L such that the reflection in L carries C into itself. If x is on C then its image x', under reflection, is also on C and the line $x \times x'$ is perpendicular to L. Hence L is the locus of the centers of a family of parallel chords in C which are perpendicular to L. Conversely, if L is the locus of the centers of a family of parallel chords perpendicular to L, then the reflection in L will carry C into itself.

The argument leading to (16.4) shows that an axis L of a conic C must pass through z, the pole with respect to C of the line ζ at infinity. When C is a parabola, z is on ζ and L belongs to the family of parallels through z. If p is a general point on C, a line M, perpendicular to any line through z, cuts C again at q, and it is easily seen that the line through z and the midpoint of p and q is an axis L, and is unique. Hence:

(27.1)   *A parabola has one axis.*

For an ellipse or hyperbola, the diameter conjugate to a line L through the center z is always a member of the family of parallel chords bisected by L. Hence L is an axis if and only if it is perpendicular to its conjugate diameter. All pairs of conjugate diameters in a circle are perpendicular so for this special type of ellipse all diameters are axes.[14] Since the center of a circle is a no-tangent point the involution of conjugate (or perpendicular)

---

[13] In this section ζ is the line at infinity. Except for ζ there is no confusion in using either ξ or L to designate a line.
[14] This fact and (16.7) imply that an angle inscribed in a semi-circle is a right angle.

diameters is elliptic (see (11.4)). When C is a hyperbola or an ellipse (but not a circle), with center z, then by (13.17) the involution of conjugate diameters and the involution of perpendiculars at z have exactly one pair of elements in common. This yields:

(27.2)   *An ellipse (not a circle) or a hyperbola has two axes characterized as the pair of conjugate diameters which are perpendicular.*

The intersection of a conic with an axis is called a *vertex* of the conic. The tangent at a vertex is perpendicular to the axis through the vertex since the tangent is parallel to the conjugate diameter of the axis. For a parabola there is only one axis and one vertex (at a finite distance). If the axis of the parabola and the tangent at the vertex are taken for the x_1 and x_2 coordinate axes respectively, then, as in (16.5),

$$x_2^2 = 2px_1$$

is a standard form of the parabola. Similarly, when the axes of an ellipse or a hyperbola are taken for coordinate axes, with the same choice of units, the representations of these conics have the familiar forms $\frac{x_1^2}{a^2} + \frac{x_2^2}{b^2} = 1$ and $\frac{x_1^2}{a^2} - \frac{x_2^2}{b^2} = 1$ respectively. We use, in the usual sense, the terms "major" and "minor" axis for an ellipse and "transverse" and "conjugate" axis for hyperbola.

The projective approach to the foci of a conic is not quite as direct.[15] *A focal point, or focus, of a conic is defined to be a point f such that conjugate lines through f are perpendicular.* In other words, in the pencil through a focus the involution of conjugate lines in the pencil coincides with the involution of perpendicular lines. Since the latter involution is elliptic, the former is also, and a focus must be a no-tangent point, provided it exists. This existence is not obvious, except for a circle where clearly the center is a focus.

Suppose, however, that C (not a circle) has a focus f and let ξ indicate the line $f \times z$, where z is the pole of ζ. Since $f \neq z$, ξ is defined. Its pole is on ζ, and the parallel chords, which lie on lines conjugate to ξ, are bisected by ξ. By assumption the conjugate of ξ, which contains f, is perpendicular to ξ, hence ξ is perpendicular to all its conjugates and is therefore an axis of C. In the case of a parabola, then, a focus would have to lie on the unique axis, and for a hyperbola it would have to be on the transverse axis since the conjugate axis consists of two-tangent points. It is also clear that if f is a point on an axis ξ and if one pair of

---

[15]The treatment of foci which follows is taken from Coxeter, "The Real Projective Plane", Section 9.7.

conjugate lines through f are perpendicular, and both are distinct from ξ, then all conjugate pairs in the pencil f are perpendicular, so f is a focus.

The actual existence of foci may be shown from the following lemma concerning conics in the projective plane.

(27.3)   *If the triangle b,a,a' is circumscribed to the conic C and if x is conjugate to b, then $x \times a$ is conjugate to $x \times a'$.*

A proof is immediately obtained by dualizing the construction for a polar following (11.8) (Figure 51). Choose a two-tangent point y on $\eta = a \times a'$, distinct from a and a', and let η' be the second tangent through y. Put $\delta = a \times b$ and $\delta' = a' \times b$. The previous polar construction amounted to this: four points on a conic determine six lines and the intersections of these, in pairs, determine three new points defining a self-polar triangle.

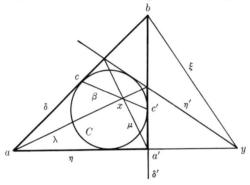

Fig. 51

Dually, four tangents, namely η, η', δ and δ' determine six points whose joins, in pairs, produce three new lines forming a self-polar triangle. The sides of this triangle are $\lambda = (\eta \times \delta) \times (\eta' \times \delta')$, $\mu = (\eta \times \delta') \times (\eta' \times \delta)$ and $\xi = b \times y$. Set $x = \lambda \times \mu$. Since x is the pole of ξ, and b is on ξ, x is conjugate to b. Being sides of a self-polar triangle, $\lambda = a \times x$ and $\mu = a' \times x$ are conjugates. The construction is reversible, starting from x conjugate to b, rather than from y, hence the theorem is established.

In the proof just given, let the contact points of δ and δ' be c and c' respectively and set $\beta = c \times c'$. For the case where C is a parabola, let δ' be taken as ζ, the line at infinity, β as the axis, and η any tangent distinct from δ (Figure 52). For any x on β, $a \times x$ and $a' \times x$ are conjugates, and since a' is on ζ the lines η and $a' \times x$ are parallel. Let the perpendicular to η at a cut β at x_0. Then $x_0 \times a'$ is perpendicular to $a \times x_0$. Since one pair of conjugate lines through x_0, both distinct from β, are perpendicular, all pairs of conjugate lines through x_0 are perpendicular, hence x_0 is a focus. There can be no other since the perpendicular to η at a is unique.

(27.4)   *A parabola has one focus. Any tangent η and the line through the focus perpendicular to η intersect on the tangent through the vertex.*

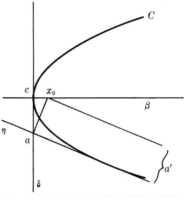

Fig. 52

When C is an ellipse, (27.3) can be applied with β chosen as one axis and η taken to be parallel to β (Figure 53). As before the lines $x \times a$ and

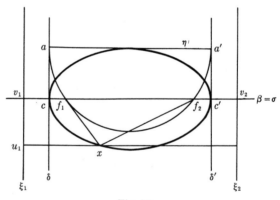

Fig. 53

$x \times a'$ are conjugate for x on β. They will be perpendicular when, and only when, x lies on the circle K having $S(a,a')$ as a diameter. When the ellipse is not a circle, K intersects β only if β is the major axis, in which case it cuts it in two points. *An ellipse therefore has two foci*, which lie on the major axis. We see also that a circle has only one focus since for that case β is tangent to K at the center of C.

Finally, for the hyperbola β may be chosen as the transverse axis, so that δ and δ' are tangents at the vertices, and η may be taken to be one of the asymptotes (Figure 54). When x is an intersection point of β and the circle K, which has $S(a,a')$ as a diameter, then $x \times a$ and $x \times a'$ are perpendicular conjugates and x is a focus. Since K and β have two intersections there are two foci. Moreover,

(27.5)   *The circle, which has a common center with a hyperbola, and which passes through the foci, also passes through the intersection points of the asymptotes with the tangents at the vertices.*

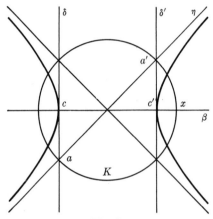

Fig. 54

The constructions for the foci make it possible to verify by direct calculation that the foci, as defined here, coincide with the foci as usually defined. However, more insight is gained through a geometric approach. If a *directrix* of a conic is defined to be the polar of a focus, it has the following well known property.

(27.6)   *As a point varies on a conic its distance from a focus is in constant ratio to its distance from the directrix corresponding to the focus. The constant ratio ε is called the eccentricity of the conic.*

PROOF: Let f be a focus of the conic C, φ the polar of f, x an arbitrary point of C, and ξ the tangent at x. Since f is a no-tangent point, φ is a nonintersector, hence between f and φ there is a vertex b. Let β indicate the tangent at b. Because β and ξ are tangents, $y = ξ \times β$ is the pole of $η = b \times x$. Then $a = η \times φ$ is the pole of $α = f \times y$. The line $γ = a \times f$ is therefore conjugate to α, and so, by the definition of a focus, α and γ are perpendicular. Since a and α are pole and polar, the points a, $α \times η$, b and x form a harmonic set, hence the lines γ, α, $f \times b = σ$, and $f \times x$ are also harmonic. The perpendicularity of $\varkappa$ and γ, together with (26.18), implies, then, that γ is one bisector of the angles formed by σ and $f \times x$. Let δ be the line through x parallel to σ and set $w = \delta \times \gamma$ and $u = \delta \times \varphi$. Take g any point of σ so that f is between b and g. Because γ is an angle bisector, $\measuredangle\, xfw = \measuredangle\, wfg$, while the parallelism of δ and σ implies $\measuredangle\, wfg = \measuredangle\, fwx$. Hence $\measuredangle\, xfw = \measuredangle\, fwx$, which gives $e(f,x) = e(w,x)$. On the other hand, for $v = \sigma \times \varphi$, the parallels δ and σ as transversals of γ, η and φ intercept proportional segments, hence $e(w,x)/e(x,u) = e(f,b)/e(b,v)$. Taking ε as the value of the last ratio, then $e(f,x)/e(x,u) = \varepsilon$ for an arbitrary point x of the conic.

The definitions usually given for an ellipse and a hyperbola, namely that the distances from a point on the conic to the foci have a constant

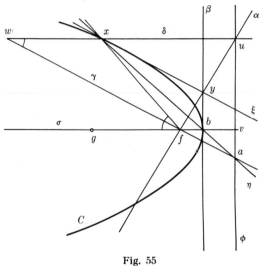

Fig. 55

sum and difference respectively, now follow as properties. Suppose, for instance, that f_1 and f_2 on σ are the foci of an ellipse C (Figure 53). Let ξ_1 and ξ_2 be the corresponding directrices, and for an arbitrary point x of C let u_i denote the foot of x on ξ_i, $i = 1,2$. Set $v_i = \sigma \times \xi_i$, $i = 1,2$. Then $e(x,f_1) + e(x,f_2) = \varepsilon \cdot [e(x,u_1) + e(x,u_2)] = \varepsilon \cdot e(v_1,v_2)$, so the sum is constant. This relation, when x is a vertex, also shows that $\varepsilon < 1$ in the case of an ellipse.

28. Hilbert's Geometry

We now turn to projective metrics which are defined in a convex domain of the affine plane A^2. As a preparation, some facts are first established concerning metrics on a straight line.

Let ξ be a line in the plane, and let I be one of the intervals on ξ with (distinct) end points x and y. We consider a metrization of I under which it becomes a metric straight line. First, if a and b are points of I, different from x and y, with b between x and a, then

$$(28.1) \qquad R(a,b,x,y) > 1 \quad and \quad R(a,a,x,y) = 1.$$

For if d is any fifth point of ξ, it was shown in (6.16) that

$$(28.2) \qquad R(a,d,x,y)R(d,b,x,y) = R(a,b,x,y).$$

In particular, if d is the point at ∞ on ξ, then (15.10) and (24.1) show that for any Euclidean metric in A^2, this reduces to

$$(28.3) \qquad R(a,b,x,y) = R(a,p_\infty,x,y)R(p_\infty,b,x,y) = \frac{e(a,x)}{e(a,y)} \cdot \frac{e(b,y)}{e(b,x)} > 1.$$

For $a = b$, (28.3) reduces to the equality in (28.1). Next, if x' is between b and x, then

$$(28.4) \qquad R(a,b,x',y) > R(a,b,x,y).$$

For

$$R(a,b,x',y) = \frac{e(a,x')}{e(b,x')} \cdot \frac{e(b,y)}{e(a,y)} \text{ and } R(a,b,x,y) = \frac{e(a,x') + e(x,x')}{e(b,x') + e(x,x')} \cdot \frac{e(b,x)}{e(a,y)}.$$

Substituting $e(a,x') = \lambda$, $e(b,x') = \mu$ and $e(x',x) = \delta$ in these relations gives

$$R(a,b,x',y) = \frac{\lambda}{\mu} \cdot \frac{e(b,y)}{e(a,y)} \text{ and } R(a,b,x,y) = \frac{\lambda + \delta}{\mu + \delta} \cdot \frac{e(b,y)}{e(a,y)}. \text{ Since } \delta > 0 \text{ and}$$

$\lambda > \mu > 0$ it follows that $\dfrac{\lambda + \delta}{\mu + \delta} < \dfrac{\lambda}{\mu}$, which, with the above expressions for $R(a,b,x',y)$ and $R(a,b,x,y)$, implies (28.4). Similarly, if y' lies between a and y, then

$$(28.5) \qquad \begin{array}{c} R(a,b,x,y') > R(a,b,x,y) \quad and \\ R(a,b,x',y') > R(a,b,x,y). \end{array}$$

For the same disposition of a, b, x and y, let the metric $h(a,b)$ on I be defined by,

$$(28.6) \qquad h(a,b) = \frac{k}{2} \log R(a,b,x,y), \qquad h(b,a) = \frac{k}{2} \log R(b,a,y,x), \qquad k > 0.^{16}$$

From (28.1), $h(a,b) > 0$ for $a \neq b$ and $h(a,b) = 0$ if and only if $a \sim b$. Also, $h(a,b) = h(b,a)$ because of (6.8). Since, from (28.2), $h(a,d) + h(d,b) = h(a,b)$, when d is between a and b, the triangle inequality holds for a,b,d in any order. All the properties of a metric are thus satisfied by $h(a,b)$. That

---

[16]Because $R(a,b,x,y) = 1/R(b,a,x,y)$, the relations in (28.6) are equivalent to $h(a,b) = \frac{k}{2} |\log R(a,b,x,y)|$ for any pair a and b on $S(x,y)$. The convenience of the choice $\frac{k}{2}$ will appear later.

I is a metric straight line in terms of $h(a,b)$ follows from the additivity of distances along *I* together with the fact that

(28.7) $R(a,b,x,y) \to \infty$ as $b \to x$ or as $a \to y$,   or as $b \to x$ and $a \to y$.

Now let D be the interior of a bounded, convex domain in the affine plane with the convex curve K as its boundary. In D a distance $h(a,b)$ is defined as follows (Figure 56). Let x be the intersection of K with the ray $R(a,b)$ and y be the intersection of K with $R(b,a)$.[17] The distance $h(a,b)$ is then defined to be $\frac{k}{2} \log R(a,b,x,y)$. The discussion of the previous paragraph shows that the first two properties of a metric are satisfied in D

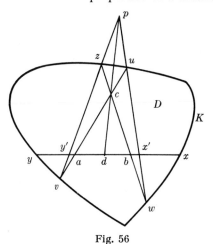

Fig. 56

by $h(a,b)$, and also that the triangle inequality holds for triples on $S(x,y)$. In D let c be a point not on $S^*(x,y)$. Take u, v, z and w for the respective intersections of K with the rays $R(a,c)$, $R(c,a)$, $R(b,c)$ and $R(c,b)$. Set $p = (v \times z) \times (w \times u)$, $y' = (v \times z) \times (x \times y)$, $x' = (w \times u) \times (x \times y)$, and $d = (p \times c) \times (x \times y)$. The perspectivity of $u \times v$ and $x \times y$ from p, together with (28.4), yields

$$R(a,c,u,v) = R(a,d,x',y') \geqslant R(a,d,x,y).$$

Similarly, (28.4) and the perspectivity of $w \times z$ and $x \times y$ from p, gives

$$R(c,b,w,z) = R(d,b,x',y') \geqslant R(d,b,x,y).$$

Multiplying the last two inequalities, and using (28.2) gives

$$R(a,c,u,v) \cdot R(c,b,w,z) \geqslant R(a,d,x,y) \cdot R(d,b,x,y) = R(a,b,x,y).$$

---

[17] $R(a,b)$ denotes the ray through b emanating from a.

Taking the logarithm of both sides, then,

$$h(a,c) + h(c,b) \geqslant h(a,b).$$

Throughout the derivation, the equality holds only if $x' = x$ and $y' = y$. This can occur only if K is not strictly convex, but contains, in fact, two non-collinear segments. Thus we have :

(28.8)   *If K does not contain two non-collinear segments, then $h(a,b)$ is a projective metric in D.*

The metric $h(a,b)$ was discovered by Hilbert.

When K does contain two non-collinear segments, $h(a,b)$ is not a projective metric. However, the metric $d(a,b) = h(a,b) + e(a,b)$ is projective in D. It will still be so, in fact, if $e(a,b)$ is replaced by any Minkowski metric $m(a,b)$.

If K is an open, instead of closed, convex curve, and if D does not contain a straight line, Hilbert's definition may still be used in the following way. For a and b in D, the line $a \times b$ intersects K in at least one point y, which lies, say on $R(b,a)$. If $R(a,b)$ also intersects K at x, then $h(a,b)$ is defined as before. When $R(a,b)$ fails to cut K, then x is taken as the intersection of $a \times b$ and the line ζ at infinity, and $h(a,b)$ is again defined by (28.6). This method amounts to closing K by adding to it the interval on ζ bounding D which is cut off by K. The interval may consist only of a point, as when K is a parabola, or may be an actual segment, as when K is a branch of a hyperbola. In the generalized case, if K is strictly convex then $h(a,b)$ is a projective metric, as before, otherwise $h(a,b) + e(a,b)$ may be taken for a projective distance.

It is easily seen that if the convex domain D contains a straight line, D is either a half-plane or else a strip between parallel lines (Ex. [17.8]). That the latter type of domain can be given a projective metric was shown in the examples following (20.6). On the other hand, if D is the interior of a half-plane, rectangular coordinates can be chosen so that D is the locus given by $x_2 > 0$. For any two points a and b in D, the second coordinates x_{2_a} and x_{2_b} are then positive and the distance $d(a,b) = e(a,b) + \left| \dfrac{1}{x_{2_a}} - \dfrac{1}{x_{2_b}} \right|$ provides D with a projective metric.
Therefore,

(28.9)   *The interior of any convex domain in the affine plane can be given a projective metric.*

The most interesting case is that where K is a strictly convex, closed curve, and D is its interior. To have a short term, we will refer to this type of domain, with the metric $h(a,b)$, as *a Hilbert geometry*. We consider some of its properties.

(28.10)   *The circles of a Hilbert geometry are strictly convex.*

According to the theory developed in Section 21 this is equivalent to the existence of perpendiculars. Hence (28.10) can be established simultaneously with the following construction for perpendiculars.

(28.11)   *In a Hilbert geometry the line P is perpendicular to the line L if and only if K has, at its intersections with P, supporting lines concurrent with L.*

PROOF: (Figure 57). Let x and y denote the intersection points of K with P and take ξ and η as the supporting lines at x and y. Assume that ξ,

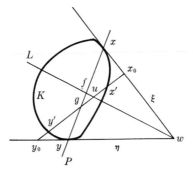

Fig. 57

η and L are concurrent at w. If f is the intersection of P and L, then for $g \neq f$ on P and $u \neq f$ on L, it must be shown that $h(g,u) > h(g,f)$. Suppose x to be on $R(g,f)$ and y to be on $R(f,g)$ and denote by x' and y' the intersections of K with $R(g,u)$ and $R(u,g)$ respectively. Finally, set $x_0 = \xi \times (u \times g)$ and $y_0 = \eta \times (u \times g)$. Then (28.4) and the perspectivity from w of $(g \times f)$ with $(g \times u)$ yields $R(g,f,x,y) = R(g,u,x_0,y_0) < R(g,u,x',y')$, implying $h(g,f) > h(g,u)$. Hence P is perpendicular to L.

The proof of the converse is less direct. Suppose g and L, non-incident, to be a point and line in D. As u traverses L let x and y be the variable intersections of K with $R(g,u)$ and $R(u,g)$ respectively. From continuity considerations it is apparent that u will assume one and only one position f for which L is concurrent with supporting lines at x and y. Hence perpendiculars exist, which implies (28.10), and P can only be perpendicular to L under the conditions of (28.11). In this construction we notice that if K is an ellipse, the points f and w are conjugate. It follows directly that:

(28.12)   *If K is an ellipse, perpendicularity is symmetric, and two intersecting lines in D are perpendicular when, and only when, they
are conjugate with respect to K.*

If K has unique supporting lines at its points of intersection with the
line P, then the lines to which P is perpendicular form a pencil. However,
in general the lines perpendicular to P do not form a pencil.

If the point p and the line L in D are non-incident, then through p
there are infinitely many lines which do not intersect L in D. The parallel
axiom, therefore, fails in Hilbert geometry. Among the non-intersectors
through p there are two extreme ones, A_1, A_2, namely the lines determined
by p and the intersections of L with K. They are called the *asymptotes*

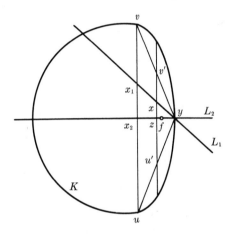

Fig. 58

through p to L. If x refers to a point in D on L and L_x is the line joining p
and x, then as x traverses L, in a given direction from a fixed point x_0,
the lines L_x tend to one of the asymptotes, A_1 or A_2. From the original
definition, it follows that asymptoticness is symmetric, that is, L is
asymptotic to A_1 and to A_2.

The following property gives a useful criterion, in terms of distance, for
two lines to be asymptotic.

(28.13)   *If x is a variable point on the line L_1 and f is its foot on the distinct
line L_2, then L_1 and L_2 are asymptotes if and only if there is a
sub-ray of L_1 such that for x on this ray the distances $\{ h(x,f) \}$
are bounded.*

PROOF: (Figure 58). Let L_1 and L_2 intersect at the point y of K, and take any
point x_i on L_i in D, $i = 1,2$. Let $R(x_1,x_2)$ and $R(x_2,x_1)$ intersect K in u and v

respectively. Let the parallel to $x_1 \times x_2$ through the point x of $S^*(x_1,y)$ intersect L_2, $S(y,u)$ and $S(y,v)$ in z,u',v'. Because K is convex, u' and v' lie in D. By (28.5) and the definition of "foot",

$$h(x,f) \leqslant h(x,z) < \frac{k}{2} \log \mathrm{R}(x,z,u',v') = h(x_1,x_2),$$

which proves that $h(x,f)$ is bounded. For the converse (Figure 59), suppose L_1 is not asymptotic to L_2 and that a and b are the intersections of K with L_1. Using the same convention as before,

$$h(x,f) = \frac{k}{2} \log \left[\frac{e(x,u)}{e(x,v)} \cdot \frac{e(f,v)}{e(f,u)} \right].$$

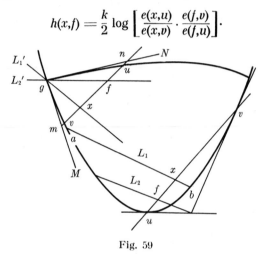

Fig. 59

As x approaches a or b, it is clear that $e(x,u)$ is bounded away from zero while $e(x,v)$ approaches zero. Because $e(f,v)/e(f,u) \geqslant \eta > o$ it follows that $h(x,f) \to \infty$ as $x \to a$ or $x \to b$.

> *Let L_1' be an asymptote to L_2' and let x on L_1' have f as foot on L_2'.*
(28.14)   *Then $h(x,f) \to 0$ for $x \to g = L_1' \times L_2'$ if and only if K is differentiable at g.*

If K is not differentiable at g then K has two extreme supporting lines, M and N, at g (Figure 59). Let $x \times f$ cut K at u and v, with x between v and f, and cut M and N in m and n respectively, with v between m and x. For a fixed position, then, $R(x_0,f_0,u_0,v_0) > R(x_0,f_0,n_0,m_0)$. For x variable between x_0 and g it follows that $R(x,f,u,v) > R(x,f,n,m) = R(x_0,f_0,n_0,m_0) > 1$. Hence $\lim\limits_{x \to g} h(x,f) \geqslant \frac{k}{2} \log R(x_0,f_0,n_0,m_0) > 0$. When K is differentiable at g, the lines $g \times u$ and $g \times v$ converge to the tangent at g as $x \to g$, and $h(x,f)$ is easily shown to approach zero.

29. Motions and Area in Hilbert Geometry
Definition of Hyperbolic Geometry

Since Hilbert distance is defined in terms of cross ratio, which is invariant under a collineation, any collineation Φ which maps D on itself will be a motion in the Hilbert geometry. Conversely, since a collineation is determined by a quadrangular set of points and their images, every motion Ψ determines a collineation Φ, which coincides with Ψ on D.

(29.1) *The motions in a Hilbert geometry are the collineations of the projective (not the affine) plane which map D on itself.*

It was shown that the reflection in a line L, when it exists, is induced by a harmonic homology Φ whose axis carries the line L. The perpendiculars to L lie on the projective lines of the pencil defined by the center of Φ, and this center is outside D. The circles of D are convex, hence by (23.9), if P is perpendicular to L, then L is perpendicular to P. The boundary K goes into itself under Φ since D does. If K is an ellipse and L is a line of D carried by the projective line ξ, then, by (13.3) and (13.6), K goes into itself under the harmonic homology whose axis is ξ and whose center is the pole of ξ with respect to K. Hence when K is an ellipse, reflections in all lines exist. As in the case of Minkowskian geometry, an ellipse is the only curve for which this is true.

(29.2) *The Hilbert geometry in the domain bounded by the convex curve K admits reflections in every line if K is an ellipse, and K is an ellipse if the reflections in all lines of one pencil (within D) exist.*

Since only the last part is not established, assume that reflections in all lines of a pencil p exist, and let L_1 and L_2 to be two lines of the pencil such that $L_1 \perp L_2$. Through a point w_i on L_i pass two supporting lines of K which cut L_j in the same points that K does, where $i \neq j$, $i,j = 1,2$. The line $w_1 \times w_2$ does not intersect K or D and may be taken to be the line at infinity. Choosing L_1 and L_2 for the x_1 and x_2 axes of an affine coordinate system, the reflections in L_1 and L_2 have the respective forms:

$$x_1' = x_1, \qquad x_2' = -x_2, \qquad \text{and} \qquad x_1' = -x_1, \qquad x_2' = x_2.$$

Each reflection carries K into itself, hence the product of the two, $x_i = -x_i$, $i = 1,2$, also carries K into itself. Therefore K has the origin p as affine center. If L is any line through p, then there are supporting lines ξ, η at the end points of L such that ξ and η are parallel. The construction in (28.11) shows that the line through p parallel to ξ is perpendicular to L. The reflection Φ in L is therefore the harmonic homology with $\xi \times \eta$ as center and with the projective line carrying L as axis. Because $\xi \times \eta$ is on the line at infinity, Φ is an affinity. If E denotes the ellipse of minimum area and center p containing K, defined in (24.15), then E goes into itself under Φ. The fact that K also goes into itself and has a common point with E shows, as in the argument of (25.3), that E and K coincide.

A Hilbert geometry in which the reflections about all lines exist is called hyperbolic. We may therefore say :

(29.3)   *A Hilbert geometry is hyperbolic if and only if the domain D is the interior of an ellipse.*

Since the reflections in every line exist, we have in hyperbolic geometry the same mobility, and therefore at least the same congruence theorems, as in Euclidean geometry (actually there are more). *The discovery of hyperbolic geometry, shortly after 1800, by Gauss (1777-1855), Bolyai (1802-1860), and Lobachevsky (1793-1856) was one of the great events in the history of mathematics.* It showed the futility of the attempts, extending over 2 000 years, to prove that the parallel axiom is a consequence of the congruence axioms. Hyperbolic geometry and elliptic geometry (the latter resulting from a projective metrization of the entire projective plane) are the so called non-Euclidean geometries, and are considered in detail in the next chapter.

However, the answer to the question of how many hyperbolic geometries there are lies immediately to hand in the following fact concerning Hilbert geometries.

Two Hilbert geometries defined in the strictly convex domains D_1 and D_2, with the metrics $h_1(a,b)$ and $h_2(a,b)$ respectively, satisfy the relation $h_1(a,b) = ch_2(a,b)$, for some constant c, if and only if a collineation of P^2 exists which carries D_1 into D_2.

PROOF : Let Φ be the collineation carrying D_1 into D_2. If a and b are points in D_1, and x and y are the intersections of $a \times b$ with K_1, then the line $a\Phi \times b\Phi$ must intersect K_2 in the points $x\Phi$ and $y\Phi$. Therefore

$$R(a,b,x,y) = R(a^\Phi,b^\Phi,x^\Phi,y^\Phi).$$

If k_i is the distance constant belonging to h_i, then clearly

$$h_2(a\Phi,b\Phi) = \frac{k_2}{k_1}h_1(a,b).$$

Conversely, let $a_1 \to a_2$ be a mapping of D_1 on D_2 under which $h_1(a_1,b_1) = ch_2(a_2,b_2)$. In spite of the factor c, it is clear that a metric straight line in D_1 goes into a metric straight line in D_2. Hence the intersection of D_1 with a projective line goes into the intersection of D_2 with a projective line. The mapping $a_1 \to a_2$ is therefore induced by a collineation, and the proof is complete.

Since any ellipse can be carried into any other ellipse by a projectivity (which is an affinity if the centers are corresponding), it follows that for any two hyperbolic metrics $h_1(x,y)$ and $h_2(x,y)$ a constant c exists such that $h_1(x,y) = ch_2(x,y)$. It will appear later that for $c \neq 1$ the corresponding spaces are not congruent.

We conclude the present chapter with a discussion of area in Hilbert geometry. Let x_1, x_2, x_3 be projective coordinates such that neither K nor D intersects the line $x_3 = 0$. In D points may then be represented in the form $(x_1, x_2, 1)$ with x_1 and x_2 regarded as affine coordinates. They may even be taken to be rectangular Euclidean coordinates with the corresponding Euclidean distance given by $e(x,y) = [(x_1 - y_1)^2 + (x_2 - y_2)^2]^{1/2}$. Suppose now that p is a point of a line L in D and that a and b are variable points which tend to p in such a way that the variable line $a \times b$ tends to L. We wish to evaluate, in this situation, the $\lim h(a,b)/e(a,b)$. Let

$$x = K \cap R(a,b), \quad y = K \cap R(b,a), \quad u = \frac{e(a,b)}{e(b,x)} \text{ and } v = \frac{e(a,b)}{e(a,y)}. \text{ [17]}$$

Then, $\dfrac{2}{k} h(a,b) = \log\left[\dfrac{e(a,x)}{e(a,y)} \cdot \dfrac{e(b,y)}{e(b,x)}\right] = \log(1 + u) + \log(1 + v)$. The Mac-Laurin expansions for the last two terms give

$$\frac{2}{k} h(a,b) = \left(u - \frac{u^2}{2} + \cdots\right) + \left(v - \frac{v^2}{2} + \cdots\right)$$

or

$$\frac{2}{k} \frac{h(a,b)}{e(a,b)} = \left[\frac{1}{e(b,x)} - \frac{1}{2}\frac{e(a,b)}{e(b,x)^2} + \cdots\right] + \left[\frac{1}{e(a,y)} - \frac{1}{2}\frac{e(a,b)}{e(a,y)^2} + \cdots\right].$$

With x_0 and y_0 as the end points of L, the last relation shows that

$$(29.4) \qquad \lim_{a,b \to p} \frac{h(a,b)}{e(a,b)} = \frac{k}{2}\left[\frac{1}{e(p,x_0)} + \frac{1}{e(p,y_0)}\right] = \Phi(p,L).$$

We show next:

(29.5) *If p is a point of D, and if on each line L of the pencil p the distinct points z_1 and z_2 are chosen so that $e(p,z_i) = 1/\Phi(p,L)$, $i = 1,2$, the locus of the points z_1 and z_2 is a strictly convex curve K_p with center p.*

PROOF: Without restriction we may assume that $k = 1$. Let L_1 and L_2 be any two lines through p, with K cutting L_1 at u_1, u_1' and L_2 at u_2, u_2'

---

[17] $M \cap N$ means the intersection of the sets M and N.

(Figure 60). Take the numbers a,b,c,d to represent $e(p,u_1)$, $e(p,u_1')$, $e(p,u_2)$ and $e(p,u_2')$ respectively. The locus point z_1 on $R(p,u_1)$ and the locus point z_2 on $R(p,u_2)$ determine a segment. Choose y any point on the open segment $S^*(z_1,z_2)$, and let $L = p \times y$ cut K at u and u', with $e = e(p,u)$ and $f = e(p,u')$. To show the strict convexity of the locus, it suffices to prove that $e(p,y) < e(p,z)$ where z is the locus point on $R(p,y)$. If $\bar{y}$ is the intersection point of L with $S(u_1,u_2)$ and $\bar{e} = e(p,\bar{y})$, convexity in the Euclidean plane of the curve K implies $e > \bar{e}$. If α is the angle between L and L_1 and β is the angle between L and L_2, then the Euclidean area of the triangle p,u_1,u_2 expressed in two ways yields

$$a\bar{e} \sin \alpha + c\bar{e} \sin \beta = ac \sin (\alpha + \beta).$$

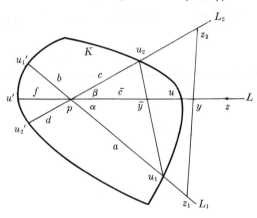

Fig. 60

Therefore,

$$\frac{\sin \alpha}{c} + \frac{\sin \beta}{a} = \frac{\sin (\alpha + \beta)}{\bar{e}} > \frac{\sin (\alpha + \beta)}{e}.$$

By the same argument,

$$\frac{\sin \alpha}{d} + \frac{\sin \beta}{b} > \frac{\sin (\alpha + \beta)}{f}.$$

Adding the two inequalities gives

$$\frac{\sin \alpha}{e(p,z_2)} + \frac{\sin \beta}{e(p,z_1)} > \frac{\sin (\alpha + \beta)}{e(p,z)}.$$

But the area of triangle p,z_1,z_2 expressed as the sum of two triangles gives

$$\frac{\sin \alpha}{e(p,z_2)} + \frac{\sin \beta}{e(p,z_1)} = \frac{\sin (\alpha + \beta)}{e(p,y)}.$$

Comparing this with the previous inequality shows that $e(p,y) < e(p,z)$.

Since K_p is strictly convex and has a center, when taken as a unit circle it defines a Minkowski metric $m(x,y)$ in the affine x_1,x_2 plane. If x and y are distinct, there is a line L_{xy} through p which is parallel to $x \times y$, and

$$(29.6) \qquad m(x,y) = e(x,y)\Phi(p,L_{xy})$$

is the Minkowski metric. Clearly $m(p,x) = 1$ is the given curve K_p, and for any collinear triple x, y and z, (29.6) implies

$$m(x,z)/m(x,y) = e(x,z)/e(x,y) = |A(x,y,z)| \qquad \text{(see footnote, page 137).}$$

From (29.4) and (29.6) we obtain:

$$(29.7) \qquad \text{If } a \neq b, \quad \lim_{a,b \to p} \frac{h(a,b)}{m(a,b)} = 1.$$

On the other hand, the relation (29.7) determines the metric $m(a,b)$ uniquely: that is, for a given choice of the line at infinity $x_3 = 0$ there is only one K_p for which (29.7) holds.

The metric $m(x,y)$ of (29.6) is called the *local Minkowski metric belonging to the point p*. The limit (29.7) shows that $m(a,b)$ is a good approximation of $h(a,b)$ in the vicinity of p. Since the Hilbert geometry is locally Minkowskian, the *Hilbert area can be defined by integration* in the following way. If B is a region in D which is bounded (in the Hilbert sense) it may be subdivided into regions B_i. If a point p_i is selected in B_i, then associated with p_i is a local Minkowski metric. By (25.14) and (25.15), the area of B_i in the local geometry is

$$(29.8) \qquad A_{p_i}(B_i) = \sigma(p_i) \iint\limits_{B_i} dx_1\, dx_2$$

where

$$(29.9) \qquad \sigma(p_i) = \pi \left[\int_{-\pi/2}^{\pi/2} \Phi^{-2}(p_i,L)d\omega \right]^{-1}.$$

Then $\sum_i A_{p_i}(B_i)$ is an approximate expression for the Hilbert area. As the number of subdivisions is increased, in such a way that the maximum diameter of all the subregions tends to zero, then, by the definition of a double integral, the sums $\sum_i A_{p_i}(B_i)$ tend to the limit

$$(29.10) \qquad A(B) = \iint\limits_{B} \sigma(x_1,x_2)dx_1\, dx_2,$$

which is the *Hilbert area* of B.

The crux of this definition is *the principle that area in a geometry is determined from the area in the local geometry by the above integration process.* A priori, there is the possibility that this principle might be inconsistent with the requirement that area in a metric space be invariant under motions. But the relation (29.7), and the fact that it determines $m(a,b)$ uniquely, implies that a motion of D taking p into p' must also take the local Minkowski metric associated with p into the local Minkowski metric belonging to p'. Therefore $A(B) = A(B')$ whenever a motion of the Hilbert geometry exists which carries B into B'. This may be confirmed by calculation.

Exercises

[19.1] If $\delta_1(x,y)$ and $\delta_2(x,y)$ are distances defined in the same set, show that $\delta_3 = \delta_1 + \delta_2$ and $\delta_4 = \max \{\delta_1, \delta_2\}$ satisfy the distance axioms.

[19.2] If δ_1 and δ_2, in [19.1], are equivalent, then any pair of the four metrics δ_i are equivalent.

[19.3] Construct an example to show that in [19.1] min $\{\delta_1, \delta_2\}$ will not in general be a metric.

[19.4] If $px < \rho$, $py < \rho$ and $xz + zy = xy$, then $pz < 2\rho$.

[19.5] Show that $\pi(x,y)$ satisfies the axioms for a metric.

[19.6] Give an example to show that, if xy is a metric, then in general $(xy)^2$ and $(xy)^4$ are not metrics. Also give an example where all three are metrics.

[20.1] If xy is a metric, then, for any fixed $\lambda > 0$, $\lambda \cdot (xy)$ is also a metric, and the segments, straight lines and great circles of the space are the same under both metrizations.

[20.2] Show that in R_3 every curve $x(t)$, for which $x_i(t)$ varies monotonically, is a segment.

[20.3] If $x(t)$ is a segment, and $x(t_1) = x(t_2)$, for $t_1 < t_2$, then $x(t) \equiv x(t_1)$ for $t_1 \leqslant t \leqslant t_2$.

[20.4] If $x(t)$, $-\infty < t < \infty$, is a straight line for the metrics $\delta_1(x,y)$ and $\delta_2(x,y)$, then it is a straight line for $\delta_3(x,y) = \delta_1(x,y) + \delta_2(x,y)$. Is the corresponding statement for a great circle also correct?

[20.5] Let μ be the subset obtained by removing a conic and its no-tangent points from the projective plane. Assuming Hamel's theorem, show that μ cannot be provided with a projective metric.

[20.6] Is P^2, under the metrization $\pi(x,y)$, a projective-metric space?

[20.7] Show that the side $x_2 > 0$ of the real x_1-axis in E^2 can be provided with a projective metric. (Hint: modify the metric of R_6.)

[20.8]* In an open, projective-metric two-space, P_1 and P_2 are two distinct, convex polygons which have a side in common. If all points of P_2 are either on or interior to P_1, show that the length of P_1 is greater than that of P_2. Derive from this the fact that a closed, convex curve has finite length.

[20.9] In an open, projective-metric space, if x lies on $S^*(a,b)$, then, except for a, $K(x,ax)$ lies in the interior of $K(b,ab)$. (The interior of $K(p,\delta)$ consists of all points y such that $py < \delta$.)

In the problems given for Section 21, all the spaces are open and two-dimensional and the metrics are projective.

[21.1] For $a \not= b$, let $B(a,b)$ denote the locus of points x such that $ax = bx$. Show that if $B(a,b)$ is always a straight line the circles are convex.

[21.2] If for every distinct pair, a and b, the locus of points x, for which $|ax - xb| = 2a < ab$, consists of two convex curves, then the circles are convex (this reduces to [21.1]).

[21.3] If the circles are convex, and $x(t)$ represents a straight line, then for any point p the function $f(t) = px(t)$ reaches its minimum for exactly one value t_0 and is decreasing for $t < t_0$ and increasing for $t > t_0$.

[21.4]* If the locus of points at a given distance from a given straight line consists of two straight lines, then the circles are convex.

[21.5] In the Euclidean x_1,x_2 plane find two continuous curves $x_2 = f_i(x_1)$, $-\infty < x_1 < \infty$, $i = 1,2$, such that all points of the first curve have the same distance from the second curve, but not conversely. Show that this cannot happen if each point of the first curve has a unique foot on the second.

[21.6]* Let $C(\eta,k)$ denote the locus of points, on one side of η, at a distance k from η. If, whenever η and a side of η are arbitrarily chosen, the points of η have a constant distance from $C(\eta,k)$, for any $k > 0$, then the circles are convex.

[21.7] Are the circles in R_6 convex?

[21.8] Every point except p has exactly one foot on the circle $K(p,\delta)$.

[21.9] Show that the metric in the example at the end of section 21 is projective.

[22.1] Find the motions of R_6. Remetrize R_6 with the distance
$$d(x,y) = d_6(x,y) + |e^{x_1} - e^{y_1}|$$
and find the motions under the new metric.

[22.2] Show that $x_1' = x_1 \cos \alpha + x_2 \sin \alpha$, $\quad x_2' = -x_1 \sin \alpha + x_2 \cos \alpha$ is a motion of E^2.

[22.3]   Show that there is no projective metric for the whole affine plane under which every affinity is a motion.

[22.4]   Let Γ be an Abelian group of motions for the metric space R with the property that for any two points, x and y in R, a motion in Γ exists which carries x into y. Prove that for any two points, x and y in R, and any motion Φ in Γ, the distance $x(x\Phi)$ equals the distance $y(y\Phi)$.

[22.5]   Under the assumptions of [22.4] show that for a given pair of points, x and y, exactly one motion in Γ exists which carries x into y.

[23.1]   Prove (23.1).

[23.2]   Use (23.7)* to show that in an open two-space three distinct points lie on at most one circle.

[23.3]   In E^2 find the reflection in the line $3x_1 + 4x_2 - 5 = 0$ and the reflection in the point $(2,9)$.

[23.4]   Find all the motions of E^1 (i. e., the real x_1-axis) with the distance $|x_1 - y_1|$.

[23.5]   In E^2, show that the product of the reflections in two distinct lines is either a rotation or a translation (of E^2, considered as an affine plane).

[23.6]   In an open, projective-metric two-space, the rotations (including the identity) about a point p form an Abelian group.

[23.7]   Find two rotations in E^2 which do not commute.

Except for [25.1], the underlying space in all problems for Sections 24 and 25 is assumed to be a Minkowski plane.

[24.1]   In a parallelogram of maximum area, for a given (Minkowski) perimeter, the adjacent sides have equal lengths and are mutually perpendicular.

[24.2]   Let a be a fixed point on a line ξ. As a point x traverses ξ in a definite direction, the circles $K(x,ax)$ tend to a straight line if and only if each (or one) of the circles $K(x,ax)$ has a tangent at a.

[24.3]   Using the preceding result, construct a Minkowski plane in which three non-collinear points exist which do not lie on a circle.

[24.4]   If $m_1(x,y)$ and $m_2(x,y)$ are two Minkowski metrics in the same affine plane, then $m_3(x,y) = \lambda_1 m_1(x,y) + \lambda_2 m_2(x,y)$ is a Minkowski metric for any fixed, positive values of λ_1 and λ_2.

[24.5]   Proceeding in a definite direction on the circle $K(p,\rho)$, from an arbitrary point g_1 of the circle, construct the sequence $g_1, g_2, g_3, \cdots$ such that $m(g_i, g_{i+1}) = \rho$. Show that $g_7 = g_1$.

[24.6]   Let a_1, a_2 and a_3 be the vertices of an equilateral triangle with sides of length k. Each of the circles $K(a_i,k)$ has an arc subtending a side of the triangle as chord, $i = 1,2,3$. Call C the convex curve formed by these three arcs. Show that the distance between any two parallel supporting lines to C is k.

[25.1]   Use the argument in the proof of (25.3) and (28.2) to prove the following affine theorem. If the affine centers of every family of parallel chords of a closed, convex curve lie on a straight line, then the curve is an ellipse.

[25.2]* If the bisector $B(a,b)$, $a \not\sim b$, is a straight line, then the reflection in $B(a,b)$ exists.

[25.3]   Deduce from [25.2]*: if all bisectors $B(a,b)$ are straight lines, then the geometry is Euclidean.

[25.4]   If $m(x,y)$ is the metric for the Minkowski plane M, then M and M_λ are congruent, where the latter Minkowski plane is metrized by $\lambda m(x,y)$, $\lambda > 0$.

[25.5]* Let E be the locus of points x for which

$$m(a,x) + m(x,b) = 2\imath > m(a,b),$$

where $a \not\sim b$. Show that E is a convex curve.

[25.6]   Show that in a general Minkowski geometry the locus of points x, such that $|m(a,x) - m(x,b)| = 2\imath < ab$, $a \not\sim b$, cannot always consist of two convex curves. (Hint: reduce to [25.3].)

[25.7]   The group of rotations about a point is isomorphic to the group of rotations about any other point.

[25.8]   If the group of rotations about a point is finite, it consists of an even number of elements. For a given even number $2n$, construct a Minkowski metric such that each rotation group consists of $2n$ elements.

[25.9]* If the group of rotations about a point has infinitely many elements, the geometry is Euclidean.

[25.10] Find the area of the ellipse $2x_1^2 + 2x_1x_2 + 3x_2^2 - 10x_1 + 6x_2 + 3 = 0$ in the Euclidean space with $\Phi(x) = 3x_1^2 - 2x_1x_2 + 7x_2^2$.

[26.1]   Find the vertices of some square in the Euclidean metric of [25.10].

[26.2]   Find the equation of $B(a,b)$, for $a = (0,0)$ and $b = (2,2)$, using the metric of [25.10].

[26.3]   With the Euclidean distance in standard form, find the rotations about the point $(1,2)$. (Hint: in (26.15) find a and b as functions of α, or use the proof of (22.7).)

[26.4]   Use motions to show that the base angles of an isosceles triangle are equal.

[26.5] If the line ξ passes through the center z of the circle K, and $x \mathbin{\text{$\not\,$}} z$ is any point of ξ which is not on K, then any circle orthogonal to K and through x intersects ξ in the conjugate point to x.

[27.1] An ellipse in the standard form $\dfrac{x^2}{a^2} + \dfrac{y^2}{b^2} = 1$ may be represented parametrically by $x = a \cos t$, $y = b \sin t$. Show that for a fixed t, $(- a \sin t, b \cos t)$ and $(a \cos t, - b \sin t)$ are points of the ellipse which lie on conjugate diameters.

[27.2] Use [27.1] to show that if the chord lengths on two conjugate diameters of an ellipse are squared and added, the sum is constant for all pairs of such diameters.

[27.3] Find analogues to [27.1] and [27.2] for hyperbolas using hyperbolic functions.

[27.4] If the tangents at g_1 and g_2 on the ellipse E intersect in the exterior point p, then the line $f \times p$, where f is a focal point, bisects the angle $\measuredangle\, g_1 f g_2$. (Hint: use the present definition of focus and (26.18).)

[27.5] Use [27.4] to show: the segment cut off on a variable tangent by two fixed tangents of an ellipse is seen from a focus under a constant angle.

[27.6] Show that the eccentricity of a conic is less than 1 for an ellipse, equal to 1 for a parabola, and greater than 1 for a hyperbola.

[27.7] The locus of points equidistant from a circle and a line is a parabola.

[27.8] The locus of the points equidistant from a circle and a point p is a branch of a hyperbola or an ellipse according as p is outside or inside the circle.

[27.9] Extend [27.8] to the locus of points equidistant from two given circles.

[28.1] Complete the proof of (28.14).

[28.2] Show that (28.13) is false for the metric $h(a,b)$ inside a closed, convex curve which contains a segment.

[28.3] In Hilbert geometry the locus of points at a constant distance from a fixed line does not consist of two straight lines. (Hint: use (28.13).)

[28.4] If the domain D of a Hilbert geometry is bounded by

$$a^{-2}\, x_1^2 + b^{-2}\, x_2^2 = 1,$$

find the equation of the circle $K(z,\rho)$, where $z = (0,0)$.

[28.5] In the geometry of [28.4] find the equation of the line perpendicular at z to the line $x_2 = \mu x_1$.

[28.6] Construct a projective metric for the domain $|x_1| < 1$, $|x_2| < 1$, in the affine plane.

[28.7] In a Hilbert geometry with differentiable K, two lines whose intersection is neither in D nor on K have exactly one common perpendicular. If the intersection is in D or on K the lines have no common perpendicular.

[28.8] Show by an example that, in general, perpendicularity of lines is not symmetric in a Hilbert geometry.

[29.1]* Give a necessary and sufficient condition for a Hilbert geometry to possess the reflection in a given point (line).

[29.2] Construct a Hilbert geometry which does not possess any rotation about one of its points.

[29.3] Construct a Hilbert geometry in which the group of rotations about one of its points consists of exactly n elements, where n is a given positive integer greater than 1.

[29.4] In the Hilbert space of [28.4], find the equation of K_z defined in (29.5).

[29.5]* Let the lines L_1 and L_2 intersect at the point p in the domain D of a Hilbert geometry. Then L_1 is perpendicular to L_2 if and only if it is perpendicular to L_2 in the sense of the Minkowski geometry defined by K_p.

Chapter V

Non–Euclidean Geometry

30. Hyperbolic Trigonometry

Hyperbolic geometry was defined as a Hilbert geometry possessing all reflections, and in (29.3) it was shown that a Hilbert geometry is hyperbolic if and only if the domain D is the interior of an ellipse E. With a suitable choice of affine coordinates x_1, x_2, E can be represented in the form

$$(30.1) \qquad x_1^2 + x_2^2 = 1.$$

If E is considered as the unit circle of the Euclidean metric

$$e(x,y) = [(x_1 - y_1)^2 + (x_2 - y_2)^2]^{\frac{1}{2}},$$

then many facts of hyperbolic geometry can be deduced immediately from our knowledge of Euclidean geometry.

The hyperbolic motions are the projectivities which carry E into itself.[1] Among such projectivities are the Euclidean rotations about the origin, $z = (0,0)$. These are also rotations in the hyperbolic metric, hence a circle with center z and a Euclidean radius $\bar{r} < 1$ is also a hyperbolic circle with center z and some radius r. The definition, (28.6), of distance in the hyperbolic space yields

$$(30.2) \qquad r = \frac{k}{2} \log \left[\frac{1}{1-\bar{r}} \cdot \frac{1+\bar{r}}{1} \right] = \frac{k}{2} \log \left[\frac{1+\bar{r}}{1-\bar{r}} \right].$$

Hence

$$(30.3) \qquad \bar{r} = \frac{e^{2r/k} - 1}{e^{2r/k} + 1} = \tanh (r/k).$$

A hyperbolic motion which carries z into p takes the hyperbolic circle $K(z,r)$ into the circle $K(p,r)$. Since the motion is a projectivity, the image locus $K(p,r)$ is again a conic, and being a closed curve it appears in the Euclidean metric $e(x,y)$ as an ellipse.

(30.4) *The circles of the hyperbolic metric are ellipses with respect to the Euclidean metric.*

---

[1] If a collineation carries E into itself, it automatically takes the interior of E into itself since it maps no-tangent points into no-tangent points.

Let η be a line through z. The Euclidean reflection in η is a projectivity which takes E into itself, hence it is a hyperbolic motion and, being involutary, is a hyperbolic reflection. Consequently:

(30.5)   *The hyperbolic perpendiculars to a line η through z coincide with the Euclidean perpendiculars to η.*

To see how facts of this type can be employed, we show:

(30.6)   *If two altitudes α_p, α_q of the triangle p, q, r intersect at a point s, then the third altitude α_r also passes through s.*

For, a motion Φ, taking s into z, and p, q and r into p', q' and r', carries α_p and α_q into the lines $\alpha_{p'}$ and $\alpha_{q'}$ which are radii of E. Because of (30.5), $\alpha_{p'}$ and $\alpha_{q'}$ are also perpendicular to $q' \times r'$ and $p' \times r'$ respectively in the Euclidean sense. It follows that $r' \times z$ is a Euclidean perpendicular to $q' \times p'$ and hence, by (30.5), is the hyperbolic perpendicular $\alpha_{r'}$. Under Φ^{-1}, then, $\alpha_{r'}$ goes into α_r through s. The existence of s must be assumed in (30.6) since it will soon appear that the altitudes of a hyperbolic triangle need not intersect.

As stated in Section 26, there is only one continuous measure of angle which is invariant (under the hyperbolic motions), has the value π for straight angles, and is additive for angles at a common vertex. Again we obtain one such measure, and therefore the only one, by taking it to be the area of the corresponding circular sector multiplied by a suitable factor of proportionality. The measure may be evaluated in the following way. For angles with center z, the Euclidean measure is also the hyperbolic measure, since Euclidean rotations about z are also hyperbolic rotations. The measure of an angle α, with arbitrary vertex p, is determined by moving p to z and finding the Euclidean measure of the image angle. It follows, in particular, that perpendicular lines form angles of measure $\pi/2$.

Now let η and ζ be two lines through z forming an angle $A < \pi/2$ (Figure 61). On η let c be a variable point with $h(z,c) = \beta$ and denote by ξ_c the line perpendicular to η at c. As c starts with $\beta = 0$ and moves away from z the line ξ_c will intersect ζ until ξ_c falls on the line ξ which is perpendicular to η and asymptotic to ζ. Let δ denote the value of β corresponding to $\xi_c = \xi$. For $\beta > \delta$, ξ_c will neither intersect ζ nor be asymptotic to it. The asymptote position occurs when $e(z,c) = \cos A$, so from (30.2),

$$\delta = \frac{k}{2} \log \left(\frac{1 + \cos A}{1 - \cos A} \right) = \frac{k}{2} \log \left(\frac{1 + \cos A}{\sin A} \right)^2 = k \log \cot \left(\frac{A}{2} \right).$$

The angle A is called the *parallel-angle*, $\pi(\delta)$, belonging to δ. It may be described as follows. At p_0 on a line ξ erect a perpendicular η, and through any point p on η let ζ be an asymptote to ξ. If $\delta = h(p,p_0)$, and $\pi(\delta)$ denotes the angle between ζ and η which is less than $\pi/2$, then

(30.7) $\qquad \delta = k \log \cot \left(\dfrac{\pi(\delta)}{2}\right), \quad or \quad \tan \left(\dfrac{\pi(\delta)}{2}\right) = e^{-\delta/k}.$

With the same conventions as before (Figure 61), consider now a position of c for which $\beta < \delta$. Let ξ_c cut ζ at b, and let $h(b,c) = \alpha$ and $h(z,b) = \gamma$. In the triangle z,b,c let B denote the hyperbolic measure of the angle at b and $\overline{B}$ its Euclidean measure. Similarly set $\overline{\alpha} = e(b,c)$, $\overline{\beta} = e(z,c)$ and

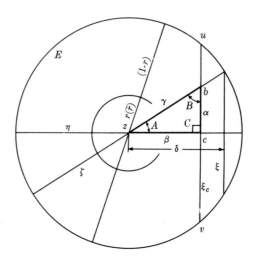

Fig. 61

$\overline{\gamma} = e(z,b)$. To develop the *trigonometry of the hyperbolic right triangle*, it is sufficient to study the triangle z,b,c, since any right triangle can be put in this position by a motion which leaves unchanged the lengths of sides and the measures of angles.

Consider first, the relation expressing:

1) *A in terms of β and γ.*

This is obtained at once from (30.3), since

(30.8) $\qquad \cos A = \cos \overline{A} = \overline{\beta}/\overline{\gamma} = \tanh (\beta/k)/\tanh (\gamma/k).$
$\qquad \textit{Similarly } \cos B = \tanh (\alpha/k)/\tanh (\gamma/k).$

The formula for $\cos B$ follows by analogy from that for $\cos A$ since the triangle can be re-oriented by a motion which interchanges the roles of

A and B. As it is placed in the figure B differs from $\overline{B}$. To see this we evaluate $\overline{\alpha}$ in terms α.[2] By definition,

$$\alpha = \frac{k}{2} \log \left[\frac{e(c,u)}{e(b,u)} \cdot \frac{e(b,v)}{e(c,v)} \right] = \frac{k}{2} \log \left[\frac{\sqrt{1 - \overline{\beta}^2}}{\sqrt{1 - \overline{\beta}^2} - \overline{\alpha}} \cdot \frac{\sqrt{1 - \overline{\beta}^2} + \overline{\alpha}}{\sqrt{1 - \overline{\beta}^2}} \right].$$

Since $\sqrt{1 - \overline{\beta}^2} = \sqrt{1 - \tanh^2 \beta'} = \text{sech } \beta'$, we have $e^{2\alpha'} = \dfrac{\text{sech } \beta' + \overline{\alpha}}{\text{sech } \beta' - \overline{\alpha}}$,

or $\overline{\alpha} = (\text{sech } \beta') \cdot \left(\dfrac{e^{2\alpha'} - 1}{e^{2\alpha'} + 1} \right)$, hence

(30.9) $$\overline{\alpha} = \tanh(\alpha/k) \text{ sech}(\beta/k).$$

Returning to the relation between B and $\overline{B}$, from (30.3), (30.8) and (30.9) it follows that

$$\cos \overline{B} = \overline{\alpha}/\overline{\gamma} = \frac{\tanh \alpha' \text{ sech } \beta'}{\tanh \gamma'} = \cos B \text{ sech } \beta' = \cos B \sqrt{1 - \overline{\beta}^2}.$$

Therefore, $\cos B > \cos \overline{B}$, and since the cosine is a decreasing function on the interval $(0, \pi)$,

(30.10) $$B < \overline{B}.$$

Moreover, from $\cos B = \dfrac{\cos \overline{B}}{\sqrt{1 - \overline{\beta}^2}}$ it is clear that $\lim\limits_{\overline{\beta} \to 0} B = \overline{B}$ and that as $\overline{\beta}$ increases from zero to δ, B decreases monotonically from $\overline{B}$ to zero. For a given value B_0, where $0 < B_0 < \overline{B} = \frac{\pi}{2} - A$, there is then exactly one position of c for which $B = B_0$. This indicates that *the angles of a hyperbolic triangle determine the sides*, a conjecture that will soon be confirmed.

Next we seek

2) *A in terms of α and β.*

From the previous results, $\tan A = \dfrac{\overline{\alpha}}{\overline{\beta}} = \dfrac{\tanh \alpha' \text{ sech } \beta'}{\tanh \beta'}$, hence,

(30.11) $$\tan A = \frac{\tanh(\alpha/k)}{\sinh(\beta/k)}, \qquad \tan B = \frac{\tanh(\beta/k)}{\sinh(\alpha/k)}.$$

A useful application of (30.11) is that:

(30. 12)   *For fixed β, as $\alpha \to \infty$ the angle B tends to 0.*

---

[2] For the remainder of this section, unless otherwise stated a prime will indicate division by k so that x', for instance, means x/k.

The expression for

3) *A in terms of α and γ*

follows from the relation between α, β and γ. The Pythagorean theorem, $\bar{\gamma}^2 = \bar{\alpha}^2 + \bar{\beta}^2$, together with (30.3) and (30.9) yields

$$\tanh^2 \gamma' = \tanh^2 \alpha' \operatorname{sech}^2 \beta' + \tanh^2 \beta',$$

or

$$1 - \operatorname{sech}^2 \gamma' = \operatorname{sech}^2 \beta' [1 - \operatorname{sech}^2 \alpha'] + 1 - \operatorname{sech}^2 \beta',$$

which reduces to

$$\operatorname{sech}^2 \gamma' = \operatorname{sech}^2 \alpha' \operatorname{sech}^2 \beta'.$$

Since the sech and cosh are positive functions, the last relation implies

(30.12) $\cosh(\gamma/k) = \cosh(\alpha/k) \cosh(\beta/k)$ *(Theorem of Pythagoras).*

Using this with (30.3) and (30.9) gives

$$\sin \Lambda = \frac{\bar{\alpha}}{\bar{\gamma}} = \frac{\tanh \alpha' \operatorname{sech} \beta'}{\tanh \gamma'} = \frac{\tanh \alpha' \operatorname{sech} \gamma' \cosh \alpha'}{\tanh \gamma'}.$$

From which it follows that:

(30.13) $\sin A = \dfrac{\sinh(\alpha/k)}{\sinh(\gamma/k)},$ $\sin B = \dfrac{\sinh(\beta/k)}{\sinh(\gamma/k)}.$

The foregoing relations are analogous to the trigonometric formulas for a Euclidean right triangle. However, in hyperbolic geometry there are additional *relations expressing the sides in terms of the angles*. For instance, the previous formulas combine to give

$$\frac{\cos A}{\sin B} = \frac{\tanh \beta'}{\tanh \gamma'} \cdot \frac{\sinh \gamma'}{\sinh \beta'} = \frac{\cosh \gamma'}{\cosh \beta'},$$

hence

(30.14) $\cosh(\alpha/k) = \dfrac{\cos A}{\sin B},$ $\cosh(\beta/k) = \dfrac{\cos B}{\sin A}.$

Then,

(30.15) $\cosh(\gamma/k) = \cosh(\alpha/k)\cosh(\beta/k) = \cot A \cot B.$

In the last chapter it was shown that if $h_1(a,b)$ and $h_2(a,b)$ are any two given hyperbolic metrics, then a constant c exists such that $h_1(a,b) = ch_2(a,b)$. For Minkowski metrics such a relation would imply that the corresponding Minkowski planes were congruent (compare exercise [25.4], Ch. IV). For hyperbolic metrics (and therefore in general for Hilbert metrics) this is not the case. Formulas (30.8), (30.11) and (30.13) show that in the right triangle z,b,c the measures of the angles A, B and C depend on k. Therefore:

(30.16)   *Hyperbolic geometries corresponding to different values of k are not congruent.*

As the proportionality constant k approaches infinity, hyperbolic formulas go over into Euclidean formulas. For instance, (30.13) becomes $\sin A = \alpha/\beta$ and (30.14) becomes $\cos A = \sin B$. This last equality expresses the fact that in Euclidean geometry $A + B = \pi/2$, whereas (30.10) shows that in hyperbolic geometry the sum is smaller.

The choice of a specific value for k in hyperbolic geometry is equivalent to selecting a unit of length. Thus a standard, or absolute unit, can be determined by intrinsic conditions, as for instance, taking unity to be the length of the hypotenuse of an isosceles right triangle with base angles $\pi/5$.[3] This is not the case in Euclidean geometry which requires some non-mathematical object, such as the standard meter at Paris, for an absolute unit.

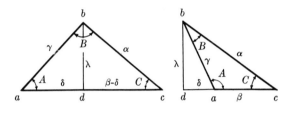

Fig. 62

The extension of hyperbolic trigonometry to general triangles is accomplished as in the Euclidean case. Let a,b,c with angles A,B,C and sides α,β,γ be an arbitrary triangle and let the altitude through b cut $a \times c$ at d with $h(b,d) = \lambda$. There are, then (see Figure 62), two cases to distinguish according as d does, or does not, lie between a and c. We will consider the former case only, leaving the latter to the reader. From

$$\frac{\sinh \lambda'}{\sinh \alpha'} = \sin C \quad \text{and} \quad \frac{\sinh \lambda'}{\sinh \gamma'} = \sin A,$$

the *hyperbolic form of the law of sines* is:

(30.17) $$\frac{\sin A}{\sinh (\alpha/k)} = \frac{\sin B}{\sinh (\beta/k)} = \frac{\sin C}{\sinh (\gamma/k)}.$$

To obtain the *law of cosines*, we observe that if $ad = \delta$,

$$\cosh \alpha' = \cosh \lambda' \cosh (\beta' - \delta') = \cosh \lambda' [\cosh \beta' \cosh \delta' - \sinh \beta' \sinh \delta'],$$
$$\text{and} \quad \cosh \gamma' = \cosh \lambda' \cosh \delta'.$$

---

[3] In this case (30.15) yields $1/k =$ area cosh $(\cot^2 \pi/5)$. More natural ways of normalizing k will be discussed later.

Substituting from the second relation in the first, and using

$$\cosh \lambda' = \cosh \gamma'/\cosh \delta'$$

yields

$$\cosh \alpha' = \cosh \beta' \cosh \gamma' - \sinh \beta' \tanh \delta'.$$

By (30.8), $\tanh \delta' = \cos A \tanh \gamma'$. This, with the last equality, gives the law of cosines:

(30.18) $\cosh (\alpha/k) = \cosh (\beta/k) \cosh (\gamma/k) - \sinh (\beta/k) \sinh (\gamma/k) \cos A.$

31. Length and Area

Let C be a continuously differentiable curve given parametrically by $x_1 = x_1(t)$, $x_2 = x_2(t)$, $a \leqslant t \leqslant b$. The *hyperbolic differential of arc length,*

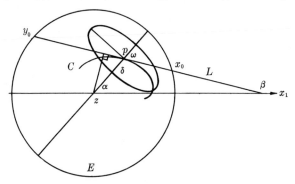

Fig. 63

dS_h, along C is determined by the function $\Phi(p,L)$ defined in (29.4). For if L is the tangent to C at $p_t = [x_1(t), x_2(t)]$, and if $S_h(t)$ and $S_e(t)$ represent the hyperbolic and Euclidean lengths of arc respectively from p_a to p_t on C, then (29.4) implies that at p

(31.1) $\dfrac{dS_h}{dS_e} = \Phi(p,L),$ *or* $dS_h = \Phi(p,L) \sqrt{dx_1^2 + dx_2^2}.$

For $p = z$, (29.4) shows that $\Phi(p,L) = k$. For $p \neq z$, let L form the angle ω with $p \times z$ and put $\delta = e(p,z) = \sqrt{x_1^2 + x_2^2}$. Finally, as in Section 29, denote the intersection of L with E by x_0 and y_0 (see Figure 63). From elementary geometry it follows that

$$e(p,x_0)\, e(p,y_0) = (1 + \delta)\,(1 - \delta) = 1 - \delta^2,$$

and

$$1/2[e(p,x_0) + e(p,y_0)] = e(x_0,y_0)/2 = \sqrt{1 - \delta^2 \sin^2 \omega},$$

hence

$$(31.2) \qquad \Phi(p,L) = \frac{k}{2}\left(\frac{e(p,x_0) + e(p,y_0)}{e(p,x_0)e(p,y_0)}\right) = k\frac{\sqrt{1 - \delta^2 \sin^2 \omega}}{1 - \delta^2}.$$

If α and β are respectively the angles of inclination of $z \times p$ and L, then $\cos \alpha = x_1/\delta$, $\sin \alpha = x_2/\delta$, $\cos \beta = x_1'/\sqrt{x_1'^2 + x_2'^2}$ and $\sin \beta = x_2'/\sqrt{x_1'^2 + x_2'^2}$, where the primes indicate differentiation with respect to t. Then from $\sin^2 \omega = \sin^2 (\beta - \alpha) = \dfrac{(x_1x_2' - x_2x_1')^2}{\delta^2(x_1'^2 + x_2'^2)}$, together with (31.1) and (31.2), we obtain,

$$(31.3) \qquad dS_h = \frac{k[x_1'^2 + x_2'^2 - (x_1x_2' - x_2x_1')^2]^{1/2}}{1 - x_1^2 - x_2^2}\,dt$$

which is usually written as

$$(31.4) \qquad dS_h^2 = k^2\frac{dx_1^2 + dx_2^2 - (x_1dx_2 - x_2dx_1)^2}{(1 - x_1^2 - x_2^2)^2}.$$

This expression is called the *hyperbolic line element in terms of x_1 and x_2*.

The hyperbolic length of an arbitrary, continuously differentiable curve may be obtained from (31.3) by integration. Frequently, however, explicit integration can be circumvented. For instance, to determine the circumference of a hyperbolic circle of radius r, it suffices, in view of (30.3), to find the hyperbolic circumference of the circle with center z and Euclidean radius $\delta = \tanh (r/k)$. For every point p on this circle, ω in (31.2) is $\pi/2$, hence $\Phi(p,L) = k/\sqrt{1 - \delta^2} = k/\mathrm{sech}\,(r/k) = k \cosh (r/k)$. Since the ratio dS_h/dS_e has the constant value $k \cosh (r/k)$, this value times $2\pi\delta$ gives the circumference, that is:

$$(31.5) \qquad \begin{array}{l}\textit{The length of the circumference of a hyperbolic circle of radius r is}\\ \textit{2πk sinh (r/k).}\end{array}$$

The locus K_p defined in (29.5) is obtained by laying off the Euclidean distance $1/\Phi(p,L)$ in both directions from p on every line L through p. Therefore K_z is the circle with Euclidean radius $1/k$. Since K_p can be obtained from K_z by a hyperbolic motion taking z into p, K_p must be an ellipse (which could be calculated from (31.2)). This expresses the fact that hyperbolic geometry is locally Euclidean.

For a fixed point p, the function $\Phi(p,L)$ obtains its maximum value $k/(1 - \delta^2)$ at $\omega = 0$ and its minimum value $k/\sqrt{1 - \delta^2}$ at $\omega = \pi/2$. Therefore the semi-major and the semi-minor axes of K_p have respectively the Euclidean lengths $\sqrt{1 - \delta^2}/k$ and $(1 - \delta^2)/k$. The minor axis, corresponding to $\omega = 0$, must lie on $p \times z$, and the Euclidean area of K_p is given by $\pi(1 - \delta^2)^{3/2}/k^2 = \dfrac{\pi}{k^2}\,\mathrm{sech}^3(r/k)$, where $r = h(p,z)$.

The function $\sigma(p)$ introduced in (29.9) was the quotient of π divided by the area of K_p (compare with π/σ in (25.14)). From (29.10), then, the hyperbolic element of area is

(31.6) $$\frac{k^2}{\sqrt{(1 - x_1^2 - x_2^2)^3}}\, dx_1\, dx_2 = k^2 \cosh^3 (r/k) dx_1\, dx_2.$$

This same result is obtained much more quickly from the fact that the area element coincides with that of the local geometry, which in this case is Euclidean. In general form, the Euclidean area element is

$$\sqrt{E G - F^2}\, dx_1\, dx_2,$$

where E, F and G are respectively the coefficients of dx_1^2, $2\, dx_1\, dx_2$, and dx_2^2 in the expression for dS_h^2. Setting $\zeta^2 = x_1^2 + x_2^2$, E, F and G are obtained from (31.4) as

(31.7) $$E = \frac{k^2(1 - x_2^2)}{(1 - \zeta^2)^2}, \qquad F = \frac{k^2 x_1 x_2}{(1 - \zeta^2)^2}, \qquad G = \frac{k^2(1 - x_1^2)}{(1 - \zeta^2)^2},$$

and $\sqrt{E G - F^2}\, dx_1\, dx_2$ is exactly the left side of (31.6). *In the future, this method will always be used to determine the area element in a geometry which is locally Euclidean.*

The hyperbolic area of any domain is determined from (31.6) by integration. To find the area of the circle with hyperbolic radius r, let the center be at z. If δ denotes the Euclidean radius, and polar coordinates ρ,ω are introduced so that $x_1 = \rho \cos \omega$, $x_2 = \rho \sin \omega$, then by (31.6) the area is

$$\int_0^{2\pi} \int_0^{\delta} \frac{k^2 \rho\, d\rho\, d\omega}{(1-\rho^2)^{3/2}} = 2\pi k^2 \left[\frac{1}{\sqrt{1-\delta^2}} - 1\right] = 2\pi k^2 [\cosh (r/k) - 1] = 4\pi k^2 \sinh^2 (r/2k).$$

(31.7)   *The area of a hyperbolic circle of radius r is $4\pi k^2 \sinh^2 (r/2k)$.*

The area of the right triangle z,b,c of Section 30 is similarly obtained. With the same convention as before the area expressed in polar coordinates is

$$k^2 \int_0^{A} \int_0^{\bar{\beta}\sec\omega} \frac{\rho\, d\rho\, d\omega}{(1-\rho^2)^{3/2}} = k^2 \int_0^{A} \left[\frac{1}{(1 - \bar{\beta}^2 \sec^2 \omega)^{1/2}} - 1\right] d\omega$$

$$= k^2 \int_0^{A} \left\{\frac{\cos \omega}{[(1 - \bar{\beta}^2) - \sin^2 \omega]^{1/2}} - 1\right\} d\omega = k^2 \left[\arc \sin \left(\frac{\sin \omega}{\sqrt{1 - \bar{\beta}^2}}\right) - \omega\right]_0^{A}$$

Using $\bar{\beta} = \tanh (\beta/k)$, and (30.14), this reduces to

$$k^2 [\arc \sin (\sin A \cosh \beta') - A] = k^2 [\arc \sin (\cos B) - A] = k^2 \left(\frac{\pi}{2} - B - A\right).$$

Therefore, the area of the triangle is

(31.8) $$k^2\left[\frac{\pi}{2} - A - B\right] = k^2\left[\pi - \left(\frac{\pi}{2} + A + B\right)\right].$$

The *excess*, $\varepsilon(a,b,c)$, of a general triangle, with vertices a,b,c and angles A,B,C, is defined by

(31.9) $$\varepsilon(a,b,c) = \pi - [A + B + C].$$

Thus (31.8) states that the area of a right triangle is k^2 times the excess. This can be generalized by observing that *excess is additive* in the sense that if a transversal through b (see Figure 64) cuts $S^*(a,c)$ in d forming the triangles a,b,d and d,b,c, then

$$\varepsilon(a,b,c) = \varepsilon(a,b,d) + \varepsilon(d,b,c).$$

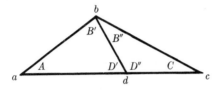

$$\Pi\text{-}(A+B+C) = \Pi - (A+B'+D') + \Pi\text{-}(C+B''+D'')$$

Fig. 64

Since area is also additive, if a general triangle is reprensented as the sum or difference of two right triangles, the application of (31.8) to the parts shows that:

(31.10)   *The area of a triangle a,b,c is $k^2\varepsilon(a,b,c)$.*

Since area is positive this relation also shows:

(31.11)   *The sum of the angles of a hyperbolic triangle is less than π.*

This could also have been deduced from (30.10).

A rather surprising implication of (31.10) is that the area of any triangle, no matter how large its sides, is less than $k^2\pi$. Also, the sum of the angles tends to π when the sides of the triangle tend to zero. Using again the right triangle z,b,c, since it was shown that B tends to zero when γ tends to infinity, the limit of the area of this triangle as $\gamma \to \infty$ is finite, namely is $k^2\left(\frac{\pi}{2} - A\right)$. If a and b are finite points and c is on E (see Figure 65) then the "triangle" a,b,c with two sides asymptotic can be written as the sum or difference of two of the above limit right "triangles", hence has the area $k^2(\pi - A - B)$. When a is finite and b and c are on E, a transversal from a will decompose the triangle into two triangles of the preceding type. Hence, $k^2(\pi - A)$ is the area. Finally, if a,b,c are all on E, that is if each pair of sides are asymptotes, the "triangle" can be subdivided into two "triangles" of the last type and the addition of their areas yields $k^2\pi$ for the area of an "asymptote triangle."

Because of this last result, specifying that an asymptote triangle have area π is equivalent to the normalization $k = 1$. In that case (31.10) shows the area of any triangle is given by its excess.

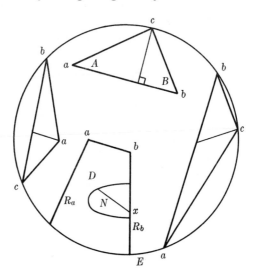

Fig. 65

It should be observed that a domain D bounded by a segment $S(a,b)$ and by two rays R_a and R_b, on the same side of $a \times b$, has infinite area if R_a and R_b do not intersect and are not asymptotes. For as x traverses R_b, away from b, the distance $h(x,b)$ and the distance from x to R_a become infinite (see Section 28). For any given N, however large, it follows that by taking $h(x,b)$ sufficiently large a semi-circle with radius N and center x lies in D. The area bounded by D is then, from (31.7), greater than $2\pi k^2 \sinh^2 (N/2k)$. Since the last expression tends to ∞ with N, the area of D is infinite.

The *area of any polygonal region* can be determined from (31.10) after decomposing the region into triangles. Thus for a four-sided region, convex or not, with vertices a,b,c,d (see Figure 66), the area is

$$k^2[\pi - (A + B_1 + D_1) + \pi - (C + B_2 + D_2)] = k^2[2\pi - (A + B + C + D)],$$

provided all angles are measured inside the region even when they exceed π. By induction the area bounded by a simple, closed n-gon, with *interior* angles $A_1, A_2, \cdots, A_n$ is

$$k^2[(n-2)\pi - (A_1 + A_2 + \cdots + A_n)].$$

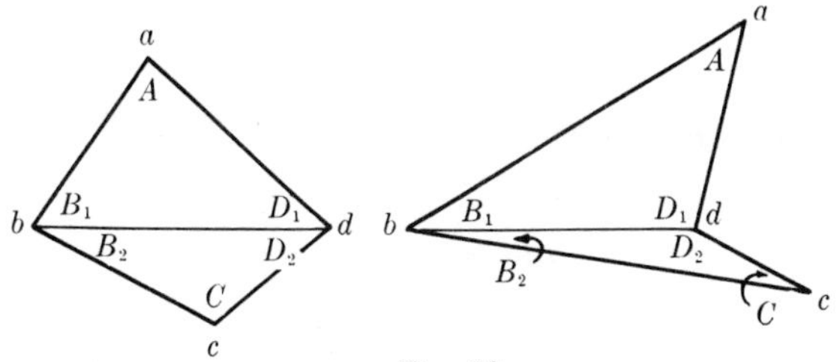

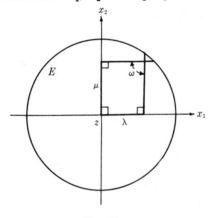

Fig. 66

The hyperbolic angle between two directions at a point can be found by the method of Section 26. For let ω be the angle with vertex (x_1, x_2) determined by the directions dx_1, dx_2 and $\delta x_1, \delta x_2$. Since the angles of the

Fig. 67

local Euclidean geometry at (x_1, x_2) coincide with the hyperbolic angles at the origin, the same will be the case everywhere.[4] Hence (26.5) can be used to obtain ω. With E, F and G given by (31.7) this yields

$$(31.12) \quad \cos \omega = \frac{E\,dx_1\delta x_1 + F(dx_1\delta x_2 + dx_2\delta x_1) + G\,dx_2\delta x_2}{\sqrt{E\,dx_1^2 + 2F\,dx_1 dx_2 + G\,dx_2^2}\ \sqrt{E\,\delta x_1^2 + 2F\,\delta x_1\delta x_2 + G\,\delta x_2^2}}$$

To see how (31.12) is applied, let us *determine the fourth angle ω in a*

---

[4] When considering area we used this principle to determine area. It would have been reasonable to do the same for angles.

quadrilateral, whose other three angles are right angles, and where the hyper-bolic lengths λ *and* μ *of the sides not adjacent to* ω *are known.* We take the vertex opposite ω at z and place the sides issuing from z along the x_1 and x_2 axes respectively (see Figure 67). The coordinates x_1 and x_2 of the vertex at ω can be expressed in the form $x_1 = \tanh(\lambda/k)$ and $x_2 = \tanh(\mu/k)$, and we can take $dx_1 = 1$, $dx_2 = 0$, $\partial x_1 = 0$, $\partial x_2 = 1$ to evaluate ω. Then (31.12) gives

$$(31.13) \qquad \cos\omega = \frac{F}{\sqrt{E}\sqrt{G}} = \frac{x_1 x_2}{\sqrt{1-x_1^2}\sqrt{1-x_2^2}}$$

$$= \frac{\tanh(\lambda/k)\tanh(\mu/k)}{\operatorname{sech}(\lambda/k)\operatorname{sech}(\mu/k)} = \sinh(\lambda/k)\sinh(\mu/k).$$

In direct developments of hyperbolic geometry, where no auxiliary Euclidean metric is used, quadrilaterals with three right angles play a fundamental role. They were first used systematically by Lambert (1728-1777).

32. Equidistant Curves and Limit Circles

The locus of a point which moves so that its hyperbolic distance α from a line η is constant is called the *equidistant curve* C_α^η. If η is taken as the x_1-axis, and (x_1, x_2) is on C_α^η, then $|x_2|$ and $|x_1|$ play the roles of $\bar{\alpha}$ and $\bar{\beta}$ in (30.9), and so are related to α by

$$|x^2| = \tanh(\alpha/k)\operatorname{sech}(\beta/k) = \sqrt{1-x_1^2}\tanh(\alpha/k).$$

Hence (see Figure 68), C_α^η is the ellipse

$$(32.1) \qquad x_1^2 + x_2^2 \coth^2(\alpha/k) = 1.$$

From the hyperbolic point of view C_α^η consists of two disconnected curves. These are not, as in Euclidean geometry, a pair of lines parallel to η, but are convex curves each of which has its concave side toward η. Both intersect η on E and are frequently referred to as the two equidistant curves at a distance α from η. Any hyperbolic motion Φ which carries the upper half-plane $x_2 > 0$ (and therefore the lower half-plane $x_2 < 0$) into itself also carries each of the curves C_α^η into itself. For such a motion takes η into itself. Hence if x is at a distance α from η, $x\Phi$ is also at distance α from η and so is on C_α^η.

The x_2-axis clearly cuts C_α^η at right angles. If now ξ is any perpendicular to η, a motion Φ exists such that $\xi\Phi$ is the x_2-axis and such that $\eta\Phi = \eta$. Because $C_\alpha^\eta\Phi = C_\alpha^\eta$ and Φ preserves angles, it follows that ξ cuts C_α^η at right angles. Hence all lines perpendicular to η are orthogonal to C_α^η.

The length $\lambda_\alpha(p_1,p_2)$ *of the arc on* C_2^η. *whose end points* p_1 *and* p_2 *have* f_1 and f_2 *as feet on* η, *may be found from the formula for the line element* dS_h. Since $dx_2/dx_1 = 0$ at $x_1 = 0$, the relations (31.7) show that

$$\left.\frac{dS_h}{dx_1}\right|_{x_1=0} = \sqrt{E} = \frac{k\sqrt{1 - x_2^2}}{(1 - x_2^2)} = \frac{k}{\sqrt{1 - \tan^2(\alpha/k)}} = k \cosh(\alpha/k),$$

where S_h on C_2^η is supposed increasing with x_1. The corresponding hyperbolic length σ_h on η (also supposed increasing with x_1) is related to x_1 by $x_1 = \tanh(\sigma_h/k)$. Therefore

$$\left.\frac{d\sigma_h}{dx_1}\right|_{x_1=0} = k.$$

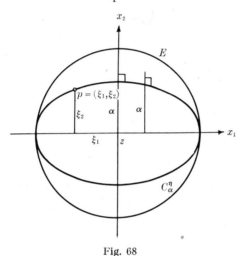

Fig. 68

With the previous relation, then,

$$\left.\frac{dS_h}{d\sigma_h}\right|_{x_1=0} = \cosh(\alpha/k).$$

But S_h and σ_h are invariant under the motions Φ considered above, hence

$$\frac{dS_h}{d\sigma_h} = \cosh(\alpha/k), \qquad -1 < x_1 < 1,$$

which implies

(32.2) $$\lambda_\alpha(p_1, p_2) = h(f_1, f_2) \cosh(\alpha/k).$$

Substituting x_2 from (32.1) in (31.4) and integrating would, of course, also yield this result.

The curves C_α^η appear in a natural way when *geodesic parallel coordinates* ξ_1, ξ_2 are introduced in the space. These are defined as follows. On an arbitrary, oriented line η, a point z is chosen as origin. The point p, with foot f on η, is then given the coordinates ξ_1, ξ_2 where $\xi_1 = \pm h(f, z)$ according as f follows or precedes z on η, and $\xi_2 = \pm h(p, f)$ according as p lies above or below η. The coordinate curves $\xi_1 =$ constant and $\xi_2 =$ constant are then respectively the lines perpendicular to η and the equidistant curves C_α^η.

The definition of ξ_1 and ξ_2 is *intrinsic* in the sense that it does not depend on an auxiliary Euclidean metric. Such an intrinsic definition can also easily be given for x_1 and x_2. We need only take two oriented lines η and σ,

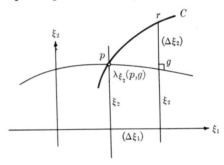

Fig. 69

which are perpendicular at z, and define the coordinates (x_1, x_2) of a general point p, by

$$x_1 = \pm \tanh [h(p, f_\eta)/k], \qquad x_2 = \pm \tanh [h(p, f_\sigma)/k],$$

where f_η, f_σ are respectively the feet of p on η and σ.

One reason for employing the coordinates ξ_1, ξ_2 is that in terms of them the line and area elements take simpler forms. Let $C : (\xi_1(t), \xi_2(t))$ be a continuously differentiable curve (see Figure 69), and consider the points $p = (\xi_1(t), \xi_2(t))$, $g = (\xi_1(t + \Delta t), \xi_2(t))$, and $r = (\xi_1(t + \Delta t), \xi_2(t + \Delta t))$. Setting $\Delta \xi_2 = h(r, g) = \xi_2(t + \Delta t) - \xi_2(t)$, and $\Delta \xi_1 = \xi_1(t + \Delta t) - \xi_1(t)$, (32.2) implies that

$$\lambda_{\xi_2}(p, g) = \Delta \xi_1 \cosh \xi_2(t)/k.$$

Then if ΔS_h represents the arc length of C from p to r, the fact that the geometry is locally Euclidean justifies the relation

$$1 = \lim_{\Delta t \to 0} \frac{(\Delta S_h)^2}{\lambda_{\xi_2}^2(p, g) + h^2(g, r)} = \lim_{\Delta t \to 0} \frac{(\Delta S_h)^2}{(\Delta \xi_1)^2 \cosh^2 (\xi_2/k) + (\Delta \xi_2)^2}$$

therefore

(32.3) $$dS_h^2 = \cosh^2 (\xi_2/k) d\xi_1^2 + d\xi_2^2.$$

The standard form $Ed\xi_1^2 + 2Fd\xi_1 d\xi_2 + Gd\xi_2^2$ for dS_h^2 yields

$$E = \cosh^2 (\xi_2/k), \ F = 0, \text{ and } G = 1.$$

The *hyperbolic area element* is therefore,

(32.4) $$\sqrt{EG - F^2}\,d\xi_1\,d\xi_2 = \cosh (\xi_2/k)d\xi_1\,d\xi_2.$$

Let the points p_1 and p_2 on an equidistant curve C_α^η have f_1 and f_2 as feet on η. We apply (32.4) to compute the area $A_\alpha(p_1 p_2)$ bounded by the segments $S(f_1,f_2)$, $S(p_1,f_1)$, $S(p_2,f_2)$ and by the arc of C_α^η from p_1 to p_2. For simplicity take the figure placed so that η is the ξ_1-axis, p_1 is the point $(0,\alpha)$, and p_2 is the point (ξ,α), with $\xi > 0$. Then,

(32.5) $$A_\alpha(p_1,p_2) = \int_0^\alpha d\xi_2 \int_0^\xi \cosh (\xi_2/k)d\xi_1 = \xi k \sinh (\alpha/k)$$
$$= kh(f_1,f_2) \sinh (\alpha/k).$$

A second type of curve which is a straight line in Euclidean geometry, but not in hyperbolic, is the limit form assumed by a circle whose radius tends to infinity. These so called *"limit circles"* are defined more precisely in the following manner. Let z be a fixed point on an oriented line η, and take x_t a variable point following z with $t = h(z,x_t)$. With C_t denoting the circle $K(x_t,t)$, the limit circle $\Lambda(z,\eta)$ is defined to be $\lim_{t \to \infty} C_t$. The existence of this limit can even be shown in more general spaces, where two points are merely assumed to lie on a metric line, hence is a fact which belongs with those marked by an asterisk in Section 21. In the present case the existence of the limit follows immediately from the model. If η is the x_1-axis and z is the origin, then C_t is an ellipse lying in E and with η as one axis (since it must go into itself in a reflection about η). For $t > t_1$, C_{t_1} is entirely inside C_t with the exception of the point z. As $t \to \infty$, or $e(z,t) \to 1$, the Euclidean curves C_t tend monotonically to an ellipse (see Figure 70), one of whose axes is the segment $S(z,(1,0))$.[5] Since $(1,0)$ is not a point of the hyperbolic plane, $\Lambda(z,\eta)$ appears in hyperbolic geometry as an open curve, but it is obviously not a straight line.

The properties of an isosceles triangle imply that a chord of a circle makes equal angles with the radii to the end points. By generalizing this notion we can obtain a *characterization of limit circles independent of the limit definition.* Let two non-intersecting, oriented lines ξ and η be called *equally oriented* if for any line ζ, intersecting both ξ and η, the positive orientations of ξ and η point to the same side of ζ. Also, if x is any point of

---

[5]It is not hard to find the equation of this ellipse (Exercise [32.8]), and to derive the properties of $\Lambda(z,\eta)$ from the equation. This would be shorter than the following discussion, but less interesting geometrically.

an oriented line ξ and y is a point not on ξ, let the angle between $S(x,y)$ and ξ, denoted by $\angle\,[S(x,y),\xi]$, mean the angle $\angle\,yxx'$ where x' is a point

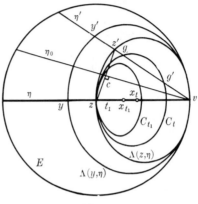

Fig. 70

of ξ following x. The generalization referred to is given, in these terms, by the lemma:

(32.6)   *If ξ and η are equally oriented, non-intersecting lines, then for any point x_0 on ξ exactly one point y_0 exists on η such that*

$$\angle\,[S(x_0,y_0),\xi] = \angle\,[S(x_0,y_0),\eta].$$

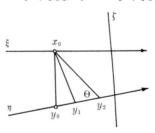

Fig. 71

PROOF: As a variable point y traverses η in the positive direction (see Figure 71) the angle $\angle\,[S(x_0,y),\eta]$ increases. For if y_2 follows y_1 on η and $\theta = \angle\,x_0y_2y_1$, then $\theta + \angle\,[S(x_0,y_2),\eta] = \pi$, from the normalization of angle measure, while $\theta + \angle\,[S(x_0,y_1),\eta] < \pi$ since the angle sum of $\triangle x_0y_1y_2$ is less than π. Because of (30.12), the angle $\angle\,[S(x_0,y),\eta]$ increases continuously from the limit zero to the limit π. At the same time it is clear that the angle $\angle\,[S(x_0,y),\xi]$ decreases continuously but stays between 0 and π. Hence, there is exactly one position y_0 on η for which

$$\angle\,[S(x_0,y_0),\eta] = \angle\,[S(x,_0y_0),\xi].$$

The point y_0 on η is sometimes called the "*corresponding point on η to x_0.*"

Referring again to Figure 70, let the asymptotes η and η' be equally oriented toward their common end point v. As before, let x_t follow z on η, with $t = h(x_t,z)$ and $C_t = K(x_t,t)$. For sufficiently large t the circle C_t intersects η' in two points g and g'. As t increases, one of these points, g', tends toward v and the other, by the definition of $\Lambda(z,\eta)$, tends toward z' the intersection point of $\Lambda(z,\eta)$ and η'. Therefore,

$$\measuredangle \, [S(z,z'),\eta] = \measuredangle \, z'zv = \measuredangle \, zz'v = \measuredangle \, [S(z,z'),\eta'],$$

hence z' is the corresponding point to z on η'. Thus $\Lambda(z,\eta)$ is characterized by:

(32.7) *If z is a point of the oriented line η, then $\Lambda(z,\eta)$ is the locus of points, corresponding to z, on the equally oriented asymptotes to η.*

These asymptotes to η are called *radii of* $\Lambda(z,\eta)$, and it is easily seen that:

(32.8) *A reflection in any radius carries $\Lambda(z,\eta)$ into itself.*

For, with the previous conventions, let z' be the point on $\eta' \cap \Lambda(z,\eta)$, where η and η' are asymptotes, and as before let g be the point of $\eta' \cap C_t$ nearest z'. Designate the reflection in η' by Φ' and the reflection in $g \times x_t$ by Φ_t. Since $C_t\Phi_t = C_t$, $\Lambda(z,\eta) = \lim_{t \to \infty} C_t\Phi_t$. But as $x_t \to v$, $g \to z'$, hence $g \times x_t \to \eta'$ and $\Phi_t \to \Phi'$. Hence $\Lambda(z,\eta) = \lim_{t \to \infty} C_t\Phi_t = \Lambda(z,\eta)\Phi'$. The notation $\Phi_t \to \Phi'$ means, precisely, that for any point p, the $\lim_{t \to \infty} p\Phi_t$ exists and equals $p\Phi'$.

We can now show:

(32.9) *If z' is the point of $\Lambda(z,\eta)$ on the radius η', then $\Lambda(z,\eta) = \Lambda(z',\eta')$.*

PROOF: Let c be the midpoint of $S(z,z')$ and let η_0 be the line perpendicular to $z \times z'$ at c. Because $S(z,z')$ makes equal angles with η and η', the reflection in η_0, which takes z into z', takes η into η'. Then η_0 cannot intersect either η or η', since the intersection would have to be on both η and η'. Therefore η_0 is an asymptote to both η and η'. Because of (32.8) the reflection Φ_0 in η_0 takes $\Lambda(z,\eta)$ into itself. On the other hand, since it is a motion it takes $\Lambda(z,\eta)$ into $\Lambda(z',\eta')$, hence $\Lambda(z,\eta) = \Lambda(z',\eta')$. Because $\Lambda(z,\eta)$ depends only on z and the common point v of all the asymptotes to η, the point v is often called *the center of* $\Lambda(z,\eta)$.

Now let y (Figure 70) be any point of η distinct from z and let y' be defined by $\eta' \cap \Lambda(y,\eta)$. The argument of the preceding proof shows that Φ_0 maps $\Lambda(y,\eta)$ into itself. Since $\eta'\Phi_0 = \eta$, then $h(z,y) = h(z',y')$, or:

(32.10) *$\Lambda(z,\eta)$ and $\Lambda(y,\eta)$ intercept a segment of length $h(z,y)$ on every radius.*

This can be seen also from the original definitions of $\Lambda(z,\eta)$ and $\Lambda(y,\eta)$.

We now establish some metric properties of limit circles. With the previous convention, let q, provided it exists, be the point in which the perpendicular to η at z cuts the radius η' of $\Lambda(z,\eta)$. (See Figure 72.) To obtain $\lambda = h(q,z)$ in terms of $\mu = h(q,z')$, let p_t be the point in which C_t cuts $S(q,x_t)$, and set $h(q,p_t) = \mu_t$. Then the hyperbolic Pythagorean theorem gives

$$\cosh(\lambda/k) = \frac{\cosh(\mu_t + t)/k}{\cosh(t/k)}$$

$$= \frac{\cosh(\mu_t/k)\cosh(t/k) + \sinh(\mu_t/k)\sinh(t/k)}{\cosh(t/k)}$$

$$= \cosh(\mu_t/k) + \sinh(\mu_t/k)\tanh(t/k).$$

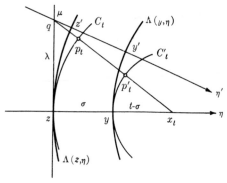

Fig. 72

As $t \to \infty$, $\mu_t \to \mu$, $\tanh(t/k) \to 1$, and the limit expression gives $\cosh(\lambda/k) = \cosh(\mu/k) + \sinh(\mu/k) = e^{\mu/k}$. Therefore,

(32.11) $$\cosh(\lambda/k) = e^{\mu/k}.$$

If η and $q \times z$ are taken for the ξ_1 and ξ_2 axes respectively, (32.11) yields at once the equation of $\Lambda(z,\eta)$ in these geodesic parallel coordinates. For if $p : (\xi_1,\xi_2)$ is an arbitrary point of $\Lambda(z,\eta)$, let f be its foot on η (Figure 73) and let g be the point in which $\Lambda(f,\eta)$ cuts the radius through p. From (32.10), $\xi_1 = h(z,f) = h(p,g)$. Hence, with respect to $\Lambda(f,\eta)$, ξ_1 plays the role of μ in (32.11) and $\xi_2 = h(p,f)$ plays the role of λ. Therefore,

(32.12) $\cosh(\xi_2/k) = e^{\xi_1/k}$ *is the equation* $\Lambda(z,\eta)$.

Since this implies $d\xi_1 = \tanh(\xi_2/k)d\xi_2$ the arc length S of $\Lambda(z,\eta)$ from z to p can be obtained from (32.3) in the form

$$S = \int_0^{\xi_2} \sqrt{1 + \sinh^2(\xi_2/k)}\, d\xi_2 = k\sinh(\xi_2/k).$$

Therefore,

(32.13) *The arc of a limit circle subtending a chord of length γ has length* $2k \sinh (\gamma/2k)$.

Now let y be a point following z on η, with $h(z,y) = \sigma$ (Figure 72). We wish to find the ratio of the arcs intercepted on $\Lambda(z,\eta)$ and $\Lambda(y,\eta)$ by two radii. Let z' and y' denote respectively the point of $\Lambda(z,\eta)$ and that of $\Lambda(y,\eta)$ on the radius η'. As before, let the perpendicular to η at z cut η' in q, and take x_t as a variable point on η tending in the positive direction to ∞. Let C_t and C_t' indicate the circles with center x_t and with radii

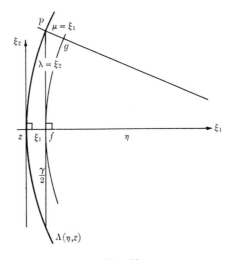

Fig. 73

$t = h(x_t,z)$ and $h(x_t,y) = t - \sigma$ respectively. The points at which C_t and C_t' cut $S(q,x_t)$ are denoted by p_t and p_t'. Since the ratio of the arc $\widehat{yp_t'}$ to $\widehat{zp_t}$ is the same as the ratio of the length of C_t' to that of C_t, (31.5) gives the ratio of the arc lengths as

$$\frac{\sinh (t - \sigma)/k}{\sinh (t/k)} = \frac{\sinh (t/k) \cosh (\sigma/k) - \cosh (t/k) \sinh (\sigma/k)}{\sinh (t/k)}$$
$$= \cosh (\sigma/k) - \sinh (\sigma/k) \coth (t/k).$$

Since $\coth (t/k) \to 1$ as $t \to \infty$ this ratio approaches the limit

$$\cosh (\sigma/k) - \sinh (\sigma/k) = e^{-\sigma/k}.$$

But as $t \to \infty$, $p_t \to y'$ and $p_t \to z'$ and the limit ratio is the value of arc length $\widehat{yy'}$ divided by the arc length $\widehat{zz'}$.[6] Since the limit depends only on σ, we have the theorem:

(32.14)   *When y follows z on η, the arcs of $\Lambda(y,\eta)$ and $\Lambda(z,\eta)$, intercepted by any pair of radii, are in the ratio $e^{-h(y,z)/k}$.*

Choosing $k = 1$, therefore, amounts to defining the unit of length in such a way that for $h(y,z) = 1$ the ratio of the larger to the smaller arc in (32.14) becomes e. Many books use this to define unit length.

Since the reflection Φ_i in the radius $\tau_{,i}$, $i = 1,2$, carries $\Lambda(z,\eta)$ into itself, the product $\Phi = \Phi_1 \Phi_2$ also maps $\Lambda(\tau_{,}z)$ into itself and preserves the orientation. The motion Φ is called a rotation about the end point of $\tau_{,}$. It can be obtained as a limit of rotations as follows. Let η be any radius, and z be the point of the limit circle on this radius. Take any point g of $\tau_{,}' = \tau_{,}\Phi$. As in the previous notation let x_t on η tend to v with C_t, the defining circles of $\Lambda(\eta,z)$. The ray $R(x_t,g)$ cuts C_t in a point p_t. If Φ_t is defined to be the rotation about x_t which takes z into p_t, then as x_t tends to v, Φ_t tends to Φ in the sense previously defined.

The rotations about the endpoint of $\tau_{,}$ form an Abelian subgroup of the group of all motions. This can be seen from the fact that Φ is the limit of the rotations Φ_t.

33. Some Synthetic Properties of Hyperbolic Geometry

The free mobility of the hyperbolic plane implies the *triangle congruence theorems* just as in the Euclidean case. In addition, *two hyperbolic triangles are congruent if their corresponding angles are equal.* In the following, brief discussion of some synthetic constructions, these theorems will be used without further proof.

Two lines ξ and ξ' which do not intersect (in D) and which are not asymptotes will be called *hyperparallels*. In the model, they intersect outside E in a point through which there are exactly two tangents to E. From the discussion of general Hilbert geometries, the line joining the points of tangency is perpendicular to both ξ and ξ' (see (28.11)), and it is the only line with this property (as is also evident from angle considerations).

---

[6]This assumes implicitly that the length of the arc $\widehat{zp_t}$ tends to the length of the arc $\widehat{zz'}$. In general the limit of length does not equal the length of the limit because the analytic expression for length depends also on the first derivitive. In the present case the tangent at a point of $\widehat{zp_t}$ tends uniformly to the tangent of $\Lambda(z,\eta)$ at the limit of the point, which justifies the above procedure.

Two hyperparallels have a unique common perpendicular. Two lines η and η' which have a common perpendicular σ are hyperparallels.

To establish the second part, η and η' cannot intersect in D, for the triangle with sides η, η' and σ would have an angle sum exceeding π, and, because of (30.7), η and η' cannot be asymptotes.

As in Euclidean geometry:

(33.2)   *If a,b,c are any three distinct points, and α, β and γ are perpendicular bisectors of the segments $S(b,c)$, $S(c,a)$ and $S(a,b)$ respectively, then α, β and γ are concurrent at a point p.*

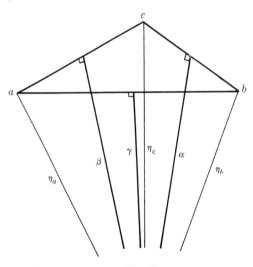

Fig. 74

However, the following possibilities occur:

If p lies in D, then a, b and c lie on a circle with center p and with α, β and γ as radii.

If p lies on E, then a, b and c lie on a limit circle with center p and with α, β and γ as radii.

If p lies outside E, then a, b and c lie on an equidistant curve to a line η, and α, β and γ are perpendicular to η. This includes the special case where a, b and c are collinear.

PROOF: Since in both hyperbolic and Euclidean geometry the perpendicular bisector of $S(x,y)$ is the locus of points equidistant from x and y, the first case follows as in Euclidean geometry. Next, suppose α and β are asymptotes (see Figure 74). Let η_a, η_b, η_c denote the asymptotes through a, b and c respectively which belong to the pencil defined by α and β. Since a reflection about α interchanges η_b and η_c, $S(b,c)$ makes equal angles with these lines, so b is on $\Lambda(c,\eta_c)$. Because the reflection about β interchanges η_c and η_a, the same argument shows that a lies on $\Lambda(c,\eta_c)$. Hence a, b and c are on $\Lambda(c,\eta_c)$ and γ is a radius of this limit circle.

Finally, let α and β be hyperparallels. Then they have a common perpendicular η (Figure 75). Let f_a, f_b and f_c denote the feet of a, b and c on η. The reflection in α carries b into c and η into itself, hence it takes $b \times f_b$ into $c \times f_c$. Therefore $h(b,f_b) = h(c,f_c)$. A similar argument shows that $h(a,f_a) = h(c,f_c)$, hence a, b and c lie on an equidistant curve to η at distance $h(a,f_a)$. Moreover, because $h(a,f_a) = h(b,f_b)$, the reflection in γ', the perpendicular bisector of $S(f_a,f_b)$, carries a into b. Hence γ' coincides with γ.

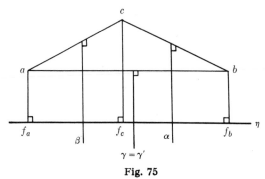

Fig. 75

Because of their many common properties, circles, limit circles and equidistant curves (including lines as the case C_0^η), are often given the common name "*cycles*". Cycle, itself, may be defined by:

(33.3)   *A cycle is the orthogonal trajectory (in the hyperbolic sense) of a pencil of straight lines.*

According as the center of the pencil is inside, on, or outside the boundary E, the cycle is a circle, limit circle, or equidistant curve.

Since the altitudes of a triangle are concurrent, as are the bisectors of the angles and the perpendicular bisectors of the sides, it is natural to inquire about the *medians*. In the triangle a,b,c, let $a \times b$ be taken as the x_1 – axis with the origin z at the midpoint (in both the Euclidean and hyperbolic sense) of $S(a,b)$. (See Figure 76.) Let a line parallel to $a \times b$ intersect $S(a,c)$ in b' and $S(b,c)$ in a'. Then $a' \times b'$ passes through the point at ∞ of $a \times b$ which is the harmonic conjugate of z with respect to a and b. The construction for the fourth harmonic then implies that $a \times a'$, $b \times b'$ and $c \times z \sim c \times c'$ are concurrent at a point u.

In order to express this last result in hyperbolic terms we observe that the x_2-axis, σ, is the perpendicular bisector of $S(a,b)$, and is also perpen-

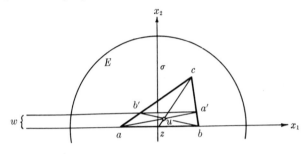

Fig. 76

dicular to $a' \times b'$ (in both the hyperbolic and Euclidean sense). Thus we have the hyperbolic (and Euclidean) theorem:

(33.4)   *In the triangle a,b,c if σ is the perpendicular bisector at z of the side S(a,b), then any perpendicular to σ which cuts S(a,c) in a point b' will also cut S(b,c) in a point a' such that $a \times a'$, $b \times b'$ and $c \times z$ are concurrent.*

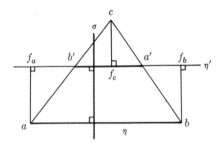

Fig. 77

Since $c \times z$ is a median, the concurrency of the medians will follow if it can be shown that whenever b' in (33.4) is the midpoint of $S(a,c)$, then a' is the midpoint of $S(b,c)$. But this is a consequence of:

(33.5)   *The perpendicular bisector of a side of a triangle is perpendicular to the line joining the midpoints of the other two sides.*

PROOF: Let η' be the join of a' and b', the midpoints of $S(b,c)$ and $S(a,c)$ in triangle a,b,c (Figure 77). Designate by f_a, f_b and f_c the feet of a,b,c on η'. The congruence of triangles b,f_b,a' and c,f_c,a' gives $h(c,f_c) = h(b,f_b)$, and the congruence of triangles a,f_a,b' and c,f_c,b' yields $h(c,f_c) = h(a,f_a)$.

Then since $h(a,f_a) = h(b,f_b)$, the reflection Φ about σ, the perpendicular bisector of $S(f_a,f_b)$, interchanges a and b. Thus Φ carries $a \times b$ into itself and σ is also the perpendicular bisector of $S(a,b)$.

The more elementary constructions of hyperbolic geometry, such as bisecting a segment (with ruler and compass) or erecting a perpendicular at a given point on a line, etc., are exactly the same as in Euclidean geometry. There are, however, some constructions which are new and essentially hyperbolic. One of these is *the construction of a line ζ' through a given point p and asymptotic, in a specified sense, to a given line ζ.*

For simplicity, take p to be the origin z. Let u and v be the intersections of ζ with E, and take $p \times u$, cutting E again at v', as the line ζ' whose construction is to be determined (Figure 78). If f is the foot of p on ζ, then

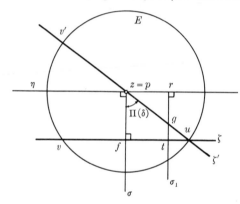

Fig. 78

$\sigma \sim p \times f$ is perpendicular to ζ in both the hyperbolic and the Euclidean sense. Similarly, if η is the perpendicular to σ at p, and g is any point of ζ' (other than z), with foot r on η, then $\sigma_1 \sim g \times r$ is perpendicular to η in both senses. Take $t \sim \zeta \times \sigma_1$. Because, in the Euclidean sense, the lines σ_1, σ, and $v \times v'$ are parallel, $R(p,g,u,v') = R(f,t,u,v)$, hence $h(p,g) = h(f,t)$. Therefore g lies on $K(p,h(f,t))$. Reversing the argument gives the following construction. *From p drop a perpendicular σ to ζ, cutting ζ at f. At p construct η perpendicular to σ, and through any point t on ζ (other than f) take σ_1 perpendicular to η. The circle with center p and radius $h(f,t)$ cuts σ_1 in a point g such that $p \times g$ is the desired asymptote ζ'.*

If $\delta = h(f,p)$ in the above figure, then $\measuredangle\, fpu = \pi(\delta)$, the parallel-angle corresponding to δ. *For a given value of δ, then, $\pi(\delta)$ may be found by erecting a perpendicular σ at a point f on a line ζ, taking p on σ so that $h(p,f) = \delta$, and continuing the foregoing construction.*

The converse problem of constructing δ, given $\pi(\delta)$, is more involved.

To solve it, we first prove a theorem of Hjelmslev (1873-1950), which is rather surprising in hyperbolic geometry.

(33.6) *If $x \to x'$ is a congruent mapping of the line ξ on the line $\xi' \not\nmid \xi$, then the centers of the segments $S(x,x')$ are distinct and collinear or else they all coincide.*

Suppose first (Figure 79) that $x \to x''$ is a congruent mapping of ξ on the line ξ'' in which $S(x,x'')$ and $S(y,y'')$ have a common center p. Then Δxyp is congruent to $\Delta x''y''p$. If now z is any other point of ξ, the congruence of Δxzp to $\Delta x''z''p$ follows from $h(x,z)=h(x'',z'')$, $h(x,p)=h(x'',p)$ and $\sphericalangle zxp = \sphericalangle z''x''p$. But because of this congruence, z, p and z'' are collinear, and p is the center of $S(z,z'')$. It has thus been shown that if two centers coincide then all do.

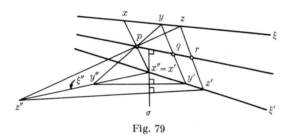

Fig. 79

Now consider the congruent mapping $x \to x'$ of ξ on ξ' in which two segments $S(x,x')$ and $S(y,y')$ have distinct centers p and q. Let r (necessarily distinct from p and q) be the center of a third pair z and z'. Extend the segments $S(y,p)$ and $S(z,p)$ through p to points y'' and z'' so that p is the center of $S(y,y'')$ and $S(z,z'')$. The correspondence $y \to y''$ and $z \to z''$ determines a congruence between ξ and $\xi'' = y'' \times z''$ in which, from the first part of the proof, all centers fall on p. Because x'' must be on $x \times p$, with p the center of $S(x,x'')$, it follows that $x'' = x'$, that is, ξ'' passes through x'. Now let σ be the bisector of $\sphericalangle y''x'y'$. It is then the perpendicular bisector of the base $S(y',y'')$ in the isosceles $\Delta y''x'y'$ and of the base $S(z',z'')$ in the isosceles $\Delta z''x'z'$. From (33.5), then, σ is perpendicular to the line $p \times q$ which joins the midpoints of sides in $\Delta yy'y''$. It is also perpendicular to $p \times r$ which joins the midpoints of the sides in $\Delta zz'z''$. If $p \times q$ and $p \times r$ were distinct, they would form with σ a triangle whose angle sum exceeds π, hence they are identical and p, q and r are collinear. The line $p \times q$ is called the *Hjelmslev line* for the mapping $x \to x'$.

If ξ is oriented then a congruent mapping of ξ on ξ' induces an orientation on ξ'. When ξ and ξ' are non-intersecting and equally oriented, as defined in Section 32, the mapping is called equally directed. For an equally directed congruence between ξ and ξ', the Hjelmslev line certainly exists, that is the centers of $S(x,x')$ cannot coincide.

If the common perpendicular σ to two hyperparallels ξ and ξ' intersects them at f and g' respectively, then the midpoint c of $S(f,g')$ is called the *symmetry point* of ξ and ξ'.

(33.7)   *If $x \to x'$ is an equally-directed congruence between the hyper-parallels ξ and ξ', then the Hjelmslev line passes through the symmetry point c of ξ and ξ'.*

The theorem is trivial if the image f' of f coincides with g'. Assume therefore that $f' + g'$, and let p and q denote the centers of $S(f,f')$ and $S(g,g')$ (Figure 80). The reflection Φ in c (equivalent to the rotation through π) is a motion which interchanges f and g' and preserves perpendicularity, hence it interchanges ξ and ξ'. From $h(f,g) = h(f',g')$ it follows that g and

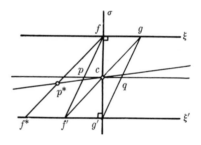

Fig. 80

f' are also interchanged. Therefore Φ interchanges $S(f,f')$ and $S(g,g')$, so $p = q\Phi$, $q = p\Phi$, and the Hjelmslev line $p \times q$ passes through c.

(33.8)   *If $x \to x'$ and $x \to x^*$ are different, equally-directed congruences between the hyperparallels ξ and ξ', then the corresponding Hjelmslev lines are distinct.*

The given conditions imply that $f^* + f'$. (See Figure 80.) If p^*, the midpoint of $S(f,f^*)$, were on $q \times p$ then, by (33.5), the perpendicular bisector of $S(f',f^*)$ would also be perpendicular to $g, \times p$. This bisector and that of $S(f',g')$ would each be perpendicular to both ξ' and $p \times g$, which is impossible. The Hjelmslev lines $c \times p$ and $c \times p^*$ are therefore distinct.

Because the *symmetry point* of two hyperparallels ξ and ξ' is the intersection of the Hjelmslev lines corresponding to two equally-directed congruences, it *may be obtained* as follows. Let a and b be arbitrary on ξ and choose a' and b' on ξ' so that b and b' are on the same side of $a \times a'$ and $h(a,b) = h(a',b')$ (Figure 81). The correspondence $a \to a'$, $b \to b'$ determines the Hjelmslev line $\eta \sim p \times q$, where p and q are the centers of $S(a,a')$ and $S(b,b')$. Now on ξ' take another pair a^*,b^* so that b and b^* are on the same side of $a \times a^*$ and $h(a,b) = h(a^*,b^*)$. The midpoints of $S(a,a^*)$ and $S(b,b^*)$, namely p^* and q^* determine a Hjelmslev line η^* which intersects η in the desired point of symmetry c.

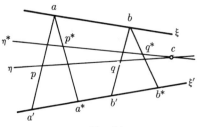

Fig. 81

To construct the common perpendicular to two hyperparallels ξ and ξ' we only have to find the symmetry point c and then construct the perpendicular to ξ from c.

Given two distinct lines $\bar{\xi}_1$ and ξ_2 it now is easy to find the line ζ which is asymptotic both to $\bar{\xi}_1$ in a given sense and to ξ_2 in a given sense, where $\bar{\xi}_1$ and ξ_2 are not themselves asymptotic in either of the given directions. Take q as any point of ξ_2, let $\bar{\xi}_1$ be the line through q asymptotic to $\bar{\xi}_1$ in

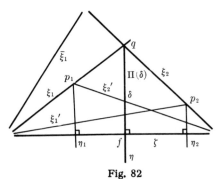

Fig. 82

the sense desired for ζ. Then clearly the common asymptote ζ of $\bar{\xi}_1$ and ξ_2 is the same as that to ξ_1 and ξ_2 (see Figure 82). Let p_i follow q on ξ_i, $i = 1,2$, and let ξ_1' and ξ_2' be the asymptotes to ξ_1 and ξ_2 from p_2 and p_1 respectively. Since ζ is also the common asymptote to ξ_1' and ξ_2 (ξ_1 and ξ_2') the perpendicular to ζ from p_2 (from p_1) will bisect the angle formed by ξ_1' and ξ_2 (ξ_1 and ξ_2'). Hence these angle bisectors, η_2 and η_1, are hyperparallels and ζ is their common perpendicular. Therefore, to construct ζ

it is only necessary to find η_1 and η_2 by bisecting the appropriate angles at p_2 and p_1 respectively and to then follow the previous construction for ζ as the common perpendicular to η_1 and η_2.

In this figure it is also clear that η, the perpendicular to ζ from q, bisects the angle between ξ_1 and ξ_2. If $f \sim \eta \times \zeta$, then the angle between η and ξ_1 is $\pi(\delta)$ for $\delta = h(q,f)$. *To find δ, given $\pi(\delta)$*, the sides of $\pi(\delta)$ need only be taken as η and ξ_1. Laying off the angle $\pi(\delta)$ on the other side of η yields ξ_2. By constructing ζ, the common asymptote to ξ_1 and ξ_2, then $f = \eta \times \zeta$ is obtained, and $\delta = h(q,f)$.

Other construction problems are now readily solved. For instance given two equally-oriented hyperparallels, ξ and ξ', and a point x on ξ, we can easily find the corresponding point x' to x on ξ' (that is, the point x' such that $S(x,x')$ makes equal angles with ξ and ξ'). For if σ is the common perpendicular to ξ and ξ', then x' is obviously the point which lies on the same side of σ as x and has the same distance from $\sigma \times \xi'$ as x does from $\sigma \times \xi$. The same problem, where ξ and ξ' are asymptotes, is left as an exercise.

34. **The Group of Hyperbolic Motions**

It was shown in (22.6) that the algebraic structure of a geometry's group of motions does not depend on the specific representation of the geometry. To determine Γ', the group of motions in hyperbolic geometry, we may therefore use the representation of the space interior to the unit circle $E : x_1^2 + x_2^2 = 1$. By (29.1), Γ' consists of the collineations which carry E into itself.[7]

In homogeneous coordinates E has the form $x_1^2 + x_2^2 - x_3^2 = 0$, and the hyperbolic polarity γ_h which defines E is

$$\gamma_h : \xi_1 = x_1, \qquad \xi_2 = x_2, \qquad \xi_3 = -x_3.$$

The motions Φ in Γ', by (13.3), are the collineations Φ for which $\Phi\gamma_h = \gamma_h\Phi$, hence Γ' is the group Γ_{γ_h} of (13.4). We wish to find the elements in Γ_{γ_h} and, for a later application, we will simultaneously determine Γ_{γ_e} for the elliptic polarity:

$$\gamma_e : \xi_1 = x_1, \qquad \xi_2 = x_2, \qquad \xi_3 = x_3.$$

The two polarities may be written in the common form:

$$\gamma : \xi_1 = x_1, \qquad \xi_2 = x_2, \qquad \xi_3 = \varepsilon x_3, \qquad \text{where } \varepsilon = \pm 1.$$

The corresponding group will be denoted by Γ_γ.

---

[7]See footnote on page 174.

The inverse of the general collineation:

$$\Phi : x_i' = \sum_{k=1}^{3} a_{ik}x_k, \qquad |a_{ik}| \neq 0, \qquad i = 1,2,3,$$

in line coordinates is

$$\Phi^{-1} : \xi_i = \sum_{k=1}^{3} a_{ki}\xi_k', \qquad i = 1,2,3.$$

The correlation $\Phi^{-1}\gamma$ is

$$\xi_i = a_{1i}x_1' + a_{2i}x_2' + \varepsilon a_{3i}x_3', \qquad i = 1,2,3,$$

hence $\Phi^{-1}\gamma\Phi$ is given by

$$\xi_i = a_{1i}\left(\sum_k a_{1k}x_k\right) + a_{2i}\left(\sum_k a_{2k}x_k\right) + \varepsilon a_{3i}\left(\sum_k a_{3k}x_k\right), \qquad i = 1,2,3.$$

That $\gamma\Phi = \Phi\gamma$, or that $\gamma = \Phi^{-1}\gamma\Phi$, means that, after a suitable normalization of the a_{ik}, the following relations hold for all x_i:

$$x_1 = x_1(a_{11}^2 + a_{21}^2 + \varepsilon a_{31}^2) + x_2(a_{11}a_{12} + a_{21}a_{22} + \varepsilon a_{31}a_{32})$$
$$+ x_3(a_{11}a_{13} + a_{21}a_{23} + \varepsilon a_{31}a_{33}),$$
$$x_2 = x_1(a_{11}a_{12} + a_{21}a_{22} + \varepsilon a_{31}a_{32}) + x_2(a_{12}^2 + a_{22}^2 + \varepsilon a_{32}^2)$$
$$+ x_3(a_{12}a_{13} + a_{22}a_{23} + \varepsilon a_{32}a_{33}),$$
$$\varepsilon x_3 = x_1(a_{11}a_{13} + a_{21}a_{23} + \varepsilon a_{31}a_{33}) + x_2(a_{12}a_{13} + a_{22}a_{23} + \varepsilon a_{32}a_{33})$$
$$+ x_3(a_{13}^2 + a_{23}^2 + \varepsilon a_{33}^2).$$

These identities imply that:

$$a_{11}^2 + a_{21}^2 + \varepsilon a_{31}^2 = 1, \quad a_{12}^2 + a_{22}^2 + \varepsilon a_{32}^2 = 1, \quad a_{13}^2 + a_{23}^2 + \varepsilon a_{33}^2 = \varepsilon,$$
(34.1) $$\text{and}$$
$$a_{1i}a_{1k} + a_{2i}a_{2k} + \varepsilon a_{3i}a_{3k} = 0, \quad i,k = 1,2,3, \quad i \neq k.$$

To find the normalization of the coefficients which leads to (34.1), Φ^{-1} is determined by employing (34.1). Substituting from $x_i' = \sum_k a_{ik}x_k$ in the expression $a_{11}x_1' + a_{21}x_2' + \varepsilon a_{31}x_3'$ reduces it, by (34.1), to x_1. Similarly x_2 results from $a_{12}x_1' + a_{22}x_2' + \varepsilon a_{32}x_3'$. In this way, Φ^{-1} is found to be

(34.2) $$\Phi^{-1} : \begin{array}{l} x_1 = a_{11}x_1' + a_{21}x_2' + \varepsilon a_{31}x_3' \\ x_2 = a_{12}x_1' + a_{22}x_2' + \varepsilon a_{32}x_3' \\ x_3 = \varepsilon a_{13}x_1' + \varepsilon a_{23}x_2' + a_{33}x_3'. \end{array}$$

Next, Γ_γ, being a group, contains Φ^{-1}. Therefore Φ^{-1} commutes with γ and satisfies the conditions (34.1), that is :

If $x_i' = \sum a_{ik}x_k$ satisfies (34.1), then

$$(34.3) \quad a_{11}^2 + a_{12}^2 + \varepsilon a_{13}^2 = 1, \quad a_{21}^2 + a_{22}^2 + \varepsilon a_{23}^2 = 1, \quad a_{31}^2 + a_{32}^2 + \varepsilon a_{33}^2 = \varepsilon,$$

and

$$a_{i1}a_{k1} + a_{i2}a_{k2} + \varepsilon a_{i3}a_{k3} = 0, \quad i,k = 1,2,3, \quad i \neq k.$$

Because $\varepsilon^2 = 1$, the determinant of Φ^{-1} satisfies

$$\begin{vmatrix} a_{11} & a_{21} & \varepsilon a_{31} \\ a_{12} & a_{22} & \varepsilon a_{32} \\ \varepsilon a_{13} & \varepsilon a_{23} & \varepsilon^2 a_{33} \end{vmatrix} = \varepsilon^2 \begin{vmatrix} a_{11} & a_{21} & a_{31} \\ a_{12} & a_{22} & a_{32} \\ a_{13} & a_{23} & a_{33} \end{vmatrix} = |\, a_{ik}\,| = |\, a_{ki}\,|.$$

On the other hand the determinant of Φ^{-1} is $|\, a_{ik}\,|^{-1}$, hence $|\, a_{ik}\,| = \pm 1$. Changing the sign of all the coefficients a_{ik} does not affect the collineation geometrically, but does change the sign of $|\, a_{ik}\,|$, hence $|\, a_{ik}\,| = 1$ may be stipulated for the elements of Γ_γ without affecting the group. Since this condition determines the algebraic form of a group element uniquely we have the theorem:

If γ is the polarity $\xi_1 = x_1$, $\xi_2 = x_2$, $\xi_3 = \varepsilon x_3$, where $\varepsilon = \pm 1$, then a collineation Φ lies in Γ_γ if and only if it can be represented in the form

$$(34.4) \qquad x_i' = \sum_k a_{ik}x_k, \quad i = 1,2,3,$$

where the coefficients a_{ik} satisfy (34.1) and $|\, a_{ik}\,| = 1$. Formally different Φ, satisfying these conditions, represent different elements of Γ_γ.

Taking $\varepsilon = -1$, the answer to our original problem is given by:

The group of motions of the hyperbolic plane, $\Gamma_{\gamma h}$, is represented in a one-to-one manner by the transformations

$$x_i' = \sum_k a_{ik}x_k, \quad \text{where } |\, a_{ik}\,| = 1,$$

$$(34.5)$$

$$a_{11}^2 + a_{21}^2 - a_{31}^2 = 1, \quad a_{12}^2 + a_{22}^2 - a_{32}^2 = 1, \quad a_{13}^2 + a_{23}^2 - a_{33}^2 = -1,$$

and

$$a_{1i}a_{1k} + a_{2i}a_{2k} - a_{3i}a_{3k} = 0, \quad i,k = 1,2,3, \quad i \neq k.$$

Particular motions are obtained by specializing the coefficients. For instance, the reflection in the x_1-axis is given by

$$x_1' = x_1, \quad x_2' = -x_2, \quad x_3' = -x_3,$$

and rotations about the origin have the form

$$x_1' = x_1 \cos \alpha + x_2 \sin \alpha, \qquad x_2' = -x_1 \sin \alpha + x_2 \cos \alpha, \qquad x_3' = x_3.$$

To interpret (34.4) for $\varepsilon = 1$ (and for other purposes) we observe that:

(34.6)   *The transformation $x_i' = \sum_k a_{ik} x_k$ leaves the form*
$$\Omega_\varepsilon(x,x) = x_1^2 + x_2^2 + \varepsilon x_3^2$$
invariant if and only if the coefficients a_{ik} satisfy (34.1).

Here invariance means that $\Omega_\varepsilon(x,x) = \Omega_\varepsilon(x',x')$. "Transformation" is used, instead of "collineation," because the value of $\Omega_\varepsilon(x,x)$ depends on the representation of the point x and so has no projective meaning. That the conditions of (34.6) are sufficient can be verified by direct computation from (34.1) which yields

$$x_1'^2 + x_2'^2 + \varepsilon x_3'^2 = \left(\sum_k a_{1k} x_k\right)^2 + \left(\sum_k a_{2k} x_k\right)^2 + \varepsilon \left(\sum_k a_{3k} x_k\right)^2$$
$$= x_1^2 + x_2^2 + \varepsilon x_3^2.$$

Conversely, the condition that the first and last of these expressions be equal for all points x leads back to (34.1).

A little more generally, if the coefficients a_{ik} satisfy (34.1), then $\Omega_\varepsilon(x-y,x-y)$ is also invariant, where $x_i' = \sum_k a_{ik} x_k$, and $y_i' = \sum_k a_{ik} y_k$. The invariance follows from (34.6) and the fact that

$$x_i' - y_i' = \sum_k a_{ik}(x_k - y_k), \qquad i = 1,2,3.$$

Putting $\Omega_\varepsilon(x,y) = x_1 y_1 + x_2 y_2 + \varepsilon x_3 y_3$, $\Omega_\varepsilon(x-y,x-y)$ can be expressed in the form

$$\Omega_\varepsilon(x-y,x-y) = \Omega_\varepsilon(x,x) - 2\Omega_\varepsilon(x,y) + \Omega_\varepsilon(y,y).$$

Since all the terms except $\Omega_\varepsilon(x,y)$ are known to be invariant, it follows that $\Omega_\varepsilon(x,y)$ is also invariant.

(34.7)   *The transformations $x_i' = \sum_k a_{ik} x_k$ which satisfy (34.1) leave $\Omega_\varepsilon(x,y)$ invariant, and conversely.*[8]

If ε is taken as $+1$ and x_1,x_2,x_3 are interpreted as rectangular coordinates in Euclidean three-space, then $[\Omega_1(x-y,x-y)]^{1/2}$ is simply the ordinary Euclidean distance between the points x and y. Since the

---

[8]Since the invariance of $\Omega_\varepsilon(x,y)$ implies that of $\Omega_\varepsilon(x,x)$, the converse follows from (34.6).

transformations which satisfy (34.1) leave $\Omega_1(x - y, x - y)$ invariant, they are motions of the Euclidean space. Indeed, the equations in (34.1) become the well known conditions for a motion of E^3 which leaves the origin fixed, namely:

(34.8) $$\sum_k a_{ki}^2 = 1, \ i = 1,2,3 \quad and \quad \sum_k a_{ki} a_{kj} = 0, \ i,j = 1,2,3, \ i \neq j.$$

Such a motion carries every Euclidean sphere $S_r^2 : \sum_i x_i^2 = r^2, \ r > 0$, into itself. Since x_i/r and $y_i/r, \ i = 1,2,3$, are the direction cosines of the rays R_x and R_y from the origin to the points x and y respectively on S_r^2, then

$$\Omega_1(x,y) = r^2 \cos \ \sphericalangle \ (R_x, R_y).$$

Hence the spherical distance on S_r^2, mentioned in Section 19, can be expressed by

$$d(x,y) = r \ \text{Arc} \cos \ [r^{-2}\Omega_1(x,y)].$$

If attention is restricted to *one* sphere S_r^2 then the ratios of x_1, x_2, x_3 determine a point on S_r^2 uniquely. In particular, if the coordinates are normalized by $\sum_i x_i^2 = 1$, instead of by $\sum_i x_i^2 = r^2$, the spherical distance takes the form

(34.9) $$d(x,y) = r \ \text{Arc} \cos \Omega_1(x,y).$$

35. Weierstrass Coordinates

The foregoing results make it natural to expect that the form $\Omega_{-1}(x,y)$ is related to hyperbolic distance. Since $x_1^2 + x_2^2 - x_3^2 < 0$ is the domain of hyperbolic geometry, we introduce

(35.1) $$x_1^2 + x_2^2 - x_3^2 = \Omega_{-1}(x,x) = -1$$

as a normalization of the projective coordinates. Then every point of D (the interior of E) has exactly two representations, one with $x_3 > 0$ and a second with $x_3 < 0$. With the present representation, (34.5), for the hyperbolic motions, where $|a_{ik}| = 1$, it is not advisable to stipulate $x_3 > 0$.[9] For the great advantage of the normalization $\Omega_{-1}(x,x) = -1$ over the previous one with $x_3 = 1$ is that, by (34.6), it is invariant under Γ_{γ_h}. But the condition $x_3 > 0$ is not invariant in the present representation

---

[9]The normalization $x_3 > 0$ leads naturally to the normalization $a_{33} > 0$. For odd dimensions the condition $|a_{ik}| = 1$ cannot be required, whereas $a_{nn} > 0$ still makes sense (compare Section 52). Therefore as long as only hyperbolic geometry is studied, this normalization is preferable. In even dimensions close analogy to the elliptic case is obtained by requiring $|a_{ik}| = 1$.

of $\Gamma_{\gamma h}$, as can be seen from the reflection in the x_1-axis. Since changing x (or y) to its second normalized representation alters the sign of $\Omega_{-1}(x,y)$ it is clear that only $|\Omega_{-1}(x,y)|$ can have a geometric meaning. To find this meaning, the invariance of $|\Omega_{-1}(x,y)|$ under $\Gamma_{\gamma h}$ implies that there is no loss of generality in taking $x = z = (0,0,1)$ and choosing the second point on the positive x_1-axis so that it has the form $u = (u_1, 0, \sqrt{1 + u_1^2})$, $u_1 > 0$. Then $|\Omega_{-1}(z,u)| = +\sqrt{1 + u_1^2}$. From the form $(u_1/\sqrt{1 + u_1^2}, 0, 1)$ for u, the Euclidean distance from z to u is given by $u_1/\sqrt{1 + u_1^2}$. Because of (30.3), then

$$\tanh h(z,u)/k = \frac{u_1}{[1 + u_1^2]^{1/2}},$$

hence

$$\sqrt{1 + u_1^2} = |\Omega_{-1}(z,u)| = \cosh h(z,u)/k.$$

The equation $\cosh v = w$ has at most two solutions, differing only in sign. Denoting the non-negative solution by Area $\cosh w$, the foregoing can be put in the general form:

(35.2)
$$h(x,y) = k \text{ Area cosh } |\Omega_{-1}(x,y)|, \quad or$$
$$|\Omega_{-1}(x,y)| = \cosh h(x,y)/k.$$

The coordinates x_1, x_2, x_3, normalized by $\Omega_{-1}(x,x) = -1$, were invented by Weierstrass (1815-1897) and are called *Weierstrass coordinates*. They are of the greatest importance in all advanced developments of hyperbolic geometry. It is important, therefore, to have an intrinsic definition for them, which may be obtained as follows. If $(\bar{x}_1, \bar{x}_2, 1)$ are the coordinates of c in the form previously considered, then with $\bar{r} = \sqrt{\bar{x}_1^2 + \bar{x}_2^2}$,

(35.3) $\qquad x_1 = \dfrac{\bar{x}_1}{\sqrt{1 - \bar{r}^2}}, \qquad x_2 = \dfrac{\bar{x}_2}{\sqrt{1 - \bar{r}^2}}, \qquad x_3 = \dfrac{1}{\sqrt{1 - \bar{r}^2}}$

are Weierstrass coordinates of c.

If f_1 and f_2 are the feet of c on the $\bar{x}_2$ and $\bar{x}_1$-axes respectively (see Figure 83), put $u_i = h(c, f_i)$, $i = 1, 2$, $r = h(c,z)$, $\lambda = h(f_2, z)$, and $\alpha = \sphericalangle czf_2$. Taking the space constant k as 1 for the remainder of this section, (30.3) gives the relation $\bar{r} = \tanh r$, hence $x_3 = \dfrac{1}{\sqrt{1 - \bar{r}^2}} = \cosh r$. From (30.13), $\cos \alpha = \sin \sphericalangle f_1 zc = \sinh u_1/\sinh r$. Since $\cos \alpha = \tanh \lambda/\tanh r$, by (30.8), then

$$\sinh u_i = \cos \alpha \sinh r = \frac{\tanh \lambda}{\tanh r} \cosh r \tanh r = \cosh r \tanh \lambda$$
$$= \bar{x}_1 \cosh r = \bar{x}_1/\sqrt{1 - \tanh^2 r} = \bar{x}_1/\sqrt{1 - \bar{r}^2}.$$

Therefore:

(35.4)　　　$\sinh u_1 = \dfrac{\bar{x}_1}{\sqrt{1 - \bar{r}^2}} = x_1.$ *Similarly,* $\sinh u_2 = x_2.$

Thus we have shown:

(35.5)　*If η_1, η_2 are oriented lines perpendicular at a point z, and if u_1, u_2 are the signed hyperbolic distances (in the usual way) of a point c from the η_2 and η_1-axes, and $r = h(c,z)$, then $x_1 = \sinh u_1$, $x_2 = \sinh u_2$, $x_3 = \cosh r$ are the Weierstrass coordinates of the point c for which $k = 1$ and $x_3 > 0$.*

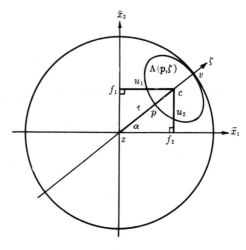

Fig. 83

Nearly all the important formulas of hyperbolic geometry assume much simpler form when expressed in Weierstrass coordinates. Distance in the space is given by (35.2), and the motions by (34.5). Because of the invariance of $\Omega_{-1}(x,y)$ under (34.5) these transformations may also be regarded as the formulas for changing from one system of Weierstrass coordinates to a second.

Since the x_i coordinates are projective, a line in the space has a representation of the form $\Omega_{-1}(x,\xi) = x_1\xi_1 + x_2\xi_2 - x_3\xi_3 = 0$, where ξ_1, ξ_2 and ξ_3 are constants. However, not every equation $\Omega_{-1}(x,\xi) = 0$ is a line of the space. To be so its distance (in the Euclidean sense) from the origin must be less than 1. This means that ξ must satisfy $|\xi_3|/\sqrt{\xi_1^2 + \xi_2^2} < 1$, or $\xi_1^2 + \xi_2^2 - \xi_3^2 > 0$. Hence a normalized representation of the line can be defined in setting $\xi_1^2 + \xi_2^2 - \xi_3^2 = 1$.

A hyperbolic line in Weierstrass coordinates has the form

(35.6) $$\Omega_{-1}(x,\xi) = 0, \text{ with } \Omega_{-1}(\xi,\xi) = 1.$$

In the form (35.6) the coefficients ξ_1, ξ_2, ξ_3 may be taken for (*Weierstrass*) *coordinates of the line.* For a given line, the condition $\Omega_{-1}(\xi,\xi) = 1$ then determines the coordinates ξ_i, $i = 1,2,3$, up to a factor of ± 1. Since a point transformation of (34.5) takes lines into lines, it induces a line transformation $\xi \to \xi'$. To find the transformation formulas for the Weierstrass line coordinates we observe that $\bar{\xi}_1 = \xi_1$, $\bar{\xi}_2 = \xi_2$, $\bar{\xi}_3 = \epsilon\xi_3$ are ordinary line coordinates, because $x \cdot \bar{\xi} = 0$ is then the equation of a line. According to (5.15) the coordinates $\bar{\xi}_i$ are transformed by

$$\bar{\xi}_i' = \sum A_{ik}\bar{\xi}_k.$$

Comparing (5.15) and (34.2) shows that $a_{ik} = A_{ik}$ for $i,k = 1,2$ and $i = k = 3$ while $a_{ik} = -A_{ik}$ when either i or k is 3 and $i \neq k$. Passing from the coordinates $\bar{\xi}_i$ and $\bar{\xi}_i'$ to ξ_i and ξ_i' it follows that

$$\xi_i' = \sum a_{ik}\xi_k, \qquad i = 1,2,3,$$

is the transformation for the Weierstrass line coordinates which is induced by the point coordinate transformation (34.5). In fact, (34.6) shows that $\Omega_{-1}(\xi',\xi') = 1$.

From (34.7) it follows that $\Omega_{-1}(x,\xi)$ is invariant, but again only $|\Omega_{-1}(x,\xi)|$ can have a geometric meaning. To determine what this is, we can suppose ξ is the $\bar{x}_1$-axis, and take x as a point on the postive $\bar{x}_2$-axis. The Weierstrass coordinates of the point and line are then $\left(0, x_2, 1/\sqrt{1 - \bar{x}_2^2}\right)$ and $(0,1,0)$, respectively. Therefore $\Omega_{-1}(x,\xi) = x_2 = \sinh h(x,z)$.

(35.7) *The hyperbolic distance of the point x from the line ξ is given by* $h(x,\xi) = \text{Area sinh } |\Omega_{-1}(x,\xi)|.$

The previous transformation also has $|\Omega_{-1}(\xi,\eta)|$ as an invariant, and various cases occur. If ξ and η are intersecting lines, ξ may be taken for the axis $(0,1,0)$ with $z \sim \xi \times \eta$. Then if α denotes the smaller angle between ξ and η, $(-\sin\alpha, \pm\cos\alpha, 0)$ are Weierstrass coordinates of η. Therefore

(35.8) $$|\Omega_{-1}(\xi,\eta)| = \cos\alpha, \text{ where } \alpha = \measuredangle(\xi,\eta) \leqslant \pi/2.$$

The case where ξ and η are asymptotes may be considered as a limit position of the intersecting case since the smaller angle between the lines tends to zero. It follows that

(35.9) $$|\Omega_{-1}(\xi,\eta)| = 1 \text{ when } \xi \text{ and } \eta \text{ are asymptotes.}$$

Finally, if ξ and η are hyperparallels, let ξ be the $\bar{x}_1$-axis and take the $\bar{x}_2$-axis as the common perpendicular to ξ and η, with η cutting the positive half of this axis at p. If $\beta = h(p,z)$, the Euclidean equation of η, namely $\bar{x}_2 = e(p,z) = \tanh \beta$, yields $x_2 \cosh \beta - x_3 \sinh \beta = 0$, whence $(0, \cosh \beta, -\sinh \beta)$ are Weierstrass coordinates of η satisfying $\Omega_{-1}(\eta,\eta) = 1$. The coordinates of ξ are $(0,1,0)$, so $|\Omega_{-1}(\xi,\eta)| = \cosh \beta$. That is:

(35.10) *If ξ and η are hyperparallels at a distance β (measured on the common perpendicular), then $|\Omega_{-1}(\xi,\eta)| = \cosh \beta$.*

As a corollary of these results we observe that $|\Omega_{-1}(\xi,\eta)| < 1$ is the condition for lines ξ and η to intersect, and that $|\Omega_{-1}(\xi,\eta)| = 0$ is the condition for perpendicularity.

As a further illustration, we derive the equations, in Weierstrass coordinates, for the various types of cycles. First we observe that since the x_i, $i = 1,2,3$, are homogeneous, it must be possible to express any geometric locus in homogeneous form, and that, analytically, any representation can be made homogeneous by multiplying the initial terms by suitable powers of $x_3^2 - x_1^2 - x_2^2 = 1$. For instance (35.2) gives $\Omega_{-1}^2(x,c) = \cosh^2 \rho$ as *the equation of a circle with center c and radius ρ*, and we may write this:

(35.11) $(x_1 c_1 + x_2 c_2 - x_3 c_3)^2 = \cosh^2 \rho = \cosh^2 \rho(x_3^2 - x_1^2 - x_2^2).$

This result, in turn, can be applied in finding *the equation of a limit circle*. For let ζ, which passes through the origin z, be the (oriented) radius of a limit circle, and designate by α the angle between ζ and the positive $\bar{x}_1$-axis (Figure 83). Let p be the point of the limit circle on ζ, and take $c = (c_1,c_2,c_3)$ as a variable point on ζ, where the c_i are Weierstrass coordinates with $c_3 > 0$. Now let $r = h(z,c) \to \infty$, that is, in the model let c approach v, the center of $\Lambda(p,\zeta)$ on E. If $(\bar{c}_1,\bar{c}_2)$ are the Euclidean coordinates of c, then:

(35.12) $\lim_{c \to v} \frac{c_1}{c_3} = \lim_{c \to v} \bar{c}_1 = \cos \alpha,$

and

(35.13) $\lim_{c \to v} \frac{c_2}{c_3} = \lim_{c \to v} \bar{c}_2 = \sin \alpha.$

Setting $s = h(z,p)$, the variable circle with center c and passing through p has radius $\lambda = r \pm s$ according as p precedes or follows z on ζ. Using the fact that $c_3 = \cosh r$, (35.5), and that $\lim_{r \to \infty} \tanh r = 1$, we then find

(35.14) $\lim_{c \to v} \frac{\cosh \lambda}{c_3} = \lim_{c \to v} \frac{\cosh r \cosh s \pm \sinh r \sinh s}{\cosh r}$

$$= \cosh s \pm \sinh s = e^{\pm s}.$$

From (35.11), the circle $K(c,\lambda)$ can be expressed by

$$\left(x_1 \frac{c_1}{c_3} + x_2 \frac{c_2}{c_3} - x^3\right)^2 = \left(\frac{\cosh \lambda}{c_3}\right)^2.$$

Since, by definition, $\Lambda(p,\zeta) = \lim_{c \to v} K(c,\lambda)$, the limits established in (35.12), (35.13) and (35.14), applied in the last equation, yield:

(35.15)   *The equation of $\Lambda(p,\zeta)$ is $(x_1 \cos \alpha + x_2 \sin \alpha - x_3)^2 = e^{\pm 2s}$, where $s = h(p,z)$, and the $+$ or $-$ sign is chosen according as p precedes or follows z on ζ.*

The equation of the equidistant curve C_α^η follows directly from (35.7) in the form:

(35.16) $$(x_1\eta_1 + x_2\eta_2 - x_3\eta_3)^2 = \sinh^2 \alpha.$$

Thus the equation in Weierstrass coordinates of any type of cycle, in general position, is readily found and has a rather simple form.

The complete theory of conic sections in hyperbolic geometry is very involved due to the many types possible. A convenient definition covering all types (including the straight line) is given by: "A conic section is the locus of points having equal distance from two cycles," where cycle is defined as in (33.3).[10] We cannot discuss the theory here, but simply remark that the non-degenerate conics are the curves whose homogeneous equations in Weierstrass coordinates are non-degenerate and of the second degree. For details the reader is referred to:

Story, On non-Euclidean Properties of Conics. *American Journal of Mathematics*, vol. 5 (1882), pp. 358-381;

Liebmann, Nichteuklidische Geometrie. Leipzig, 1905, pp. 182-196. (The later editions of the same book do not discuss conics as thoroughly as this first one does.)

Killing, Die Mechanik in den nichteuklidischen Raumformen. *Journal für Mathematik*, vol. 98 (1885), pp. 1-48.

36. Definition of Elliptic Geometry

The preceding discussion suggests how a metric, similar to the hyperbolic one, might be defined for the whole projective plane by using $\Omega_1(x,y)$. Since every point x can be normalized in exactly two ways so that

(36.1) $$\Omega_1(x,x) = x_1^2 + x_2^2 + x_3^2 = 1,$$

---

[10]We observe that this definition is a generalization of the Euclidean facts in exercises [27.8] and [27.9].

we expect full duality and seek to maintain it algebraically. We choose line coordinates ξ_1, ξ_2, ξ_3 so that also

(36.2) $$\Omega_1(\xi,\xi) = \xi_1^2 + \xi_2^2 + \xi_3^2 = 1.$$

The expression $|\Omega_1(x,y)|$ is invariant under the collineations of $\Gamma_{\gamma e}$:

(36.3)
$$x_i' = \sum_k a_{ik}x_k, \qquad |a_{ik}| = 1, \qquad \sum_i a_{ik}^2 = 1, \qquad k = 1,2,3,$$
$$\sum_k a_{ki}a_{kj} = 0, \qquad i,j = 1,2,3, \qquad i \neq j.$$

Again, (5.15) and (34.2) imply that

(36.4) $\quad \xi_i' = \sum_k a_{ik}\xi_k$ *is the line transformation induced by* (36.3).

The expressions $|\Omega_1(x,y)|$, $|\Omega_1(x,\xi)|$ and $|\Omega_1(\xi,\eta)|$ are thus all invariants of (36.3). As (34.9) implies, and as was proved in Section 19, the Cauchy inequality (19.1) justifies the relation

(36.5) $$|\Omega_1(x,y)| \leqslant \Omega_1^{1/2}(x,x)\Omega_1^{1/2}(y,y) = 1.$$

We may therefore define a metric, called *elliptic,* by means of

(36.6) $$\varepsilon(x,y) = k \text{ Arc cos } |\Omega_1(x,y)|.$$

Any pair of lines ξ,η are now intersecting and the *smaller angle* which they form, $\sphericalangle(\xi,\eta)$, is defined by

(36.7) $$\sphericalangle(\xi,\eta) = \text{Arc cos } |\Omega_1(\xi,\eta)|.$$

Here complete duality with (36.6) prevails only if $k = 1$. The factor k cannot be introduced in (36.7) if $k \neq 1$ without destroying the normalization giving π for the measure of straight angles.

To see that $\varepsilon(x,y)$ actually is a projective metric we recall (see (19.1)) that the equality sign holds in the inequality (36.5) if and only if the triples x_i and y_i are proportional. Therefore $\varepsilon(x,y) = 0$ if and only if x and y are the same point. That $\varepsilon(x,y) = \varepsilon(y,x)$ follows at once from $\Omega_1(x,y) = \Omega_1(y,x)$. The remaining facts are most easily obtained by using the invariance of $\varepsilon(x,y)$ under the transformations of $\Gamma_{\gamma e}$. We show first:

(36.8) *There are collineations in $\Gamma_{\gamma e}$ which carry a given point x (line ξ) into a given point $x' \neq x$ (line $\xi' \neq \xi$).*

Observe that for any point y the polar ξ_y under γ has the equation

(36.9) $$\Omega_1(x,y) = 0, \quad or \quad \varepsilon(x,y) = k\pi/2.$$

The polar is therefore the locus of points x at a maximum distance from y. By (13.6), the harmonic homology Φ, with y as center and ξ_y as axis, commutes with γ. It belongs therefore to Γ_{γ_e} and leaves distance invariant. Let z, not on ξ_y, be any point distinct from y and take $z' = z\Phi$. Then for $z^* \sim \xi_y \times (z \times z')$ we have

$$(36.10) \qquad \varepsilon(z,z^*) = \varepsilon(z',z^*) \quad and \quad \varepsilon(y,z) = \varepsilon(y,z').$$

Because of this, Φ is called both a *reflection in y and in ξ_y*.

Clearly y, z^*, z and z' form a harmonic set. If now x and x' are given distinct points, the hyperbolic involution on $x \times x'$ with x and x' as fixed points has by (13.17) a corresponding pair y_1, y_2 in common with the elliptic involution induced on $x \times x'$ by γ_e. The reflection in y_i or ξ_{y_i}, $i = 1,2$, lies in Γ_{γ_e} and takes x into x'.

To establish the triangle inequality,

$$(36.11) \qquad \varepsilon(a,b) + \varepsilon(b,c) \geqslant \varepsilon(a,c),$$

it may be assumed, because of (36.8), that b is $(0,0,1)$. Also, a, b and c may be supposed distinct (otherwise (36.11) is trivial). If a and c are normalized so that a_3 and c_3 are not negative, then

$$|\Omega_1(a,b)| = a_3, \qquad |\Omega_1(b,c)| = c_3, \qquad |\Omega_1(a,c)| = \left|\sum_i a_i c_i\right|,$$

and (36.11) is equivalent to establishing

$$\text{Arc cos } a_3 + \text{Arc cos } c_3 \geqslant \text{Arc cos}\left|\sum_i a_i c_i\right|.$$

Since between 0 and π the cosine decreases, the last relation is valid if and only if taking the cosine of both sides reverses the inequality. That is, the problem is reduced to justifying that

$$(36.12) \quad a_3 c_3 - \sqrt{1 - a_3^2}\sqrt{1 - c_3^2} = a_3 c_3 - \sqrt{a_1^2 + a_2^2}\sqrt{c_1^2 + c_2^2} \leqslant \left|\sum_i a_i c_i\right|.$$

But the Cauchy inequality,

$$(36.13) \qquad (a_1 c_1 + a_2 c_2)^2 \leqslant (a_1^2 + a_2^2)(c_1^2 + c_2^2)$$

implies

$$-\sqrt{a_1^2 + a_2^2}\sqrt{c_1^2 + c_2^2} \leqslant -|a_1 c_1 + a_2 c_2|,$$

hence

$$a_3 c_3 - \sqrt{a_1^2 + a_2^2}\sqrt{c_1^2 + c_2^2} \leqslant a_3 c_3 - |a_1 c_1 + a_2 c_2| \leqslant \left|\sum_i a_i c_i\right|.$$

Thus (36.12) is established which implies (36.11). Moreover, the triangle equality can hold only when (36.12) is an equality, hence only when a_1 and a_2 are proportional to c_1 and c_2. But in that case c satisfies the equation

$$\begin{vmatrix} x_1 & x_2 & x_3 \\ a_1 & a_2 & a_3 \\ 0 & 0 & 1 \end{vmatrix} = 0,$$

which represents the line $a \times b$, so the three points are collinear.

To show that the projective lines are also metric lines it suffices, in virtue of (36.8), to prove this for the line $x_3 = 0$. Any point of $x_3 = 0$ has a unique representation in the form

$$x_\omega = (\cos \omega/k, \sin \omega/k, 0) \qquad 0 \leqslant \omega < k\pi.$$

If x_α is a second point of the line, then $(1/k) \, | \, \omega - \alpha \, | < \pi$ and

$$\Omega_1(x_\omega, x_\alpha) = \cos (\omega/k) \cos (\alpha/k) + \sin (\omega/k) \sin (\alpha/k) = \cos | \, \omega - \alpha \, |/k.$$

Therefore,

$$\text{arc } \cos | \, \Omega_1(x_\omega, x_\alpha) \, | = \min \{ \, | \, \omega - \alpha \, |/k, \, \pi - | \, \omega - \alpha \, |/k \, \},$$

so

$$\varepsilon(x_\omega, x_\alpha) = \min \{ \, | \, \omega - \alpha \, |, \, k\pi - | \, \omega - \alpha \, | \, \}, \qquad 0 \leqslant \omega, \qquad \alpha < k\pi.$$

This shows (compare (20.6)) that $x_3 = 0$, and hence that every projective line, is congruent to the circle $S^1_{k/2}$, which completes the details of *proving* $\varepsilon(x,y)$ *to be a projective metric*.

If $\measuredangle (\xi, \eta)$, as defined in (36.7), is taken for a metric it follows, by duality, that every pencil of lines with this metric is congruent to the circles $S^1_{1/2}$, that is, is congruent to an ordinary pencil of lines with the angle $\leqslant \pi/2$ as measure.

We now consider the *elliptic motions* more closely. If we adopt the former definition of a rotation about p, namely a motion having only p as a fixed point, then the reflection in p is not a rotation since it leaves all points of ξ_p invariant. But it is natural to think of this reflection as a rotation through π. Therefore a *rotation about* p is defined to be a motion which has p as a fixed point and which has no other fixed points save possibly points of ξ_p. This definition is self-dual since a rotation about p has ξ_p and possibly lines through p, as fixed lines. This may be expressed by saying that a rotation about p is a translation along ξ_p.

Because of theorem (9.2), *every motion of the elliptic plane has at least one fixed point*. The motions which leave a point p fixed form a subgroup Γ_p of the group $\Gamma_{\gamma e}$, and since a motion always exists carrying p into a point g, then by (22.6) the groups Γ_p and Γ_g are isomorphic. To study Γ_p, therefore, it suffices to take $p = (0,0,1)$.

For the point $p = (0,0,1)$ to be fixed under (36.3), we must have $a_{13} = a_{23} = 0$. The polar of p, namely $x_3 = 0$ must map into itself and, as was shown in Section 15, this implies $a_{32} = a_{31} = 0$. The condition $\sum_i a_{i3}^2 = 1$, reduces then to $a_{33} = \pm 1$, and the remaining conditions simplify to:

$$a_{11}^2 + a_{21}^2 = 1, \qquad a_{21}^2 + a_{22}^2 = 1, \qquad a_{11}a_{12} + a_{21}a_{22} = 0.$$

These equations were solved in Section 26 (compare (26.14)) where it was found that

$$\begin{pmatrix} a_{11} & a_{12} \\ a_{21} & a_{22} \end{pmatrix} = \begin{pmatrix} \cos \alpha & -\sin \alpha \\ \sin \alpha & \cos \alpha \end{pmatrix}, \text{ or } \begin{pmatrix} a_{11} & a_{12} \\ a_{21} & a_{22} \end{pmatrix} = \begin{pmatrix} \cos \alpha & \sin \alpha \\ \sin \alpha & -\cos \alpha \end{pmatrix}.$$

In the latter case all points of the line $x_2 = x_1 \tan (\alpha/2)$ were shown to be fixed, hence the motion is the reflection in this line. The former case, together with $|a_{ik}| = 1$ yields $a_{33} = 1$, hence the mapping is given by

(36.14)
$$\begin{aligned} x_1' &= \quad x_1 \cos \alpha + x_2 \sin \alpha \\ x_2' &= -x_1 \sin \alpha + x_2 \cos \alpha \\ x_3' &= \quad x_3. \end{aligned}$$

This is either the identity ($\alpha = 0$) or a rotation about p. For, by (36.4), the line $\xi = (\xi_1, \xi_2, 0)$ through p goes into

$$\xi' = (\xi_1 \cos \alpha - \xi_2 \sin \alpha, \ \xi_1 \sin \alpha + \xi_2 \cos \alpha, \ 0).$$

But $\xi' \sim \xi$ only for $\alpha = 0$ and $\alpha = \pi$. Hence for $\alpha \neq 0$ the motion (36.14) has only p as a fixed point if $\alpha \neq \pi$ and is the reflection in p (or in $x_3 = 0$) when $\alpha = \pi$. Since a reflection in a line is also a rotation, we have the remarkable fact:

(36.15)   *Every elliptic motion, which is not the identity, is a rotation.*

The smaller angle between ξ and ξ' above is

$$\begin{aligned} \text{Arc cos } |\Omega_1(\xi, \xi')| &= \text{Arc cos } |\xi_1^2 \cos \alpha - \xi_1 \xi_2 \sin \alpha + \xi_1 \xi_2 \sin \alpha + \xi_2^2 \cos \alpha | \\ &= \text{Arc cos } |(\xi_1^2 + \xi_2^2) \cos \alpha | = \text{Arc cos } |\cos \alpha | \\ &= \min \{ \alpha, \pi - \alpha \}, \end{aligned}$$

which shows that *the number α in (36.14) has the same interpretation as in the Euclidean case, namely the angle through which the rotation revolves a line on p.* That min $\{ \alpha, \pi - \alpha \}$ is obtained instead of α is due to the fact that all definitions were made in terms of the smaller angle. This can be remedied in an obvious way by defining the measure of the other angle between two lines to be the supplement of the first. We will use this notion in the future whenever it is convenient to do so.

The circle about $p = (0,0,1)$, with radius ρ, has the equation

$$|\Omega_1(p,x)| = \cos (\rho/k),$$

hence

$$x_3^2 = \cos^2 (\rho/k), \quad \text{or} \quad x_1^2 + x_2^2 = \sin^2 (\rho/k), \quad 0 < \rho \leqslant k\pi/2.$$

The polar of the center, in this case $x_3 = 0$, is itself a circle about p. It is the circle of maximum radius. When $\rho < k\pi/2$ any line through the center intersects the circle in two, diametrically opposite points g_α and $g_{\pi+\alpha}$. For $p = (0,0,1)$ they are expressed by

$$g_\alpha = (\cos\alpha\,\sin\rho/k,\ \sin\alpha\,\sin\rho/k,\ \cos\rho/k),$$
$$g_{\alpha+\pi} = (-\cos\alpha\,\sin\rho/k,\ -\sin\alpha\,\sin\rho/k,\ \cos\rho/k).$$

As $\rho \to \pi k/2$ the two points g_α and $g_{\alpha+\pi}$ tend to the same point

$$r_\alpha = r_{\alpha+\pi} = (\cos\alpha,\ \sin\alpha,\ 0)\ \text{on } \xi_p.$$

To better understand this strange phenomenon, orient ξ_p so that increasing α corresponds to the positive direction. For a fixed ρ between 0 and $k\pi/2$ denote by S_α the oriented segment from r_α to g_α (where "segment" between two points means the shorter of the two intervals which they define on the line they determine). If now r_α starts from r_0 and traverses ξ_p in the positive direction, then it returns for $\alpha = \pi$ to $r_\pi = r_0$. The variable segment S_α, however, does not return to its original position S_0 but to the segment S_π which connects $r_\pi = r_0$ and $g_\pi(\neq g_0)$. The segments S_0 and S_π are oppositely directed from r_0 and only after r_α traverse ξ_p a second time does S_α return to S_0. This means that we cannot distinguish opposite sides of the line ξ_p. *A person walking in the elliptic plane along ξ_p in the positive direction, having in the beginning his head in the direction of S_0, returns, after traversing ξ_p, to his original position but with his head in the opposite direction.* (Thus the elliptic plane has, so to speak, only one side.) This is not true of a general circle $K(p,\rho)$, $\rho < k\pi/2$. If ξ_p is replaced by a circle $K(p,\rho')$, $\rho' < k\pi/2$, then, with the same conventions, S_α returns to its original position when r_α traverses $K(p,\rho')$ only once.

A second way of understanding this difference is to observe that an ordinary circle $K(p,\rho)$ *decomposes the plane into two parts, whereas ξ_p does not.* More precisely, all points not on $K(p,\rho)$ can be expressed as the union of two disjoint sets μ_1 and μ_2 such that it is impossible to connect any point of one set to any point of the other by a continuous curve which does not intersect $K(p,\rho)$. For ξ_p, however, no such sets exist and any two points not on ξ_p can be connected without crossing ξ_p. This is obvious since $p_1 \times p_2$ cuts ξ_p only once and one of the two intervals, defined by p_1 and p_2 on $p_1 \times p_2$, does not contain the intersection point. For the case of $K(p,\rho)$, μ_1 may be defined as the set of points x for which $\varepsilon(p,x) < \rho$ and μ_2 as those points for which $\varepsilon(p,x) > \rho$. If $z(t) = (z_1(t), z_2(t), z_3(t))$, $0 \leqslant t \leqslant 1$, is a continuous curve from $p_1 = z(0)$ to $p_2 = z(1)$, where p_i is in μ_i, $i = 1,2$, then $\varepsilon(p,z(t))$ varies continuously with t. Since $\varepsilon(p,z(0)) < \rho$ and $\varepsilon(p,z(1)) > \rho$ it follows that for some value $\bar{t}$, $0 \leqslant \bar{t} \leqslant 1$, $\varepsilon(p,z(\bar{t})) = \rho$, hence $z(t)$ crosses $K(p,\rho)$.

Unfortunately there is no real model of the projective plane in ordinary space (one three-dimensional way of interpreting it will be given later).

However, *there are surfaces in E^3 on which the phenomenon of one-sidedness occurs*. The simplest of these is the so called Moebius strip. If a rectangular strip of paper, having consecutive vertices a,a',b',b (see diagram) is folded so that a falls on a' and b on b', a simple loop results, and the former segments $S(a,a')$, $S(b,b')$ become the boundaries of the loop surface, and these are distinct. The surface clearly has two sides. We can make two pencil marks which cannot be joined, by any curve drawn on the paper, which does not cross one of the boundary curves. However, if the original strip is twisted as it is folded, so that a falls on b' and b on a' (Figure 84), the resulting surface, called a Moebius strip (after Moebius (1790-1868)), is seen to have a single boundary curve. Two dots placed anywhere on the paper can be joined by a continuous curve which does not cross the bound-

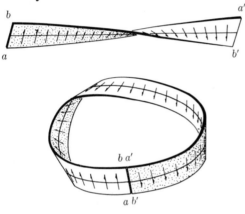

Fig. 84

ary. For this reason the Moebius strip, like the elliptic plane, is said to be one-sided, and many analogies between the two surfaces are easily discovered. It should be noted, however, that the existence on a surface of closed curves which do not decompose the surface is not in general related to the one or two-sidedness of the surface. A torus, that is a surface shaped like a doughnut (pictured in Figure 96), is a two-sided surface though it carries closed curves which do not decompose the surface.

37. Elliptic Trigonometry
Relation of Elliptic and Spherical Geometry

In developing the trigonometry of the right triangle we will leave the constant k out of the calculations, but add it in the final formulas. Because motions exist, a general right triangle a,b,c may be placed so that the vertex c of the right angle is $(0,0,1)$ with the sides of the angle along $\delta_1(x_1 = 0)$ and $\delta_2(x_2 = 0)$. (See Figure 85.) The other two vertices then have the form $a = (0, a_2, a_3)$ and $b = (b_1, 0, b_3)$, and we choose the representations so that a_3 and b_3 are non-negative. If $\alpha = \varepsilon(b,c)$, $\beta = \varepsilon(a,c)$ and $\gamma = \varepsilon(a,b)$, then $\cos \alpha = |\Omega_1(b,c)| = b_3$ and $\cos \beta = |\Omega_1(a,c)| = a_3$. Hence with the proper orientation of the axes, $a = (0, \sin \beta, \cos \beta)$ and $b = (\sin \alpha, 0, \cos \alpha)$. Thus, $\cos \gamma = |\Omega_1(a,b)| = \cos \alpha \cos \beta$. Therefore

$$(37.1) \qquad \cos \gamma/k = \cos \alpha/k \cos \beta/k$$

is the *elliptic form of the Pythagorean theorem*.

For the other trigonometric formulas we need the line coordinates of $a \times b$. In non-normalized form these can be expressed by

$$(\sin \beta \cos \alpha, \cos \beta \sin \alpha, -\sin \alpha \sin \beta).$$

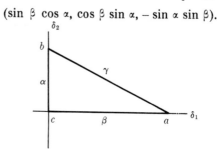

Fig. 85

Setting

$$(37.2) \qquad \begin{aligned} \lambda^{-1} &= [\sin^2 \beta \cos^2 \alpha + \cos^2 \beta \sin^2 \alpha + \sin^2 \alpha \sin^2 \beta]^{1/2} \\ &= [\sin^2 \beta + \cos^2 \beta \sin^2 \alpha]^{1/2} = [\sin^2 \beta \cos^2 \alpha + \sin^2 \alpha]^{1/2}, \end{aligned}$$

elliptic line coordinates of $a \times b$ are $(\lambda \sin \beta \cos \alpha, \lambda \cos \beta \sin \alpha, -\lambda \sin \alpha \sin \beta)$.

Let A and B denote the angles of the triangle at a and b respectively. If $A = \pi/2$ then $a \times c$ is the polar of b and $\alpha = \gamma = k\pi/2$. Therefore, $\alpha < k\pi/2$ implies $A < \pi/2$ and similarly $\beta < k\pi/2$ implies $B < \pi/2$. In any case, $A \leqslant \pi/2$ and $B \leqslant \pi/2$. The relation

$$(37.3) \qquad \cos^2 A = \Omega_1^2(\delta_1, a \times b) = \lambda^2 \sin^2 \beta \cos^2 \alpha$$

expresses A *in terms of* α *and* β. From (37.2), $\sin^2 \beta \cos^2 \alpha = \lambda^{-2} - \sin^2 \alpha$. Hence

$$\sin^2 A = 1 - \cos^2 A = 1 - \lambda^2 \sin^2 \beta \cos^2 \alpha = 1 - \lambda^2(\lambda^{-2} - \sin^2 \alpha) = \lambda^2 \sin^2 \alpha.$$

Combining this with (37.3) gives the equivalent, but simpler, form:

$$(37.4) \qquad \tan A = \frac{\tan (\alpha/k)}{\sin (\beta/k)}, \qquad \tan B = \frac{\tan (\beta/k)}{\sin (\alpha/k)}. \quad [11]$$

---

[11] As before, the symmetry of the situation gives simultaneous results for A and B.

To obtain A *in terms of* β *and* γ, we first use (37.2) and (37.3) to produce

$$\cos^2 A = \frac{\sin^2 \beta \cos^2 \alpha}{\sin^2 \beta + \cos^2 \beta \sin^2 \alpha}.$$

Replacing $\cos \alpha$ by $\cos \gamma/\cos \beta$, from (37.1), this becomes

$$\cos^2 A = \frac{\tan^2 \beta \cos^2 \gamma}{\sin^2 \beta + \cos^2 \beta(1 - \cos^2 \gamma/\cos^2 \beta)} = \frac{\tan^2 \beta \cos^2 \gamma}{1 - \cos^2 \gamma},$$

so

(37.5) $$\cos A = \frac{\tan (\beta/k)}{\tan (\gamma/k)}, \qquad \cos B = \frac{\tan (\alpha/k)}{\tan (\gamma/k)}.$$

Other standard relations now follow by simple calculations. For example:

(37.6) $$\sin A = \frac{\sin (\alpha/k)}{\sin (\gamma/k)}, \qquad \sin B = \frac{\sin (\beta/k)}{\sin (\gamma/k)}.$$

(37.7) $$\cos (\alpha/k) = \frac{\cos A}{\sin B}, \qquad \cos (\beta/k) = \frac{\cos B}{\sin A}.$$

(37.8) $$\cos (\gamma/k) = \cot A \cot B.$$

As in hyperbolic geometry, the angles of the right triangle determine the sides.

If b and c are distinct, a consequence of the Pythagorean theorem, $\cos \gamma/k = \cos \alpha/k \cos \beta/k$, is that $\cos \gamma/k < \cos \beta/k$, that is $\gamma > \beta$, unless $\cos \beta/k = 0$. If $\beta < k\pi/2$, then c is the unique foot of a on δ_2. Moreover, $|\Omega_1(a, \delta_2)| = \sin \beta$. Thus the term *"perpendicular"* can be used *in elliptic geometry* with the same sense defined for open spaces (Section 21), except that "foot" and "perpendicular" cease to be unique when a is the pole of δ_2. Formally:

> *if p is not the pole of η, then p has a unique foot f on η, and*
(37.9) $\angle (\eta, p \times f) = \pi/2$. *The distance of p from η is given by*
$$\varepsilon(p, \eta) = \varepsilon(p, f) = k \text{ Arc sin } |\Omega_1(p, \eta)| < k\pi/2.$$

Using the name "circular functions" for the sin, cos, etc., we can standardize as follows the *close analogy between the formulas of hyperbolic and elliptic trigonometry*. If, in elliptic formulas for a right triangle, the circular functions of lengths are replaced by the corresponding hyperbolic functions (sinh for sin, etc.), and the circular functions of angles are unaltered, the new formulas are hyperbolic. Conversely, replacing the hyperbolic functions by the corresponding circular ones changes hyperbolic trigonometry to elliptic trigonometry.

The identity of the above elliptic formulas with those of spherical trigonometry has probably been noticed. This phenomenon, of course, is not surprising in view of the similarity between the spherical distance, (34.9):

$$d(x,y) = r \text{ Arc cos } \Omega_1(x,y),$$

and the elliptic distance, (36.6):

$$\varepsilon(x,y) = k \text{ Arc } \cos |\Omega_1(x,y)|.$$

Since $\Omega_1(x,-y) = \Omega_1(-x,y) = -\Omega_1(x,y) = -\Omega_1(-x,-y)$, the number $|\Omega_1(x,y)|$ is either $\Omega_1(x,y)$ or $\Omega_1(x,-y)$. Therefore,

(37.10) $\varepsilon(x,y) = kr^{-1} \min \{ d(x,y), d(x,-y) \}.$

We may look on the situation as follows. Associated with the point x on the sphere S_r^2 is the point x in the elliptic plane P_e^2, which has the same coordinates. The points (x_1,x_2,x_3) and $(-x_1,-x_2,-x_3)$ on the sphere then map into the same point on the plane. *The correspondence is thus two-to-one.* If Ψ represents the mapping and x,y on S_r^2 are such that $d(x,y) \leqslant \pi/2$ then $d(x,y) = \varepsilon(x\Psi,y\Psi)$. When $d(x,y) \geqslant \pi/2$ then the point $-y$, antipodal to y, has distance $d(x,-y) \leqslant \pi/2$ from x, and

$$d(x,-y) = \varepsilon(x\Psi,-y\Psi) = \varepsilon(x\Psi,y\Psi).$$

To Ψ corresponds also a two-to-one mapping of the group, Γ_{sp}, of spherical motions on the group, $\Gamma_{\gamma e}$, of elliptic motions. An element of Γ_{sp} has the form

$$x_i' = \sum_k a_{ik}x_k, \text{ where } \sum_k a_{ki}^2 = 1, \quad i = 1,2,3, \quad \text{and}$$

$$\sum_j a_{ji}a_{jk} = 0, \qquad i,k = 1,2,3, \qquad i \neq k,$$

hence it differs from an elliptic motion by omitting the condition $|a_{ik}| = 1$ (instead $|a_{ik}| = \pm 1$ in Γ_{sp} (see Section 34)). The mapping Φ_{S_0}: $x_i' = -x^i$ in Γ_{sp}, which interchanges antipodal points, corresponds to the identity in $\Gamma_{\gamma e}$. Of the two distinct elements in Γ_{sp}, Φ_S and $\Phi_S\Phi_{S_0}$, one has determinant $+1$ and is algebraically identical with an element Φ_e of $\Gamma_{\gamma e}$. Since the other spherical motion is associated with $1 \cdot \Phi_e$, which is again Φ_e, there is a two-to-one pairing of the motions in Γ_{sp} with those of $\Gamma_{\gamma e}$.

We return to the mapping Ψ of S_r^2 on P_e^2. If x moves on a great circle η of S_r^2, $x\Psi$ moves on a straight line $\eta\Psi$ in P_e^2, but $x\Psi$ traverses $\eta\Psi$ twice as x circles η once. The smaller angle between two great circles η and ζ equals the smaller angle between $\eta\Psi$ and $\zeta\Psi$, so the mapping does not alter the angular metric. If a,b,c are vertices on S_r^2 of a triangle whose sides do not exceed $\pi/2$, then the corresponding sides in the triangle $a\Psi,b\Psi,c\Psi$ have the same lengths. The relations of sides and angles in the two triangles are thus the same, which explains why the formulas for the elliptic and spherical right triangle are identical.

There are, however, great differences, as well as similarities, between the two geometries. Two points on a sphere do not always determine a great circle uniquely, and on the sphere every simple, closed curve has an inside and an outside. Beside such obvious differences there are some unexpected ones. If the sphere is considered imbedded in Euclidean space and r is taken as unity, the relation between the spherical distance

$$d(x,y) = \text{Arc cos } \Omega_1(x,y)$$

and the Euclidean distance $e(x,y) = \Omega_1^{1/2}(x-y, x-y)$ is given by

$$e^2(x,y) = \Omega_1(x,x) - 2\Omega_1(x,y) + \Omega_1(y,y) = 2 - 2 \cos d(x,y).$$

Therefore $e(x,y)$ is an increasing function of $d(x,y)$, so $e(x_1,y_1) = e(x_2,y_2)$ implies $d(x_1,y_1) = d(x_2,y_2)$ and conversely.

Because of (23.12), two-dimensional Euclidean geometry has the property that any isometry between two subsets can be extended to the entire two-space, that is to a motion. It will be seen later that Euclidean

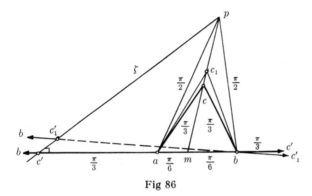

Fig 86

three-space also has this property. If a,b,c and a',b',c' are two isometric triples on S_r^2, then, because of these remarks, they are also isometric triples in E^3. The quadruples $z = (0,0,0), a,b,c$ and z, a',b',c' are then isometric in E^3. Accepting the stated property of E^3, there is then a motion of Euclidean three-space taking one quadruple into the other, and clearly this must induce a motion of S_r^2 on itself. *It follows that S_r^2 also has the property that an isometry between two triples (or any two subsets) can be extended to a motion. One of the less easily anticipated differences between S_r^2 and P_e^2 is that P_e^2 does not have this property.*

To see this, let $k = 1$ and take a,b in P_e^2 so that $\varepsilon(a,b) = \pi/3$ (Figure 86). Let m be the center of $S(a,b)$, so $\varepsilon(a,m) = \varepsilon(m,b) = \pi/6$, and take c' the midpoint of the complementary interval on $a \times b$ so that. $\varepsilon(c',a) = \varepsilon(c',b) = \pi/3$. If p denotes the pole of $a \times b$, as a variable point x travels from m to p on $p \times m$ (along either segment $S(p,m)$) the distance $\varepsilon(a,x) = \varepsilon(b,x)$ varies continuously from $\pi/6$ to $\pi/2$. Because of continuity, x passes through a position c for which

$\varepsilon(a,c) = \varepsilon(b,c) = \pi/3$. The triples a,b,c' and a,b,c are isometric, since every distance in each is $\pi/3$. But clearly there cannot be a motion of P_e^2 carrying a,b,c' into a,b,c since every motion which leaves a and b fixed leaves $a \times b$ pointwise invariant and so leaves c' fixed.

If the same construction is carried out on a sphere (Figure 87), yielding a triangle $\bar{a},\bar{b},\bar{c}$ isometric to the elliptic triangle a,b,c, then there are *two* points, $\bar{c}'$ and $-\bar{c}'$, on the great circle through $\bar{a},\bar{b}$, such that $d(\bar{a},\bar{c}') = d(\bar{b},-\bar{c}') = \pi/3$ and $d(\bar{a},-\bar{c}') = d(\bar{b},\bar{c}') = 2\pi/3$. Hence neither the triple $\bar{a},\bar{b},\bar{c}'$ or $\bar{a},\bar{b},-\bar{c}'$ is isometric to $\bar{a},\bar{b},\bar{c}$.

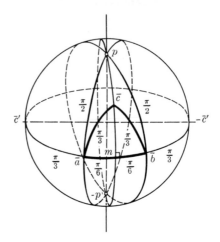

Fig. 87

It might be thought that this difference occurred because one triple in P_e^2 was a degenerate triangle. But this is not the case. If ζ is the perpendicular to $a \times b$ at c' and c_1' is chosen close to c' on ζ, then $\varepsilon(a,c_1')$, which equals $\varepsilon(b,c_1')$, will be slightly larger than $\varepsilon(a,c')$ because c' is the foot of a on ζ (Figure 86). In the construction, as x travels from m to c the distance $\varepsilon(a,x)$ increases from $\pi/6$ to $\pi/3$, hence for x at some point c_1, close to c but slightly beyond, $\varepsilon(a,c_1) = \varepsilon(b,c_1) = \varepsilon(a,c_1') = \varepsilon(b,c_1')$. The triples a,b,c_1 and a,b,c_1' are then isometric. Both triangles are non-degenerate. They are not congruent, however, since the angle $\measuredangle\ ac_1'b$ is nearly π while the angle $\measuredangle\ ac_1b$ is less than $\pi/2$.

Thus we see that the congruence theorems do not all hold unrestrictedly in the elliptic plane (though they are valid in spherical geometry). As a consequence, it is not unreservedly true that one can obtain the trigonometric formulas for oblique, elliptic triangles purely by analogy with the hyperbolic case. For instance, for a triangle with angles A, B and C and with opposite sides α, β, γ, one might expect the elliptic cosine law, paralleling the hyperbolic one (30.18), to be

(37.11) $\qquad \cos \alpha = \cos \beta \cos \gamma + \sin \beta \sin \gamma \cos A.$[12]

This formula is correct, and well known, for spherical trigonometry and is derived from the right triangle formulas in the same manner as we obtained the hyperbolic law. Since it implies that the sides determine the angles, and hence that isometric triples are congruent, it cannot possibly be true for all elliptic triangles.

Calculation shows the formula is correct for the elliptic triangle a,b,c in our example. From the right triangle a,m,c, we know that

$$\cos A = \frac{\tan \varepsilon(a,m)}{\tan \varepsilon(a,c)} = \frac{\tan \pi/6}{\tan \pi/3} = 1/3,$$

and the above law gives

$$\cos A = \frac{\cos \alpha - \cos \beta \cos \gamma}{\sin \beta \sin \gamma} = \frac{\cos \pi/3 - \cos^2 \pi/3}{\sin^2 \pi/3} = 1/3.$$

The law is clearly not valid for the degenerate triangle a,b,c' since it again gives $\cos A = 1/3$, when A is known to be a straight angle. Because of continuity the law cannot be true for $\Delta abc_1'$ either, which differs only slightly from $\Delta abc'$. The reason for this can also be seen from the figure. The line ζ is a perpendicular from the vertex c_1' to the base line $a \times b$ of the triangle a,b,c_1'. But neither the sum nor the difference of the triangles c_1',c',a and c_1',c',b is the triangle a,b,c_1', as would be the case in both the hyperbolic and spherical situations, and which is used in deriving the law of cosines.

It would not be difficult to establish the precise conditions for the validity of spherical formulas in the elliptic plane. But the results are not important enough to warrant the effort.

38. The Elliptic Line Element. Length and Area

To obtain the *elliptic line element*, we consider, as in previous cases, a parametrized curve $C : x = x(t)$. The points

$$(x_1 x_2, x_3) \text{ and } (x_1 + \Delta x_1, x_2 + \Delta x_2, x_3 + \Delta x_3)$$

on C both satisfy $\Omega_1(y,y) = 1$, so from

$$\Omega_1(x + \Delta x, x + \Delta x) = \Omega_1(x,x) + 2\Omega_1(x, \Delta x) + \Omega_1(\Delta x, \Delta x) = 1$$

---

[12]The plus sign on the right is to be expected rather than the minus sign due to the difference in the middle signs in the expansions of cosh $(x - y)$ and cos $(x - y)$ which enter in the proof of the law of cosines.

we obtain

$$\Omega_1(x, \Delta x) = - (1/2)\Omega_1(\Delta x, \Delta x)$$

Then (for $k = 1$)

$$\cos \varepsilon(x, x + \Delta x) = \Omega_1(x, x + \Delta x) = \Omega_1(x, x) + \Omega_1(x, \Delta x)$$
$$= 1 - (1/2)\Omega_1(\Delta x, \Delta x).$$

Therefore

$$\sin^2 \varepsilon(x, x + \Delta x) = 1 - \cos^2 \varepsilon(x, x + \Delta x)$$
$$= \Omega_1(\Delta x, \Delta x) \cdot [1 - (1/4)\Omega_1(\Delta x, \Delta x)].$$

Since $\Omega_1(\Delta x, \Delta x) \to 0$ as $\Delta t \to 0$,

$$\lim_{\Delta t \to 0} \frac{\sin^2 \varepsilon(x, x + \Delta x)}{\Omega_1(\Delta x, \Delta x)} = 1.$$

This, with

$$\lim_{\Delta t \to 0} \frac{\sin^2 \varepsilon(x, x + \Delta x)}{\varepsilon^2(x, x + \Delta x)} = 1 \text{ gives } \lim_{\Delta t \to 0} \frac{\varepsilon^2(x, x + \Delta x)}{\Omega_1(\Delta x, \Delta x)} = 1.$$

If C is continuously differentiable, then $\lim_{\Delta t \to 0} \dfrac{\Delta S}{\Delta C} = 1$, where ΔS is arc length and $\Delta C = \varepsilon(x, x + \Delta x)$ is the corresponding chord length. Therefore,

$$1 = \lim_{\Delta t \to 0} \frac{\Delta S^2}{\Delta C^2} = \lim \frac{\Delta S^2}{\varepsilon^2(x, x + \Delta x)} = \lim \frac{\Delta S^2}{\Omega_1(\Delta x, \Delta x)}$$

yields the elliptic line element (for general k) in the form:

$$(38.1) \qquad dS_e^2 = k^2\Omega_1(dx, dx) = k^2 \sum_{i=1}^{3} dx_i^2,$$

where the coordinates satisfy $\Omega_1(x,x) = 1$.

As an example, we use (38.1) to find the circumference of a circle with radius $r < k\pi/2$. If the center is taken at $(0,0,1)$, the circle has the parametric representation

$$(38.2) \quad x_1 = \cos \omega \sin (r/k), \qquad x_2 = \sin \omega \sin (r/k), \qquad x_3 = \cos (r/k),$$

already used in Section 36. Then,

$$k^2 \sum_{i=1}^{3} dx_i^2 = k^2 \sin^2 (r/k)d\omega^2,$$

hence $k \displaystyle\int_0^{2\pi} \sin (r/k)d\omega = 2\pi k \sin (r/k)$ is the circumference of the circle. More generally,

$$(38.3) \qquad k \int_0^{\theta} \sin (r/k)d\omega = k\theta \sin (r/k) \text{ is the length of the circular arc,}$$

with central angle θ and radius r.

If both ω and r are variable in (38.2), then

$$dx_1 = -\sin\omega\,\sin(r/k)d\omega + (1/k)\cos\omega\,\cos(r/k)dr$$
$$dx_2 = \cos\omega\,\sin(r/k)d\omega + (1/k)\sin\omega\,\cos(r/k)dr$$
$$dx_3 = -(1/k)\sin(r/k)dr.$$

Therefore,

(38.4) $$dS^2 = k^2\sum_i dx_i^2 = dr^2 + k^2\sin^2(r/k)d\omega^2.$$

Since *this representation of the line element, in terms of the polar coordinates r, ω, is a quadratic form in dr and $d\omega$*, it follows that elliptic geo-

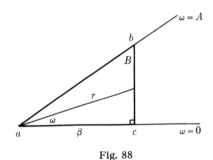

Fig. 88

metry is locally Euclidean. The values of E, F, G in polar coordinates are $E = 1$, $F = 0$ and $G = k^2\sin^2(r/k)$, so:

(38.5) $W\,dr\,d\omega = \sqrt{EG - F^2}\,dr\,d\omega = k\sin(r/k)\,dr\,d\omega$ *is the polar coordinate form of the elliptic area element.*

The area of a circle with radius r is then

$$k\int_0^{2\pi}\int_0^r \sin(r/k)dr\,d\omega = 2\pi k^2[1 - \cos(r/k)] = 4\pi k^2\sin^2(r/2k).$$

As $r \to k\pi/2$, $4\pi k^2\sin^2(r/2k) \to 2\pi k^2$, the area of the entire elliptic plane.

Virtually the same calculations as in the hyperbolic case can now be applied to find the area of an elliptic right triangle. We take the vertex a of $\triangle abc$ at the origin and the vertex c of the right angle on $\omega = 0$ so that b lies on the line $\omega = A$ (Figure 88). With $k\alpha$, $k\beta$, $k\gamma$ as the sides opposite the angles A, B, C, (37.5) gives the equation of $b \times c$ in the form

(38.6) $$\tan r/k = \tan\beta\,\sec\omega.$$

With $r = k \tan^{-1}(\tan \beta \sec \omega)$, the area of the triangle can thus be expressed as

$$k \int_0^A \int_0^r \sin (t/k) dt \, d\omega = k^2 \int_0^A [1 - \cos (r/k)] d\omega$$

$$= k^2 A - k^2 \int_0^A \frac{\cos \omega \, d\omega}{[\sec^2 \beta - \sin^2 \omega]^{\frac{1}{2}}} = k^2 A - k^2 [\sin^{-1}(\cos \beta \sin \omega)]_0^A.$$

From (37.7), $\cos \beta \sin A = \cos B$, therefore

$$[\sin^{-1}(\cos \beta \sin \omega)]_0^A = \sin^{-1}(\cos \beta \sin A) = \sin^{-1}(\cos B) = \frac{\pi}{2} - B.$$

Substituting this result above yields:

The area of $\triangle abc$, with right angle C, is

(38.7)
$$k^2\left(A + B - \frac{\pi}{2}\right) = k^2(A + B + C - \pi).$$

For any triangle the quantity $(A + B + C - \pi)$, denoted by $\varepsilon(a,b,c)$, is called the *excess* of the triangle. As in the hyperbolic case, the result in (38.7) extends to any triangle, that is:

(38.8)   *The area of any triangle a,b,c is $k^2\varepsilon(a,b,c)$.*

The triangles of maximal area are the degenerate ones, all of whose angles are π, which again gives $2\pi k^2$ as the area of the elliptic plane. Another consequence of (38.8) is:

(38.9)   *The sum of the angles in an elliptic triangle is greater than π.*

We note briefly that $\xi_1 = k\omega/2$, $\xi_2 = \frac{\pi k}{2} - r$ are the coordinates which correspond to the equidistant-curve coordinates in the hyperbolic plane. The curves $\xi_2 = $ constant are equidistant to $x_3 = 0$ and the lines $\xi_1 = $ constant cut the curves $\xi_2 = $ constant orthogonally. The polar coordinates, r,ω, have (as do all polar coordinates) a singularity at $r = 0$ in the sense that all lines $\omega = $ constant pass through that point. In both the Euclidean and hyperbolic geometries we encountered regular coordinate systems. That is, each point of the space was the intersection of a unique pair of coordinate curves. The standard coordinates

(38.10) $y_1 = x_1/x_3, \qquad y_2 = x_2/x_3,$

were regular under either a Euclidean or hyperbolic interpretation since in both cases the line $x_3 = 0$ was outside the space. It is a theorem (which we will not prove) that *no coordinate system in the projective plane (or on the sphere) is regular everywhere.* The coordinates are either singular at some points, as polar coordinates are at the origin, or else they fail to be defined for some points, as is the case for $x_3 = 0$ in (38.10).

39. Cayley's Method

Using the forms Ω_1 and Ω_{-1} established a close analogy between hyperbolic and elliptic geometry. It is natural to ask, however, whether or not $\varepsilon(x,y)$ *can be obtained independently* in the same way that we originally defined $h(x,y)$, that is, *in terms of a conic with distance taken as the logarithm of a cross ratio*. We will see that this is possible provided that an imaginary conic, and consequently imaginary numbers, are used.

To obtain this result we will make use of a method due to Cayley (1821-1895) which employs general polarities instead of the special form $\xi_1 = x_1$, $\xi_2 = x_2$, $\xi_3 = \pm x_3$. This will also give us the opportunity to see how the formulas of both geometries appear in this more general approach. According to (10.9) any polarity may be expressed in the form

$$(39.1) \quad \begin{aligned} \xi_i &= \sum_k a_{ik}x_k, \quad |a_{ik}| \neq 0, \quad a_{ik} = a_{ki}, \quad i = 1,2,3 \\ x_i &= \sum_k A_{ik}\xi_k, \quad |A_{ik}| \neq 0, \quad A_{ik} = A_{ki}, \quad i = 1,2,3, \end{aligned}$$

where A_{ik} is the quotient of the cofactor of a_{ik} and the determinant $|a_{ik}|$. The locus E of self-conjugate points in the polarity is given analytically by the equation

$$\Omega(x,x) = \sum_{i,k} a_{ik}x_ix_k = 0,$$

which represents a real conic if the polarity is hyperbolic. Similarly, the self-conjugate lines satisfy the equation

$$\Omega'(\xi,\xi) = \sum_{i,k} A_{ik}\xi_i\xi_k = 0.$$

When the conic is real it defines a hyperbolic geometry in the set of no-tangent points of E. For if x and y are two such points and u and v are the intersections of $x \times y$ with E, then the metric

$$h(x,y) = \frac{k}{2} \, |\log R(x,y,u,v)|$$

is hyperbolic.

For the arbitrary polarity (39.1), let y and z be two non-self-conjugate points such that $y \times z$ is not a self-conjugate line. Then $y \times z$ intersects $\Omega(x,x) = 0$ in two points $y - \lambda_i z$, $i = 1,2$, which may be complex. If

$$\Omega(x,y) = \sum_{i,k} a_{ik}x_iy_k, \text{ and } \Omega'(\xi,\eta) = \sum_{i,k} A_{ik}\xi_i\eta_k,$$

then λ_1 and λ_2 are the roots of the equations

$$\Omega(y - \lambda z, y - \lambda z) = \Omega(y,y) - 2\lambda\Omega(y,z) + \lambda^2\Omega(z,z) = 0.$$

From the theory of equations, then:

(39.2) $\qquad \lambda_1 + \lambda_2 = 2\Omega(y,z)/\Omega(z,z), \qquad \lambda_1\lambda_2 = \Omega(y,y)/\Omega(z,z).$

We now generalize (6.2) by taking

$$R(y,z,y + \lambda z, y + \mu z) = \lambda/\mu$$

still to be the definition of cross ratio when either λ or μ fails to be real. In particular,

$$R(y,z,y - \lambda_1 z, y - \lambda_2 z) = \lambda_1/\lambda_2,$$

for real or complex values of λ_i. By making use of

(39.3) $\qquad\qquad \cos(t/i) = \cos(i \cdot t) = \cosh t$

and

(39.4) $\qquad\qquad \cosh(\pm (1/2) \log t) = \pm (1/2)(\sqrt{t} + 1/\sqrt{t})$

in conjunction with (39.2), we obtain

$$
\begin{aligned}
&\cosh[\pm (1/2) \log R(y,z,y - \lambda_1 z, y - \lambda_2 z)] \\
&= \cos\;\;[\pm (1/2i) \log R(y,z,y - \lambda_1 z, y - \lambda_2 z)] \\
\text{(39.5)}\quad &= \pm (1/2)[(\lambda_1/\lambda_2)^{1/2} + (\lambda_2/\lambda_1)^{1/2}] \\
&= \pm \frac{\lambda_1 + \lambda_2}{2\sqrt{\lambda_1 \cdot \lambda_2}} = \pm \frac{\Omega(y,z)}{\sqrt{\Omega(y,y)\Omega(z,z)}}.
\end{aligned}
$$

We can now answer the question raised at the beginning of this section. Taking the special case $\Omega(x,x) = \Omega_1(x,x)$, with u,v as the (imaginary) intersections of $y \times z$ with $\Omega_1(x,x) = 0$, we obtain from (36.6) and (39.5):

(39.6) $\qquad \varepsilon(y,z) = |(k/2i) \log R(y,z,u,v)| = \frac{k}{2}| \log R(y,z,u,v)|$[13]

For the general polarity (39.1) we have similarly:

(39.7) $\qquad\qquad h(y,z) = k \text{ Area cosh} \left| \dfrac{\Omega(y,z)}{\sqrt{\Omega(y,y)\Omega(z,z)}} \right|$

or

(39.8) $\qquad\qquad \varepsilon(y,z) = k \text{ Arc cos} \left| \dfrac{\Omega(y,z)}{\sqrt{\Omega(y,y)\Omega(z,z)}} \right|,$

according as the polarity is hyperbolic or elliptic.

Analogous results hold for angular measure. If ξ and η are two, non-

---

[13]The complex logarithm is a multi-valued function. In the present case we mean the determination which has the smallest absolute value.

self-conjugate lines which intersect at a non-self-conjugate point, then $\Omega'(\xi,\xi) \neq 0$ and $\Omega'(\eta,\eta) \neq 0$. Through $\xi \times \eta$ there are then two self-conjugate lines, $\xi - \lambda_i \eta$, $i = 1,2$, which may be complex. Since λ_1, λ_2 are the roots of $\Omega'(\xi - \lambda\eta,\ \xi - \lambda\eta) = 0$, the same analysis as before leads to the relations:

$$\cos\left[\pm (1/2i) \log R(\xi,\eta,\ \xi - \lambda_1\eta,\ \xi - \lambda_2\eta)\right]$$
$$= \cosh\left[\pm (1/2) \log R(\xi,\eta,\ \xi - \lambda_1\eta,\ \xi - \lambda_2\eta)\right]$$
$$= \pm\ \frac{\Omega'(\xi,\eta)}{\sqrt{\Omega'(\xi,\xi)\Omega'(\eta,\eta)}}\ .$$

In the hyperbolic case ξ and η form an angle only if $\xi \times \eta$ is a no-tangent point, hence only if λ_1 and λ_2 are imaginary. We can therefore say, for either type of polarity, that if ζ and σ are the (imaginary) tangents from $\xi \times \eta$ to $\Omega(x,x) = 0$ (or the self-conjugate lines through $\xi \times \eta$), then

(39.9) $$\angle\ (\xi,\eta) = |(1/2i) \log R(\xi,\eta,\zeta,\sigma)|$$

and

(39.10) $$\cos\angle\ (\xi,\eta) = \left|\frac{\Omega'(\xi,\eta)}{\sqrt{\Omega'(\xi,\xi)\Omega'(\eta,\eta)}}\right|$$

The first of these formulas shows that angles, as well as distance, can be expressed in terms of cross ratio. For $\Omega = \Omega_{\pm 1}$, $a_{ik} = A_{ik}$ hence Ω and Ω' have the same coefficients and (39.10) coincides with (35.8) and (36.7).

If, in the hyperbolic case, $\xi \times \eta$ is a two-tangent point, λ_1 and λ_2 are real and $\xi - \lambda_1\eta$ and $\xi - \lambda_2\eta$ are real tangents to $\Omega(x,x) = 0$. We then have

(39.11) $$\cosh\left[\pm (1/2) \log R(\xi,\eta,\zeta,\sigma)\right] = \left|\frac{\Omega'(\xi,\eta)}{\sqrt{\Omega'(\xi,\xi)\Omega'(\eta,\eta)}}\right|.$$

Comparing this with (35.10) we see that if ξ and η are hyperparallels and ζ and σ are the tangents to E from the point $\xi \times \eta$, outside the model, then the length of the common perpendicular segment between ξ and η is $|(1/2 \log R(\xi,\eta,\zeta,\sigma)|$.

It is worth investigating how these ideas can be used *to obtain Euclidean geometry*. Since the geometry on a sphere becomes more nearly Euclidean as the radius increases, the problem may be approached in the following way. With x_1, x_2, x_3 indicating homogeneous rectangular coordinates (so $\bar{x}_1 = x_1/x_3$, $\bar{x}_2 = x_2/x_3$ are ordinary rectangular coordinates), define

$$\Omega_{\varepsilon R}(x,x) = x_1^2 + x_2^2 + \varepsilon R^2 x_3^2, \quad \text{where} \quad \varepsilon = \pm 1.$$

The corresponding form in line coordinates is

$$\Omega'_{\varepsilon R}(\xi,\xi) = \xi_1^2 + \xi_2^2 + \varepsilon R^{-2}\xi_3^2.$$

If ξ and η are two intersecting lines, and if, corresponding to $\Omega_{\varepsilon R}$, $\alpha_{\varepsilon R}(\xi,\eta)$ indicates the measure of the smaller angle between ξ and η, then (39.10) implies

$$\cos \alpha_{\varepsilon R}(\xi,\eta) = \left| \frac{\xi_1\eta_1 + \xi_2\eta_2 + \varepsilon R^{-2}\xi_3\eta_3}{(\xi_1^2 + \xi_2^2 + \varepsilon R^{-2}\xi_3^2)^{1/2}(\eta_1^2 + \eta_2^2 + \varepsilon R^{-2}\eta_3^2)^{1/2}} \right|,$$

hence:

(39.12) $$\lim_{R \to \infty} \cos \alpha_{\varepsilon R}(\xi,\eta) = \left| \frac{\xi_1\eta_1 + \xi_2\eta_2}{\sqrt{\xi_1^2 + \xi_2^2}\sqrt{\eta_1^2 + \eta_2^2}} \right|.$$

Since $\sum x_i \xi_i = 0$ is the equation of the line ξ, the numbers $\xi_1/\sqrt{\xi_1^2 + \xi_2^2}$ and $\xi_2/\sqrt{\xi_1^2 + \xi_2^2}$ are the sine and cosine respectively of the Euclidean angle between ξ and the $\bar{x}_1$ – axis. A similar interpretation holds for $\eta_1/\sqrt{\eta_1^2 + \eta_2^2}$ and $\eta_2/\sqrt{\eta_1^2 + \eta_2^2}$. The right side of (39,12) is thus the cosine of the smaller Euclidean angle, $\alpha_e(\xi,\eta)$, between ξ and η. That is:

(39.13) $$\lim_{R \to \infty} \alpha_{\varepsilon R}(\xi,\eta) = \alpha_e(\xi,\eta).$$

A similar result can be obtained for distance between points provided the constant k is chosen appropriately. Let $h_R(x,y)$ and $\varepsilon_R(x,y)$ be the distances corresponding to Ω_{-R} and Ω_{+R}, with $k = R$. Then, from (39.7),

$$\sinh^2\left(\frac{h_R(x,y)}{R}\right) = \frac{-\Omega_{-R}(x,x)\Omega_{-R}(y,y) + \Omega_{-R}^2(x,y)}{\Omega_{-R}(x,x)\Omega_{-R}(y,y)}$$

$$= \frac{-[R^{-2}(\bar{x}_1^2 + \bar{x}_2^2) - 1][R^{-2}(\bar{y}_1^2 + \bar{y}_2^2) - 1] + [R^{-2}(\bar{x}_1\bar{y}_1 + \bar{x}_2\bar{y}_2) - 1]^2}{[R^{-2}(\bar{x}_1^2 + \bar{x}_2^2) - 1][R^{-2}(\bar{y}_1^2 + \bar{y}_2^2) - 1]}$$

$$= \frac{R^{-2}[\bar{x}_1^2 + \bar{x}_2^2 + \bar{y}_1^2 + \bar{y}_2^2 - 2\bar{x}_1\bar{y}_1 - 2\bar{x}_2\bar{y}_2] + R^{-4}f(\bar{x}_1,\bar{x}_2,\bar{y}_1,\bar{y}_2)}{[R^{-2}(\bar{x}_1^2 + \bar{x}_2^2) - 1][R^{-2}(\bar{y}_1^2 + \bar{y}_2^2) - 1]},$$

where f is some function whose exact expression is not needed. For whatever the value of f is, as $R \to \infty$, $R^{-2}f \to 0$, hence:

$$\lim_{R \to \infty} R^2 \sinh^2\left(\frac{h_R(x,y)}{R}\right) = (\bar{x}_1 - \bar{y}_1)^2 + (\bar{x}_2 - \bar{y}_2)^2 = e^2(\bar{x},\bar{y}).$$

Setting $u = h_R(x,y)/R$, as $R \to \infty$, $u \to 0$. Making use of $\lim_{u \to 0} \dfrac{\sinh u}{u} = 1$, the previous result can be expressed by

$$e^2(\bar{x},\bar{y}) = \lim_{R \to \infty} R^2 \sinh^2(h_R(x,y)/R) = \lim_{R \to \infty} h_R^2(x,y)(\sinh u/u)^2$$
$$= \lim_{R \to \infty} h_R^2(x, y).$$

Practically the same calculation shows that as $R \to \infty$, $\varepsilon_R^2(x,y)$ tends also to $e^2(\bar{x},\bar{y})$. Thus,

(39.14) $$\lim_{R \to \infty} h_R(x,y) = \lim_{R \to \infty} \varepsilon_R(x,y) = e(\bar{x},\bar{y}).$$

Except that the calculations are more complicated, there is no difference in applying the same argument, using the general form Ω, to obtain the general forms (25.13) and (26.5) for Euclidean angle and distance.

There remains the question *whether Euclidean distance and angle can be expressed directly as the logarithms of cross ratios instead of as the limit of such ratios.* This is not possible for distance, but the following heuristic argument shows how it can be done for angles. The line $x_3 = 0$ cuts the locus $\Omega_{\varepsilon R}(x,x) = 0$ in the points $(i,1,0)$ and $(-i,1,0)$, independent of ε and R. These are called the *"circular points"* because not only $\Omega_{\varepsilon R}(x,x) = 0$, but any Euclidean circle

$$x_1^2 + x_2^2 + dx_1x_3 + ex_2x_3 + fx_3^2 = 0$$

(whether real or imaginary) passes through both of them.[14] Any point p (not on $x_3 = 0$) determines with the circular points, in turn, two lines ζ and σ, called the isotropic lines through p. Now let ξ and η be two ordinary lines through p, and let ζ_R and σ_R denote the lines through p self-conjugate with respect to $\Omega_{\varepsilon R}(x,x) = 0$. Since it is easily verified that ζ_R and σ_R tend to the isotropic lines ζ and σ as $R \to \infty$, it is natural to expect the relation:

(39.15) $\qquad \alpha_e(\xi,\eta) = |\,(1/2i)\log\ R(\xi,\eta,\zeta,\sigma)\,|.$

It was the discovery of this famous result by Laguerre (1834-1886) which gave this whole direction of research its impetus.

The equality (39.15) is easily verified by calculation. Since neither ξ nor η is the line $x_3 = 0$ we may take their equations as $\bar{x}_1 \sin \alpha - \bar{x}_2 \cos \alpha + a = 0$ and $\bar{x}_1 \sin \beta - \bar{x}_2 \cos \beta + b = 0$, where the $\bar{x}_1$–axis makes the angles α and β respectively with ξ and η. Since ξ,η,ζ,σ, cut $x_3 = 0$, in the points $(\cos \alpha, \sin \alpha, 0)$, $(\cos \beta, \sin \beta, 0)$, $(i,1,0)$ and $(-i,1,0)$, (6.12) gives us

$$R(\xi,\eta,\zeta,\sigma) = \frac{\begin{vmatrix} \cos \alpha & \sin \alpha \\ i & 1 \end{vmatrix} \cdot \begin{vmatrix} \cos \beta & \sin \beta \\ -i & 1 \end{vmatrix}}{\begin{vmatrix} \cos \alpha & \sin \alpha \\ -i & 1 \end{vmatrix} \cdot \begin{vmatrix} \cos \beta & \sin \beta \\ i & 1 \end{vmatrix}}$$

$$= \frac{(\cos \alpha - i \sin \alpha)(\cos \beta + i \sin \beta)}{(\cos \alpha + i \sin \alpha)(\cos \beta - i \sin \beta)} = \frac{e^{-i\alpha} \cdot e^{i\beta}}{e^{i\alpha} \cdot e^{-i\beta}} = e^{2i(\beta - \alpha)}.$$

Hence $|\,(1/2i)\log\ R(\xi,\eta,\zeta,\sigma)\,| = |\,\beta - \alpha\,| = \alpha_e(\xi,\eta).$

This relation shows us how to *define Euclidean geometry from affine geometry by singling out a polarity.* The lines ξ and η are perpen-

---

[14] The circular points have no projective meaning: analytically, because they are not invariant under general collineations; geometrically, because they depend on the interpretation of x_1, x_2, x_3 as homogeneous rectangular coordinates. As the intersection of all Euclidean circles with the line at infinity they are invariant under all Euclidean motions. This is easily confirmed by calculation.

dicular if and only if their intersections with the line at infinity, namely the points $z^\xi = (x_1^\xi, x_2^\xi, 0)$ and $z^\eta = (x_1^\eta, x_2^\eta, 0)$, satisfy the relation $x_1^\xi x_1^\eta + x_2^\xi x_2^\eta = 0$. That is, the points z^ξ and z^η must be paired in the involution $x_1 x_1' + x_2 x_2' = 0$ on the line $x_3 = 0$. This involution is elliptic, since it has the (imaginary) circular points for its fixed elements.

A Euclidean motion is certainly an affinity (see (15.3)),

$$\bar{x}_1' = a_{11}\bar{x}_1 + a_{12}\bar{x}_2 + a_{13}$$
$$\bar{x}_2' = a_{21}\bar{x}_1 + a_{22}\bar{x}_2 + a_{23},$$

which preserves area and perpendicularity. It is therefore an equi-affinity which takes any corresponding pair in the involution $x_1 x_1' + x_2 x_2' = 0$ on $x_3 = 0$ again into a corresponding pair on this line. More briefly, the affinity must leave the involution invariant. The converse is also true, that is:

An equi-affinity,

$$\bar{x}_1' = a_{11}\bar{x}_1 + a_{12}\bar{x}_2 + a_{13}$$
$$\bar{x}_2' = a_{21}\bar{x}_1 + a_{22}\bar{x}_2 + a_{23},$$

(39.16) *is a Euclidean motion, hence satisfies*

$$a_{11}^2 + a_{12}^2 = a_{21}^2 + a_{22}^2 = 1, \qquad a_{11}a_{21} + a_{12}a_{22} = 0,$$

if and only if it leaves the involution

$$x_1 x_1' + x_2 x_2' = 0 \text{ on } x_3 = 0 \text{ invariant.}$$

For the affinity in homogeneous coordinates has the form

$$x_i' = \sum_k a_{ik}x_k, \quad i = 1,2, \quad x_3' = x_3.$$

A simple calculation shows that

$$A_{11} = \varepsilon a_{22}, \quad A_{12} = -\varepsilon a_{21}, \quad A_{21} = -\varepsilon a_{12}, \quad A_{22} = \varepsilon a_{11} \quad \text{and} \quad A_{33} = \varepsilon,$$

(where $\varepsilon = \pm 1$) hence the homogeneous line transformation has the form

$$\xi_1' = \varepsilon(a_{22}\xi_1 - a_{12}\xi_2), \quad \xi_2' = \varepsilon(-a_{21}\xi_1 + a_{11}\xi_2), \quad \xi_3' = \varepsilon\xi_3.$$

By hypothesis, $\xi_1\eta_1 + \xi_2\eta_2 = 0$ implies $\xi_1'\eta_1' + \xi_2'\eta_2' = 0$ which is equivalent to

$$0 = (a_{22}\xi_1 - a_{12}\xi_2)(a_{22}\eta_1 - a_{12}\eta_2) + (-a_{21}\xi_1 + a_{11}\xi_2)(-a_{21}\eta_1 + a_{11}\eta_2)$$
$$= \xi_1\eta_1(a_{21}^2 + a_{22}^2) + \xi_2\eta_2(a_{11}^2 + a_{12}^2) - (\xi_1\eta_2 + \xi_2\eta_1)(a_{11}a_{21} + a_{12}a_{22}).$$

Particular lines satisfying $\xi_1\eta_1 + \xi_2\eta_2 = 0$ are $\xi = (1,0,0)$ and $\eta = (0,1,0)$. Substituting these in the last relation shows that $a_{11}a_{21} + a_{12}a_{22} = 0$. If now a perpendicular pair, ξ, η, is chosen, for which $\xi_1\eta_1 \neq 0$, then $\xi_2\eta_2 = -\xi_1\eta_1 \neq 0$ and substituting this pair in the above relation yields $a_{11}^2 + a_{12}^2 = a_{21}^2 + a_{22}^2$. Then, as we've seen before, the condition $a_{11}a_{12} - a_{21}a_{22} = \varepsilon$ shows that $a_{11}^2 + a_{12}^2 = 1$.

Thus Euclidean geometry can be built up by starting with equi-affine geometry and distinguishing on the line at infinity an elliptic involution. In terms of this involution perpendicularity can be defined, and from this one concept all other Euclidean concepts can be derived. For instance, angular measure may be obtained as follows. If α' and β' denote the perpendiculars to α and β respectively at $\alpha \times \beta$, and if (γ, δ) is the notation to indicate the smaller angle between two lines γ and δ, then (compare exercise [6.2]):

$$|R(\alpha,\beta,\alpha',\beta')| = \frac{\sin(\alpha,\alpha')}{\sin(\alpha,\beta')} \cdot \frac{\sin(\beta,\beta')}{\sin(\beta,\alpha')}$$

$$= \frac{1}{\sin\left[\frac{\pi}{2} - (\alpha,\beta)\right] \sin\left[\frac{\pi}{2} - (\alpha,\beta)\right]} = \sec^2(\alpha,\beta).$$

If the involution is taken in the general form:

$$Ex_1x_1' + F(x_1x_2' + x_1'x_2) + Gx_2x_2' = 0, \qquad EG - F^2 > 0,$$

the facts of Euclidean geometry are obtained in the general forms discussed in Section 26. The reader interested in this approach to the subject is referred to "The Real Projective Plane" by Coxeter.

It is well to be aware of the limitations of the method used in this section. It cannot be extended to apply in a Hilbert geometry, for example, since for a general convex curve no adequate analogue to the polar form $\Omega(x,y)$ exists.

Exercises

[30.1]   Show that the length δ of the side of a regular n-gon inscribed in a circle of radius r is given by

$$\sinh\frac{\delta}{2k} = \sin\frac{\pi}{n}\sinh\frac{r}{k}.\text{[15]}$$

By a limit process find the length of the circle's circumference.

[30.2]   If the radius of a circle is greater than $k\log\cot\frac{\pi}{2n}$ it is impossible to circumscribe a regular n-gon about the circle.

[30.3]   The radii, R and r, of the circles respectively circumscribed and inscribed to a regular n-gon satisfy the relation

$$\tanh rk^{-1} = \tanh Rk^{-1}\cos\frac{\pi}{n}.$$

[30.4]   In the hyperbolic plane a regular n-gon exists with a given angle $A < \frac{(n-2)\pi}{n}$. Determine the length of the side as a function of A.

---

[15]A regular n-gon is a convex polygon with equal sides and equal angles.

[30.5] Let a "regular covering" of the Euclidean or hyperbolic plane
mean the decomposition of the plane into congruent, regular n-gons
such that the intersection of any pair of them is empty, or else is a
vertex or a (whole) side. Then every vertex is common to exactly
the same number, say m, of n-gons. Prove that a regular covering
of the Euclidean plane is only possible with triangles, squares or
hexagons of arbitrary size, and find m for each case.

[30.6] In the hyperbolic plane, a regular covering is possible for every
$m > 3$ if $n \geqslant 5$. Find all n belonging to $m = 3,4$. Show that m and n
determine the side of the n-gon.

[30.7] The length of the segment joining the midpoints of two sides of a
triangle is less than half the length of the third side.

For the following two exercises, recall the proof of the corresponding Euclidean formulas.

[30.8] If A, B and C are the angles of a triangle and α, β and γ are the
respective opposite sides, then for $2\sigma = \alpha + \beta + \gamma$ and $k = 1$,

$$\cos^2 \frac{A}{2} = \frac{\sinh \sigma \sinh (\sigma - \alpha)}{\sinh \beta \sinh \gamma}, \quad \sin^2 \frac{A}{2} = \frac{\sinh (\sigma - \beta) \sinh (\sigma - \gamma)}{\sinh \beta \sinh \gamma}.$$

[30.9] The radius ρ of the circle inscribed in a triangle is given (with the
conventions of [30.8]) by

$$\tanh^2 \rho = \sinh (\sigma - \alpha) \cdot \sinh (\sigma - \beta) \sinh (\sigma - \gamma) \operatorname{csch} \sigma.$$

[30.10] Find the equation of a spiral for which two consecutive intersections with a ray through $(0,0)$ have constant hyperbolic distance δ.

[30.11] The equation of a hyperbolic circle about $(0,b)$ with radius r is
$$x_1^2(1 - b) + x_2^2(1 - b^2\bar{r}^2) - 2x_2 b(1 - \bar{r}^2) - \bar{r}^2 - b^2 = 0, \text{ where } \bar{r} = \tanh \frac{r}{k}.$$

[30.12] Obtain the Euclidean theorem of Pythagoras as the limit of the
corresponding hyperbolic theorem when $k \to \infty$.

[31.1]* In terms of the sides α, β, γ and perimeter 2σ, the area Δ of a triangle is

$$\sin \frac{\Delta}{2} = \frac{[(e^{2\sigma} - 1)\ (e^{2\sigma - 2\alpha} - 1)\ (e^{2\sigma - 2\beta} - 1)\ (e^{2\sigma - 2\gamma} - 1)]^{\frac{1}{2}}}{(e^{\alpha} + 1)\ (e^{\beta} + 1)\ (e^{\gamma} + 1)}$$

[31.2] Among all triangles with a given base α and such that $\beta + \gamma = \lambda > \alpha$,
show that the isosceles triangle ($\beta = \gamma = \lambda/2$) has the greatest area.
Show that for a given perimeter the triangle of greatest area is
equilateral.

[31.3] Use (31.3) to prove (30.11).

[31.4] Show that as the vertex α of triangle a,b,c traverses the semi-circle
with $S(b,c)$ as diameter, the angle A is not constant. (Hint: evaluate
the area of the triangle.)

[31.5] In a quadrangle with three right angles, if the sides of the fourth and acute angle ω are equal, the lengths of the other two sides are equal. Find the sides in terms of ω.

[32.1] Find the transformation of x_1, x_2 into geodesic parallel coordinates ξ_1, ξ_2, when the origin is the same point in both systems and $x_2 = 0$ and $\xi_2 = 0$ coincide.

[32.2] In terms of ξ_1, ξ_2 (with $k = 1$), find the equation of a line η such that the perpendicular to η from the origin has length p and makes an angle α with ξ_1-axis.

[32.3] In the terminology of [32.2], if α, p and α_1, p_1 are defined for two distinct lines η and η_1, then η and η_1 intersect, are asymptotes, or are hyperparallels according as

$$\tanh^2 p + \tanh^2 p_1 - 2 \tanh p \tanh p_1 \cos (\alpha - \alpha_1) \gtreqless \sin^2 (\alpha - \alpha_1).$$

[32.4] The equation of the circle with radius r and center at $\xi_1 = \xi_2 = 0$ is $\sinh^2 \xi_1 + \tanh^2 \xi_2 - \tanh^2 r \cosh^2 \xi_1 = 0$.

[32.5] Let z be a point on a definite limit circle Λ and let ξ be the radius of Λ through z. Assign a positive and a negative direction from z on both $\Lambda(z,\xi)$ and ξ. For any point p in the plane define coordinates u,v as follows. If η is the radius of Λ through p, v is the signed length of arc on Λ intercepted by ξ and η, while u is the signed distance intercepted on ξ by $\Lambda(z,\xi)$ and $\Lambda(p,\eta)$. Find the line element and area element in terms of u and v.

[32.6] Two limit circles have a common center. Find the area bounded by two radii and the arcs which they intercept on the limit circles.

[32.7] Show that 5 of the 6 domains into which a limit circle and two radii decompose the plane have infinite area, and that the area of the 6th is finite and proportional to the arc length on the limit circle intercepted by the radii.

[32.8] In the x_1, x_2 coordinates, the limit circle through $(0,0)$ with center at $(0,1)$ has the equation $x_1^2 + 2x_2^2 - 2x_2 = 0$. (Hint: use [30.11].)

[32.9] Prove the statement at the end of Section 32 that the rotations about an endpoint of η are limits of rotations about points of η.

[33.1] If the midpoints of the sides of a triangle are taken for the vertices of a second triangle, the perpendicular bisectors of the sides of the first are the altitudes of the second.

[33.2] Give constructions for: the midpoint of two given points, the bisector of an angle, the perpendicular to a line through a given point.

[33.3] Given the asymptotes ξ,η, and the point x on ξ, construct the point on η corresponding to x.

[33.4] Show that the Hjelmslev line of an equally directed congruence between asymptotes ξ and ξ' is an asymptote to both ξ and ξ'.

[33.5] If $\Lambda(x,\xi)$ and $\Lambda(y,\eta)$ have the same center, and α and β are respectively the perpendicular to η through x and the perpendicular to ξ through y, then, if α and β intersect, the radius through $\alpha \times \beta$ is perpendicular to the line $x \times y$.

[33.6] Are there triangles with three acute angles which do not have a circumcircle? Are there triangles whose angle sum is arbitrarily close to π which do not have a circumcircle?

[33.7] The angle bisectors of a triangle are concurrent, and an inscribed circle exists for every triangle.

[34.1] A translation along η is a motion Φ without fixed points of the hyperbolic plane which carries η and each halfplane bounded by η into itself. Show that the $\lim\limits_{n \to \infty} x\Phi^n = x\Psi$ exists for every x in D or on E. What is the nature of the limit mapping Ψ?

[34.2] Show that the translations (including 1) along a line η form a group. Find this group for the line $x_2 = 0$ in the representation (34.5).

[34.3] Find the reflection in $x_1 = 1/2$ and in $x_1 = x_2$.

[34.4] In the form (34.5), find the reflection in the point $(1/2,0)$.

[34.5] Find the motions, in the form (34.5), which carry the pencil of asymptotes through $(0,1)$ into itself.

[34.6] In the group of problem [34.5], find the subgroup of motions which leave invariant the limit circles with center $(0,1)$.

[35.1] Let f_1 and f_2 be two fixed points. Show that in Weierstrass coordinates (with $k = 1$) the locus of a point x, such that
$$h(f_1,x) + h(x,f_2) = 2a > h(f_1,f_2),$$
has the equation
$$\Omega^2_{-1}(f_1,x) + \Omega^2_{-1}(f_2,x) - 2\Omega_{-1}(f_1,x)\Omega_{-1}(f_2,x) \cosh 2a - \Omega_{-1}(x,x) \sinh^2 2a = 0.$$

[35.2]* Using Weierstrass coordinates show that the altitudes of any triangle are concurrent (in, on, or outside E).

[36.1] If pairs of points, x,y and x',y', are given such that $\varepsilon(x,y) = \varepsilon(x',y')$ show that a motion of the elliptic plane exists which carries x into x' and y into y'.

[36.2] In the elliptic plane, the locus $B(a,b)$, $a \not\perp b$ (compare Section 23), consists of two straight lines.

[36.3] Use the preceding result to show that in the elliptic plane there are in general four circles through three given, non-collinear points.

[36.4] State and prove the elliptic analogue to theorem (33.3).

[36.5] State and prove the elliptic analogue to problem [35.1].

[37.1] Prove (36.6), (36.7) and (36.8).

[37.2]* Find a $\delta > 0$ such that for any two congruent sets μ and μ' contained in a circle of radius δ, a motion of the elliptic plane exists which carries μ into μ'.

[37.3] Show that the law of cosines holds for triangles with sides smaller than the δ of exercise [37.2].

[37.4] In a triangle for which $\sigma = (\alpha + \beta + \gamma)/2$ is sufficiently small, show that

$$\cos^2 \frac{A}{2} = \frac{\sin \sigma \sin (\sigma - \alpha)}{\sin \beta \sin \gamma}.$$

[37.5] In spherical geometry the locus $B(a,b)$, $a \not\rightarrow b$, consists of one great circle. Explain the difference between this and the situation in [36.2] by means of the two-to-one mapping of the sphere on the elliptic plane.

[38.1]* Show that in the coordinates y_1, y_2 of (38.10) the elliptic line element has the form

$$dS_e^2 = k^2 \frac{dy_1^2 + dy_2^2 + (y_1 dy_2 - y_2 dy_1)^2}{(1 + y_1^2 + y_2^2)^2}.$$

Chapter VI

Spatial Geometry

40. Three–Dimensional Projective Space

Most facts of three-dimensional geometry are simply extensions of the corresponding facts in plane geometry. However, there are also many new phenomena. Naturally, it is these which will be stressed, while the former will be treated very briefly.[1]

We first list some algebraic facts which will be constantly used. The *rank δ of the matrix*

$$\mu = \begin{pmatrix} a_1^1 & a_2^1 \cdots a_n^1 \\ a_1^2 & a_2^2 \cdots a_n^2 \\ \vdots & \vdots \quad \vdots \\ a_1^r & a_2^r \cdots a_n^r \end{pmatrix} = (a_i^k), \qquad \begin{array}{l} k = 1, \cdots r \\ i = 1, \cdots n \end{array}$$

is the largest number j for which a square subarray of μ exists, with j rows and columns, whose determinant does not vanish.

A system of r n-tuples, $b^k = (b_1^k, b_2^k, \cdots, b_n^k)$, $k = 1,2,\cdots,r$, is said to be *linearly dependent* if numbers $\lambda_1, \lambda_2 \cdots, \lambda_r$ exist, which are not all zero, such that

$$\sum_{k=1}^{r} \lambda_k b^k = 0, \text{ that is } \sum_{k=1}^{r} \lambda_k b_i^k = 0, \; i = 1,2,\cdots,n. \text{ If no such numbers}$$

λ_k, $k = 1,2,\cdots,r$, exist the system is said to be *linearly independent*. Clearly any system containing an all-zero n-tuple is linearly dependent, for if $b^1 = (0,0,\cdots,0)$, then $\lambda_1 = 1$, $\lambda_k = 0$, $k = 2,\cdots,r$ satisfy

$$\sum_{k=1}^{r} \lambda_k b_i^k = 0, \; i = 1,2,\cdots,n. \text{ In a linearly dependent system if } \lambda_1 \neq 0,$$

then $b^1 = -\sum_{k=2}^{r} (\lambda_k/\lambda_1) b^k$, that is $b_i^1 = -\sum_{k=2}^{r} (\lambda_k/\lambda_1) b_i^k$, $i = 1,2,,\cdots n$. In this case, b^1 is said to be expressed as linear combination of $b^2, b^3, \cdots, b^r$, or to be dependent upon $b^2, b^3, \cdots, b^r$.

---

[1] Frequently the extension of the plane case is merely mentioned. The proofs of such generalizations from two to three dimensions serve as exercises which are indicated by the sign § appearing behind unproved statements.

The fundamental relationship between rank of a matrix and linear dependence is given by:

(40.1)   *If μ has rank δ, then δ suitable rows $a^k = (a_1^k, a_2^k, \cdots a_n^k), k = 1, 2, \cdots \delta$, form a linearly independent system, and every system of more than δ rows is linearly dependent. If the rows $a^{k_1}, a^{k_2}, \cdots a^{k_\delta}$ are linearly independent, then every other row can be expressed as a linear combination of them.*

Consider now the system of equations:

$$(40.2) \qquad \sum_{i=1}^{n} a_i^k x_i = 0, \qquad k = 1, 2, \cdots, r.$$

The matrix $\mu = (a_i^k)_{i=1, \cdots, n}^{k=1, \cdots, r}$ is called the matrix of the system (40.2). Because the system is homogeneous it has a solution $x_i = 0$, $i = 1, 2, \cdots, n$ which is called the *trivial solution*. Also, if

$$x^j = (x_1^j, x_2^j, \cdots, x_n^j), \qquad j = 1, 2, \cdots, s$$

are solutions of the system, then any linear combination of them, $\sum_{j=1}^{s} \lambda_j x^j$, is again a solution. The number of linearly independent solutions is determined as follows.

(40.3)   *If δ is the rank of μ, there are exactly $n-\delta$ linearly independent, non-trivial solutions of the system.*

It follows from this that if $n - \delta > 0$ and if $x^1, x^2, \cdots, x^{n-\delta}$ are linearly independent solutions, then every other non-trivial solution x can be expressed in the form $x = \sum_{j=1}^{n-\delta} \lambda_j x^j$, where not all the λ_j are zero since x is non-trivial. As corollaries of (40.3) we have the following.

(40.4)   *If μ has rank n, the system has no non-trivial solution.*

(40.5)   *The equation $\sum_{i=1}^{n} a_i x_i = 0$, where not all a_i are zero and $n \geqslant 2$, has $n - 1$ (and not more) linearly independent solutions.*

(40.6)   *Two equations $\sum_{i=1}^{n} a_i^k x_i = 0$, $k = 1, 2$, where $n \geqslant 3$ and neither all a_i^1 nor all a_i^2 are zero, have $n - 2$ (and not more) linearly independent solutions unless a_i^1 and a_i^2 are proportional sets.*

(40.7) *The maximum number of linearly independent n-tuples of real numbers is n.*

We now apply these facts to the three-dimensional space P^3, which was defined in Section 19. Excluding $(0,0,0,0)$, the points of P^3 are represented by the classes of all quadruples of real numbers $x = (x_1, x_2, x_3, x_4)$, where two quadruples represent the same point if and only if they are proportional. The *points* $x^1, x^2, \ldots, x^n$ of P^3 are called *independent* if the set of quadruples representing them are linearly independent. (If this is the case for one set of representations it is true for all sets of representations.[§]) From (40.7), the maximal number of independent points in P^3 is 4.

If $p^{1*}, \cdots, p^{4*}$ are given representations of four independent points and x^* is a given representation of an arbitrary point, then the equations

$$x_k^* = \sum_{i=1}^{4} x_i' p_k^{i*}, \qquad k = 1, \cdots, 4,$$

determine the x_i' uniquely. If the representation of x is changed, say x^* is replaced by μx^*, then the x_i' are also multiplied by μ. Therefore the class $[x]$ only determines the x_i' up to a factor, and the x_i' are called *projective coordinates* of x. The numbers x_i are special projective coordinates corresponding to the choices $p^{i*} = d_i$, where

$d_1 = (1,0,0,0), \ \ d_2 = (0,1,0,0), \ \ d_3 = (0,0,1,0), \ \ d_4 = (0,0,0,1), \ \ e = (1,1,1,1).$

A *plane* is defined as the locus of the points x satisfying a linear equation

$$x \cdot \xi = \sum_{i=1}^{4} x_i \cdot \xi_i = 0, \text{ where not all } \xi_i \text{ are } 0.$$

To say that four points are independent is therefore the same as saying that they do not lie in a plane. Points which lie in a plane are called *coplanar*. The following are the analogues of (3.10) and (3.17).

(40.8) *Given five points no four of which are co-planar, there is exactly one projective coordinate system for which these points are, in a given order, the points d_1, d_2, d_3, d_4 and e.[§]*

(40.9) *If x_i' and x_i'' are two systems of projective coordinates, they are related by equations of the form*

$$x_i'' = \sum_{k=1}^{4} a_{ik} x_k', \qquad i = 1,2,3,4, \qquad |a_{ik}| \neq 0.[§]$$

(40.10) *If the numbers x_i' are projective coordinates and x_i'' are defined by (40.9) then the x_i'' are also projective coordinates.[§]*

The point locus $x \cdot \xi = 0$ is the same as the locus $x \cdot \eta = 0$ if and only if ξ_i and η_i are proportional, that is $[\xi] = [\eta]$ or $\xi \sim \eta$. Therefore the ξ_i are called the coordinates of the plane $x \cdot \xi = 0$, if $\xi \neq 0$. The linear transformation (40.9) carries the linear equation $x' \cdot \xi' = 0$ into another non-degenerate linear equation $x'' \cdot \xi'' = 0$, hence the definition of plane is independent of the choice of the projective coordinate system.[§] Since the original numbers x_i no longer play a distinguished role, they will henceforth be used for arbitrary coordinates and the following analogue to (3.20) holds.

The coordinate transformation

$$x'_i = \sum_{k=1}^{4} a_{ik} x_k, \qquad i = 1,2,3,4, \qquad |a_{ik}| \neq 0,$$

induces the transformation, in plane coordinates,

(40.11) $$\xi'_i = \sum A_{ik} \xi_k, \qquad i = 1,2,3,4, \qquad |A_{ik}| = |a_{ik}|^{-1},$$

where A_{ik} is the co-factor of a_{ik}, in the matrix (a_{ik}), divided by $|a_{ik}|$. The inverse transformations are

$$x_i = \sum_{k=1}^{4} A_{ki} x'_k, \qquad \xi_i = \sum_{k=1}^{4} a_{ki} \xi'_k.[§]$$

In space a point and plane are dual concepts. The equation $x \cdot \xi = 0$, for variable x and fixed ξ represents all points in ξ, and for variable ξ and fixed x represents all planes through x.

The *planes* $[\xi^1], \cdots, [\xi^r]$ are called *independent or dependent* according as the quadruples $(\xi^i_1, \cdots, \xi^i_4)$, $i = 1,2,\cdots,r$, are independent or dependent. Because of (40.5) a plane ξ contains three, but not more, independent points. If a^1, a^2, a^3 are three independent points in ξ, then

$$|a^1, a^2, a^3, x| = 0$$

is the equation of ξ, hence

(40.12) $$\xi \sim \left(\begin{vmatrix} a^1_2 & a^1_3 & a^1_4 \\ a^2_2 & a^2_3 & a^2_4 \\ a^3_2 & a^3_3 & a^3_4 \end{vmatrix}, \ - \begin{vmatrix} a^1_1 & a^1_3 & a^1_4 \\ a^2_1 & a^2_3 & a^2_4 \\ a^3_1 & a^3_3 & a^3_4 \end{vmatrix}, \ \begin{vmatrix} a^1_1 & a^1_2 & a^1_4 \\ a^2_1 & a^2_2 & a^2_4 \\ a^3_1 & a^3_2 & a^3_4 \end{vmatrix}, \ - \begin{vmatrix} a^1_1 & a^1_2 & a^1_3 \\ a^2_1 & a^2_2 & a^2_3 \\ a^3_1 & a^3_2 & a^3_3 \end{vmatrix} \right).$$

Similarly, there are three, and not more than three, independent planes through a point x. If $\alpha^1, \alpha^2, \alpha^3$ are three such planes, then $|\alpha^1, \alpha^2, \alpha^3, \xi| = 0$ is *the equation of x in plane coordinates*, and the coordinates of x are similar to those of ξ in (40.12).

The *concept of line may be expected to be selfdual* since it is both the connection of two points and the intersection of two planes, and we try to express this in the definition of a line. To say that two points (or planes) are distinct is equivalent to saying that they are independent. The set of points (planes) dependent on two distinct points x,y (planes ξ,η) is called a *pencil of points (planes)* and denoted by $L(x,y)$ $(L(\xi,\eta))$. Every point of $L(x,y)$ is then expressible in the form $\lambda x + \mu y$. If x' and y' are two distinct points in $L(x,y)$, then $L(x,y) = L(x',y')$ (see (40.1)). Similar statements hold for pencils of planes. Now let ξ be incident with two distinct points, x and y, that is $\xi \cdot x = 0$ and $\xi \cdot y = 0$. If z is any point of $L(x,y)$, then $\xi \cdot z = \xi \cdot (\lambda x + \mu y) = \lambda(\xi \cdot x) + \mu(\xi \cdot y) = 0$, so z is incident with ξ. *A point pencil and a plane pencil are said to be incident if each point of the point pencil is incident with every plane of the plane pencil. A straight line is now defined to be a point pencil and a plane pencil which are incident.* Of course, any two distinct points x and y (planes ξ and η) of the point (plane) pencil determine the line. Without ambiguity, therefore, the line may be denoted by $[x,y]$ or $[\xi,\eta]$. However, at present the notation $x \times y$ would be misleading since no algorithm has been given to determine the "coordinates" of the line in terms of x_i and y_i. This will be done in Section 42.

The line $[x,y] = [\xi,\eta]$ is said to be incident with, or to pass through, all points of the pencil $L(x,y)$, and also to be incident with, or to lie on, every plane of the pencil $L(\xi,\eta)$.

(40.13) *A plane ξ and a line L, not incident with ξ (considered as a point locus), have exactly one common point.*

For if L consists of the points $\lambda x + \mu y$, then

$$(\lambda x + \mu y) \cdot \xi = \lambda(x \cdot \xi) + \mu(y \cdot \xi) = 0$$

determines the ratio λ/μ uniquely since $x \cdot \xi$ and $y \cdot \xi$ do not both vanish. The dual to (40.13) is:

(40.14) *There is exactly one plane through a given line L and a given point x not incident with L.*

This plane will be denoted by $(x \wedge L)$. If M is any other line in the plane, $(M \wedge L)$ also denotes the same plane, but regarded as determined by M and L.

In the following discussion of P^3 we will want to apply the results obtained for P^2 to the planes of P^3. To do so, we must first be sure that *the points and lines of a plane in P^3 are actually the points and lines of a projective plane.* To see this, let z^1, z^2, z^3 be three independent points of a plane ξ in P^3. If z^{1*}, z^{2*}, z^{3*} indicate fixed representations of z^1, z^2, z^3, then for each point x in ξ, and a representation x^*, there is a unique set of constants $\lambda_1, \lambda_2, \lambda_3$ such that

(40.15) $x^* \equiv \lambda_1 z^{1*} + \lambda_2 z^{2*} + \lambda_3 z^{3*}.$

There must be at least one set λ_i since, as we observed previously, any point of ξ is dependent upon any independent triple of points in ξ. There is only one set of $\lambda_i's$ since $x^* \equiv \mu_1 z^{1*} + \mu_1 z^{2*} + \mu_3 z^{3*}$, together with (40.15), implies

$$(40.16) \qquad \sum_{i=1}^{3} (\mu_i - \lambda_i) z_j^{i*} = 0, \qquad j = 1,2,3.$$

Because z^1, z^2, z^3 are independent, the matrix (z_j^i) has rank 3, hence the system (40.16) has only the trivial solution $\mu_i = \lambda_i$, $i = 1,2,3$. If the representation of x is changed from x^* to σx^*, $\sigma \neq 0$, then in (40.15) the values λ_i will be changed to $\sigma \lambda_i$. To each point of ξ, then, there corresponds a class of proportional number triples $(\lambda_1, \lambda_2, \lambda_3) \neq (0,0,0)$ and this correspondence is one-to-one. By definition, the triples $(\lambda_1, \lambda_2, \lambda_3)$ form a projective plane P^2 with $\lambda_1, \lambda_2, \lambda_3$ as projective coordinates. If $y = \sum \mu_i z^{i*}$ is a point of ξ distinct from x, then the pencil $\alpha x + \beta y$ has the representation

$$(\alpha x + \beta y) = \sum_{i=1}^{3} (\alpha \lambda_i + \beta \mu_i) z^{i*}.$$

Thus to a linear combination of x and y there corresponds the same linear combination of the real number triples associated with x and y. Therefore the line $[x,y]$ in P^3 corresponds to the line $\lambda \times \mu$ in P^2.

Now in P^2, every projective concept can be defined in terms of incidence alone. For example, (6.15) characterizes, in terms of incidence, the cross ratio preserving transformations of a line on itself, and so contains implicitly a definition of cross ratio in terms of incidence. Thus the fact that the correspondence between ξ and P^2 is one-to-one and incidence preserving could be used to show that cross ratio is preserved.

However, if the cross ratio in P^3 of the points $x, y, \alpha x + \beta y, \alpha' x + \beta' y$ on $[x,y]$ is defined by

$$(40.17) \qquad R(x,y,\alpha x + \beta y, \alpha' x + \beta' y) = \beta \alpha' / \alpha \beta',$$

then the previous remarks show at once that these points have the same cross ratio as the corresponding set in P^2, since the corresponding points in P^2 are λ, μ, $\alpha \lambda + \beta \mu$, $\alpha' \lambda + \beta' \mu$.

From the duality of point and plane in P^3 it is natural to define the cross ratio of four planes in a pencil, ξ, η, $\alpha \xi + \beta \eta$, $\alpha' \xi + \beta' \eta$, by (40.17) also, that is as $\beta \alpha' / \alpha \beta'$. Analogously to (6.5) we show:

(40.18)   *If the four distinct planes $\xi^1, \xi^2, \xi^3, \xi^4$ of a pencil are cut by a line L in the points x^1, x^2, x^3, x^4 and by a plane η in the lines M^1, M^2, M^3, M^4, where the axis of the pencil does not lie on η or intersect L, then*

$$R(\xi^1, \xi^2, \xi^3, \xi^4) = R(x^1, x^2, x^3, x^4) = R(M^1, M^2, M^3, M^4).$$

PROOF: Let ξ^1 and ξ^2 and x^1 and x^2 be taken for the base elements in the plane and point pencils respectively so that

$$\xi^3 = \alpha_1\xi^1 + \alpha_2\xi^2, \qquad \xi^4 = \beta_1\xi^1 + \beta_2\xi^2,$$
$$x^3 = \lambda_1 x^1 + \lambda_2 x^2, \qquad x^4 = \mu_1 x^1 + \mu_2 x^2.$$

Because $\xi^i \cdot x^i = 0$, $i = 1,2,3,4$,

$$0 = (\alpha_1\xi^1 + \alpha_2\xi^2) \cdot (\lambda_1 x^1 + \lambda_2 x^2) = \alpha_1\lambda_2(\xi^1 \cdot x^2) + \alpha_2\lambda_1(\xi^2 \cdot x^1)$$

and

$$0 = (\beta_1\xi^1 + \beta_2\xi^2) \cdot (\mu_1 x^1 + \mu_2 x^2) = \beta_1\mu_2(\xi^1 \cdot x^2) + \beta_2\mu_1(\xi^2 \cdot x^1).$$

Because the given elements of the pencils are distinct, none of the linear combination constants are zero, hence

$$(\xi^1 \cdot x^2)/(\xi^2 \cdot x^1) = -\alpha_2\lambda_1/\alpha_1\lambda_2 = -\beta_2\mu_1/\beta_1\mu_2,$$

and the last equality implies $R(\xi^1,\xi^2,\xi^3,\xi^4) = R(x^1,x^2,x^3,x^4)$. Now in the plane η, any line N, which is not in the plane pencil of lines determined by the lines M^i, cuts the lines M^i in distinct points y^i, $i = 1,2,3,4$. By (6.5), $R(M^1,M^2,M^3,M^4) = R(y^1,y^2,y^3,y^4)$, and by the proof just given $R(y^1,y^2,y^3,y^4) = R(\xi^1,\xi^2,\xi^3,\xi^4)$, hence $R(M^1,M^2,M^3,M^4) = R(\xi^1,\xi^2,\xi^3,\xi^4)$.

The totality of lines and planes which pass through a fixed point z is called the *bundle z*. This concept is dual to that of the points and lines in a projective plane P^2. (Compare with Exercise [2.2].) Therefore, every theorem or concept in P^2 yields a dual theorem or concept in the bundle. The following are a few examples.

The lines through a fixed point of P^2 and the planes through a fixed line of z (the bundle) form a pencil. Two *plane pencils of z are perspective from the plane* η of the bundle if corresponding planes of the pencils intersect in lines on η. *Two line pencils of the bundle (each in a plane of z) are perspective from a line L of z* if the planes determined by corresponding lines of the pencils all contain L. The former relation between perspectivities and projectives still holds. That is, if the first pair of three plane pencils in z are perspective from η_1 and the second pair are perspective from η_2, then the first and third pencils are projective.[§] As formerly, any projectivity between two pencils of the bundle is the product of three, or fewer, perspectivities.

A *collineation between two bundles* may be defined in the same way as a projectivity between planes was defined in Section 5. But since theorem (9.1) has been proved, it is geometrically more satisfactory to use its dual and to define a collineation between two bundles as a one-to-one correspondence of their planes which preserves incidence. Such a collineation is induced (though not uniquely) by a collineation of P^3 on itself (see next section).

A *correlation of z on z'* is a one-to-one incidence preserving transformation of the planes of z on the lines of z'. A correlation of z on itself, whose square is the identity, is a *polarity*. If the line L and the plane ξ are corresponding elements of a polarity, then L is called *the polar line to* ξ and ξ is called *the polar plane to L*. The plane η *is conjugate* to ξ if it contains the polar line to ξ, and the *line M is conjugate to N* if it lies in the polar plane to M. If the polarity has self-conjugate elements it is called hyperbolic, otherwise it is called elliptic.

The locus of self-conjugate planes (lines) is called *a plane cone (line cone) with apex z*. Each plane of the plane cone contains its polar line and the locus of these polar lines is the line cone. The distinction between plane and line cone can therefore be dropped. The lines of the cone are called its generators and its planes are called tangent planes.

Consider a bundle z and a plane ξ not through z. A plane η of the bundle cuts ξ in a line L_η and a line L of the bundle intersects ξ in a point x_L. Conversely L_η in ξ determines with z a plane through z, and a point x_L of ξ determines with z a line L through z. Both mappings of the bundle on the plane, $\eta \rightarrow L_\eta$ and $L \rightarrow x_L$, preserve incidence, hence a collineation of the bundle $\eta \rightarrow \eta'$ induces the collineation of ξ, $L_\eta \rightarrow L_{\eta'}$, and conversely. The polarity $x_L \rightarrow L_\eta$ on ξ induces the polarity $L \rightarrow \eta$ on z, and conversely. Hence conjugate planes or lines of z correspond to conjugate lines or points of ξ. Since self-conjugate elements of z correspond to self-conjugate elements of ξ, *the intersection of a cone K of z with ξ is a conic C*. Conversely if C is any conic in ξ, the lines determined by z and the points of C form a line cone K through z whose tangent planes are the planes determined by z and the tangent lines to C.

The *equation of K as a plane or point locus in P^3* is easily obtained if the coordinates are chosen conveniently. If ξ has the coordinates $(0,0,0,1)$, i.e., is the plane δ_4, or $x_4 = 0$, and z is the point d_4, then the points d_1, d_2 and d_3 lie in ξ. Since every point $x = (x_1,x_2,x_3,0)$ of ξ can be expressed in the form $x = x_1 d_1 + x_2 d_2 + x_3 d_3$, it follows from the previous discussion that x_1,x_2,x_3 are projective coordinates in ξ of the point x. Hence C will have an equation of the form $\sum_{i,k=1}^{3} a_{ik} x_i x_k = 0$, $a_{ik} = a_{ki}$, where $|a_{ik}| \neq 0$.

Considered as a locus in P^3 the equation of C is clearly satisfied by $z = d_4$. Moreover, if $y = (y_1,y_2,y_3,y_4)$ satisfies the equation, then $\bar{y} = (y_1,y_2,y_3,0)$, in ξ, also satisfies it. Every point of the pencil $L(\bar{y},z)$ is on the locus, since by inspection $\lambda\bar{y} + \mu d_4$ satisfies the equation. Conversely, every point of $L(a,z)$ satisfies the locus, where a is a point of the conic. Hence the equation of the cone K is also $\sum_{i,k=1}^{3} a_{ik} x_i x_k = 0$. If d_1,d_2,d_3 form a self-polar triangle with respect to the conic, K takes the simpler form

$$\sum_{i=1}^{3} b_i x_i^2 = 0.$$

41. Collineations of the Projective Space

A *collineation* of the projective space P^3 is defined to be a mapping of P^3 on itself which, in suitable coordinates, has the form

$$\bar{x}_i = x_i, \qquad i = 1,2,3,4.^2$$

Under this collineation the plane with coordinates ξ_i goes into a plane with the same coordinates: $\bar{\xi}_i = \xi_i$. Therefore

(41.1)   *A collineation preserves all incidences (including those of point and line and plane and line).*§

The following facts are obtained exactly as in the plane case:

(41.2)   *In a fixed coordinate system x_i the general projectivity Ψ of P^3 takes the form*

$$x_i' = \sum_{k=1}^{4} a_{ik} x_k, \qquad i = 1,2,3,4, \qquad |a_{ik}| \neq 0.$$

In a specified order, there is exactly one projectivity which carries five given points, no four of which are co-planar, into five given points with the same property. If A_{ik} denotes the co-factor of a_{ik} divided by $|a_{ik}|$, then the inverse of Ψ is the collineation:

$$\Psi^{-1}: x_i = \sum_{k=1}^{4} A_{ki} x_k', \qquad i = 1,2,3,4, \qquad |A_{ki}| \neq 0.$$

Induced by Ψ are the plane transformations:

$$\Psi: \xi_i' = \sum_{k=1}^{4} A_{ik} \xi_k, \qquad i = 1,2,3,4.$$

$$\Psi^{-1}: \xi_i = \sum_{k=1}^{4} a_{ki} \xi_k', \qquad i = 1,2,3,4.$$

As a consequence of these facts:

(41.3)   *The cross ratio of four planes, four points in a pencil, or of four concurrent lines in a plane, is invariant under a collineation.*§

---

²Only collineations of P^3 on itself will occur, and none between different spaces.

Also, from (41.2) it is clear that:

(41.4)   *The mapping of $[x,y] = [\xi,\eta]$ on $[x',y'] = [\xi',\eta']$ induced by Ψ, regarded as either a point or plane transformation, is a projectivity.*[§]

The collineation Ψ maps all points in a plane ξ on points in the image plane ξ'. Every point x' of ξ' satisfies $x' \cdot \xi' = 0$, which implies

$$(x'\Psi^{-1}) \cdot (\xi'\Psi^{-1}) = x \cdot \xi = 0,$$

hence every point of ξ' is the image of a point in ξ. From (41.1) it follows, then, that Ψ induces a one-to-one, incidence preserving mapping of ξ on ξ'. This, with (9.1), implies:

(41.5)   *The mapping of the plane ξ on ξ', induced by Ψ, is a collineation.*[3]

Exactly as in Section 5 it is seen that:

(41.6)   *The collineations of P^3 form a (non-Abelian) group.*[§]

As in the plane case, the theorem that

(41.7)   *A one-to-one, incidence preserving transformation of the projective space P^3 on itself is a collineation*

can be reduced to establishing

(41.8)   *A one-to-one, incidence preserving transformation, Ψ, of the projective space P^3 on itself is the identity if it leaves five points fixed, no four of which are co-planar.*

Let a_i, $i = 1,2,3,4,5$, be the fixed points in (41.8). The plane η determined by a_2, a_3, a_4 goes into itself since these points are fixed. Similarly the line L determined by a_1 and a_5 maps on itself. Because no four of the five, fixed points are co-planar, the point f in which L cuts η is not a_2, a_3 or a_4. Since f is fixed under Ψ, the mapping of η on itself induced by Ψ has four fixed points. From the main theorem for the plane, Ψ induces a collineation on η, which, having four fixed elements, is the identity. By the same argument, all ten planes determined by triples from a_1, a_2, a_3, a_4, a_5 are pointwise invariant under Ψ. Any line which does not lie in any of these ten planes, intersects them in at least three different points, hence goes into itself under Ψ. Every point, being the vertex of a line bundle, must then go into itself under Ψ.

Analogous to a plane homology, a *space homology* is defined as a collineation of P^3 on itself which leaves a plane η pointwise invariant and which maps into itself every plane in a bundle a, where a is not on η. The point a is called the *center*, and η the *axial plane*, of the homology.

---

[3]An analytic proof of this fact, though longer, has the advantage that it can be extended to complex, projective space. Since (9.1) does not hold for complex, projective planes, the above proof lacks this extension.

With the same conditions, but with a on η, the mapping is called an *elation*. In either case, it follows that every line through a goes into itself, hence a point and its image are collinear with a. We have, then, as in the plane case:

(41.9)   *A homology Φ is uniquely determined by its axial plane, non-incident center a, and one corresponding pair x,x', collinear with a, neither of which is a or on η.*

PROOF : Let b_1, b_2, b_3 be three non-collinear points of η, no one of which is on the line $[a,x]$. Then no four of the points b_1, b_2, b_3, a, x are co-planar, and this is also true of the set b_1, b_2, b_3, a, x'. Hence, there is exactly one collineation Φ mapping the first set on the second in the order indicated. Since the points b_i are fixed, η goes into itself, and the line $[a,x]$ is invariant because a, x and x' are collinear. Therefore x_0, the intersection point of η and $[a,x]$, is fixed. Having four fixed points, no three collinear, η has every point as a fixed element. A plane ξ, through a, has the line $[\xi,\eta]$ and the point a as fixed elements hence $\xi\Phi \sim \xi$ and Φ is a homology. In particular, when $x' \sim x$ the homology is the identity. Also, it is clear that Φ induces on every plane ξ of the bundle a plane homology whose center is a and whose axis is the line $[\xi,\eta]$.

A *homology* is defined to be *harmonic* if every corresponding pair is separated harmonically by the center and the point in which their line cuts the axial plane.

(41.10)   *A homology Φ is harmonic if one corresponding pair, x,x', is separated harmonically by the center a and the point in which the line $[x,x']$ intersects the axial plane η.*

For let y and y' be any other point and image. Then Φ induces a homology in the plane determined by $[x,x']$ and $[y,y']$. Since by (9.6) the plane homology is harmonic, the points y,y',a and the intersection point of η and $[a,y]$ form a harmonic set. Also, as before:

(41.11)   *Given a plane η and two distinct points x and x', not on η, there is exactly one harmonic homology Φ, which has η for its axial plane and which maps x on x'.*

Let x_0 be the intersection point of η and $[x,x']$, and take a to be the harmonic conjugate of x_0 with respect to x and x'. From (41.9), there is exactly one homology Φ, with axial plane η and center a, such that $x\Phi = x'$. By (41.10), Φ is harmonic.

The discussion of the plane case suggests that a mapping of the form

(41.12) $$x_i' = x_i + a_i x_4, \qquad i = 1,2,3, \qquad x_4' = x_4,$$

is an elation. By inspection, every point of the plane $(0,0,0,1)$, that is $x_4 = 0$, is fixed. The mapping is a collineation since its determinant has the value 1. The induced plane transformation,

$$\xi_i' = \xi_i, \qquad i = 1,2,3, \qquad \xi_4' = -a_1\xi_1 - a_2\xi_2 - a_3\xi_3 + \xi_4,$$

has for fixed elements the planes ξ whose coordinates satisfy

$$a_1\xi_1 + a_2\xi_2 + a_3\xi_3 = 0.$$

But this is the equation of the point $(a_1,a_2,a_3,0)$. Hence the mapping is an elation whose axial plane is $x_4 = 0$ and whose center is the point $(a_1,a_2,a_3,0)$. If $y = (y_1,y_2,y_3,1)$ and $z = (z_1,z_2,z_3,1)$ are an arbitrary pair of points not in the axial plane $x_4 = 0$, then (41.12), with $a_i = z_i - y_i$, $i = 1,2,3$, carries y into z. The same argument as in (9.8) shows that there is only one such elation with $x_4 = 0$ as axial plane. Therefore:

(41.13)   *Given a plane η, and a distinct pair of points x,x', not on η, there is exactly one elation Φ, with axial plane η, such that $x\Phi = x'$.*

Thus far the situation has paralleled the two-dimensional case. However, differences now appear. In the plane every involution was shown to be a harmonic homology (see (9.7)). But in space an involution need not even possess fixed points.[4] For instance,

$$x_1' = x_3, \qquad x_2' = x_4, \qquad x_3' = -x_1, \qquad x_4' = -x_2$$

is an involution, but to be a fixed point $a = (a_1,a_2,a_3,a_4)$ would have to satisfy

$$\lambda a_1 = a_3 = -a_1/\lambda \qquad \text{and} \qquad \lambda a_2 = a_4 = -a_2/\lambda,$$

for some real, non-zero λ. Since $a \neq 0$, a_1 and a_2 cannot both vanish. But if either is not zero then $\lambda^2 = -1$ and λ is not real. Even when an involution in P^3 does have fixed points it need not be a harmonic homology. An example is the collineation

$$x_1' = x_1, \qquad x_2' = x_2, \qquad x_3' = -x_3, \qquad x_4' = -x_4.$$

The planes $x_3 = 0$ and $x_4 = 0$ meet in a line of fixed points, as do the planes $x_1 = 0$ and $x_2 = 0$.

An involution Φ which leaves all points fixed on two skew (i.e., non-coplanar) lines, L and M, is called *biaxial*, the fixed lines being the axes. Any point x (see Figure 89), not on L or M, lies on exactly one line N_x which intersects both L and M. For the plane $(x \wedge L)$ intersects M in

---

[4] Hence the analogue of (9.2) does not hold. In (9.2) we used the fact that the equation $\Delta(\lambda) = 0$ (see (9.4)), whose roots determined the fixed points, was a cubic and so had a real root. The corresponding equation is now a quartic and may have no real roots. The general fact is that a collineation on itself of an even dimensional projective space must have fixed elements, while for odd dimensional spaces collineations exist without fixed points (compare with the one-dimensional case in Section 8).

a point y and the line $N_x = [x,y]$ in $(x \wedge L)$ intersects L in a point z. Also, of course, $N_x = [(x \wedge L),\ (x \wedge M)]$. The involution cannot have any fixed point which is not on L or M, for it would then have a fixed quintuple, no four co-planar, and would be the identity. Because Φ leaves y and z fixed, it carries N_x into itself in an induced hyperbolic involution having y and z for fixed points. Hence, x and its image separate y and z harmonically. Summed up:

(41.14)   *For any two skew lines L and M, there is a unique biaxal involution Φ having L and M as axes. A point x, not on L or M, and its image $x\Phi$ are harmonic conjugates with respect to the points z on L, and y on M, which form with x a collinear triple. Every plane in the pencil on L and the pencil on M maps on itself.*

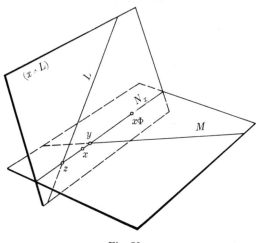

Fig. 89

The last remark is a consequence of the fact that any plane through L cuts M and so has a fixed point and a fixed line which are not incident.

(41.15)   *An involution of projective space which has a fixed point is biaxial or else is a harmonic homology.*

Let Φ be the involution and a be a fixed point. We consider two cases.

1)   *Every line through a goes into itself.*

Then every plane through a, being determined by a plane pencil of invariant lines, is also invariant. Hence Φ induces on every plane through a a homology, which, by (9.7), is harmonic. If the axes of two of these plane homologies were skew, then, by (41.14), Φ would be biaxial, which is clearly impossible. Hence the axes of all the plane homologies lie in a plane

η, and Φ is a harmonic homology whose center is a and whose axial plane is η.

2) *There is at least one point y such that its image y' does not lie on* $[a,y]$.

Since Φ carries a,y,y' into a,y',y it maps the plane η of a,y,y' on itself, and so induces in η a harmonic homology. Because $[a,y]$ is not invariant, a is not the center of the homology and so must lie on the axis L. If b denotes the plane homology center and all space lines through b are invariant, then Φ is a harmonic homology by the reasoning of case 1). On the other hand, if a line through b is not invariant, then since b is a fixed point we are back to the start of case 2). The same argument shows then that b is on the axis M of a harmonic homology induced in a plane $\bar{\eta}$ (not η) through b. Because L and M cannot be co-planar it follows that Φ is biaxial.

We consider next the notion of a *perspectivity* in space. Let η and η' be two distinct planes, with $L = [\eta,\eta']$. If z is a point not on η or η', then a line through z cuts η and η' in the points x and x' respectively and the mapping $x \to x'$ is said to define a perspectivity between the planes from the center z. Analogous to (7.1) we prove:

(41.16)   *A collineation Φ of the plane η on the plane η' is a perspectivity if it leaves fixed every point (or three points) of the line $L = [\eta,\eta']$.*

PROOF: Let x and y be two points of η which are not on L. Then $[x,y]$ intersects L in a point a. Since Φ leaves a fixed and preserves incidence, the line $[x',y']$ also contains a. Hence, the lines $[x,x']$ and $[y,y']$ are co-planar and intersect in a point z. If now w is any point of η which is not on L or $[x,y]$, the same kind of argument shows that $[w,w']$ must intersect both $[x,x']$ and $[y,y']$ and so must pass through z.

(41.17)   *Every projectivity Φ of a plane η on a plane η' is a product of four or fewer perspectivities.*

PROOF: Assume first that η and η' are different. If $L = [\eta,\eta']$, then $L\Phi$ is a line L' in η', and Φ induces a projectivity Φ' of L on L'. From (7.2) or (7.3), Φ' may be expressed as the product of line perspectivities $\Phi'_1,\cdots,\Phi'_k$, in η', where $k \leqslant 3$. Let a^i denote the center of the perspectivity Φ'_i of L_{i-1} on L_i, $i = 1,2,\cdots,k$, where $L_0 = L$ and $L_k = L'$. Through L_i take any plane η^i distinct from η', $i = 1,2,\cdots,k$, and denote by Φ_i the perspectivity from a^i of η^{i-1} on η^i, $i = 1,2,\cdots,k$, where $\eta^0 = \eta$. Then Φ_i induces the mapping Φ'_i. Hence $\Phi^{-1}\Phi_1\Phi_2\cdots\Phi_k$ is a projectivity of η' on η^k which leaves every point of L' fixed and so, by (41.16), is a perspectivity Φ_{k+1} of η' on η^k from some center w. Since Φ_{k+1}^{-1} is also a perspectivity, $\Phi_{k+1} = \Phi^{-1}\Phi_1\cdots\Phi_k$ implies that $\Phi = \Phi_1\Phi_2\cdots\Phi_k\Phi_{k+1}^{-1}$ is the product of

$k + 1$ perspectivities where $k + 1 \leqslant 4$. If $L \neq L'$ and the intersection a of L and L' corresponds to itself under Φ, then $k = 1$. This case is

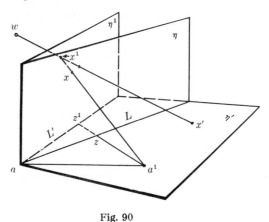

Fig. 90

illustrated in Figure 90. When $L \neq L'$ and a is not mapped on itself by Φ, then $k = 2$ as illustrated in Figure 91, where $x^1 = x\Phi_1$ and $x^2 = x\Phi_1\Phi_2$.

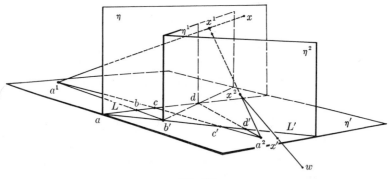

Fig. 91

Finally, if $\eta \sim \eta'$ then a suitable perspectivity Φ_0 takes η into $\eta_0 + \eta'$ such that $[\eta_0, \eta']$ does not correspond to itself under $\Phi\Phi_0$. By the preceding part of the proof, $\Phi\Phi_0$ is then expressible as the product of three (or fewer) perspectivities. Thus (41.17) is established.[5]

---

[5]The dual to a perspectivity between the plane η, η' from a point z is a perspectivity between bundles y, y' from a plane ζ. The formulation of the definition and the properties analogous to (41.16) and (41.17) are left as exercises for the reader.§

42. Line Coordinates. Linear Complexes

Coordinates have been introduced in P^3 for points and for planes but not for lines. The notion of fixing, by coordinates, a geometric element other than a point or plane and then studying the loci of equations in these coordinates is due to Pluecker (1801-1868). In particular, coordinates for lines in space, the topic of this section, are due to him.

In developing this idea the following Laplace expansion of a 4th order determinant will be useful.

$$
\begin{aligned}
(42.1) \quad
\begin{vmatrix} a_1 & a_2 & a_3 & a_4 \\ b_1 & b_2 & b_3 & b_4 \\ c_1 & c_2 & c_3 & c_4 \\ d_1 & d_2 & d_3 & d_4 \end{vmatrix}
&= \begin{vmatrix} a_1 & a_2 \\ b_1 & b_2 \end{vmatrix} \cdot \begin{vmatrix} c_3 & c_4 \\ d_3 & d_4 \end{vmatrix}
- \begin{vmatrix} a_1 & a_3 \\ b_1 & b_3 \end{vmatrix} \cdot \begin{vmatrix} c_2 & c_4 \\ d_2 & d_4 \end{vmatrix} \\
&+ \begin{vmatrix} a_1 & a_4 \\ b_1 & b_4 \end{vmatrix} \cdot \begin{vmatrix} c_2 & c_3 \\ d_2 & d_3 \end{vmatrix}
+ \begin{vmatrix} a_2 & a_3 \\ b_2 & b_3 \end{vmatrix} \cdot \begin{vmatrix} c_1 & c_4 \\ d_1 & d_4 \end{vmatrix} \\
&- \begin{vmatrix} a_2 & a_4 \\ b_2 & b_4 \end{vmatrix} \cdot \begin{vmatrix} c_1 & c_3 \\ d_1 & d_3 \end{vmatrix}
+ \begin{vmatrix} a_3 & a_4 \\ b_3 & b_4 \end{vmatrix} \cdot \begin{vmatrix} c_1 & c_2 \\ d_1 & d_2 \end{vmatrix}.
\end{aligned}
$$

This special case can easily be verified by computation.

Now, corresponding to the distinct points x and y, consider the numbers p_{ik} defined by

$$(42.2) \qquad p_{ik} = \begin{vmatrix} x_i & x_k \\ y_i & y_k \end{vmatrix}, \qquad i, k = 1,2,3,4.$$

These are not all zero since $x \not\div y$. If the representations of x and y are changed, then each of the numbers p_{ik} is multiplied by the same factor. On $[x,y]$ any point z, distinct from x and y, has a representation of the form $z = \lambda x + \mu y$, where $\lambda \neq 0$ and $\mu \neq 0$. Then

$$
\begin{vmatrix} x_i & x_k \\ z_i & z_k \end{vmatrix} = \begin{vmatrix} x_i & x_k \\ \lambda x_i + \mu y_i & \lambda x_k + \mu y_k \end{vmatrix} = \mu \begin{vmatrix} x_i & x_k \\ y_i & y_k \end{vmatrix},
$$

hence the ratios of the numbers p_{ik} are unchanged if y is replaced by z. By the same token, the ratios will be the same if x and y are replaced by any two points of $[x,y]$ which are distinct.

We wish to show that with suitable restrictions the numbers p_{ik} can be taken as coordinates of $[x,y]$. First we observe that these numbers are not arbitrary. Obviously $p_{ii} = 0$ and $p_{ik} = -p_{ki}$, hence only six of the numbers can be interesting. We follow tradition in singling out the sextuple

$$(42.3) \qquad p = (p_{12}, p_{13}, p_{14}, p_{34}, p_{42}, p_{23}).$$

Even these six numbers are not arbitrary, for in (42.1) if we set $a = c = x$ and $b = d = y$, we see that the numbers in p satisfy the relation

$$(42.4) \qquad p_{12}p_{34} + p_{13}p_{42} + p_{14}p_{23} = 0.$$

Subject to this restriction however, *the numbers p_{ik} can be taken for the coordinates of the line.* That is:

(42.5) *Given six numbers $p_{12},p_{13},p_{14},p_{34},p_{42},p_{23}$, not all zero, which satisfy (42.4), there is exactly one line $[x,y]$ such that the numbers (42.2) obtained from x and y form the given set.*

First, to see that there is not more than one line with given p_{ik} it suffices to show that the set obtained from x and y determines the intersection points of $[x,y]$ with the planes $x_i = 0$, $i = 1,2,3,4$. For $[x,y]$ can lie in at most two of these planes, hence at least two of the intersections determine $[x,y]$ uniquely. Taking the line in the form $\lambda x + \mu y$, the values $\lambda = y_1$ and $\mu = -x_1$ clearly give the point of the line in $x_1 = 0$. This point is $(0, -y_1x_2 + x_1y_2, -y_1x_3 + x_1y_3, -y_1x_4 + x_1y_4) = (0,p_{12},p_{13},p_{14})$. Similar calculations for the remaining planes show:

(42.6) *The line $[x,y]$ intersects the planes $x_i = 0$, $i = 1,2,3,4$, in the "points" $(0,p_{12},p_{13},p_{14})$, $(-p_{12},0,p_{23},-p_{42})$, $(-p_{13},-p_{23},0,p_{34})$, $(-p_{14},p_{42},-p_{34},0)$ respectively, at least two of which represent different points.[6] Conversely, if the numbers p_{ik} are not all zero, then at least two of the above quadruples have a non-zero coordinate since each number p_{ik} appears in two of them. These two quadruples, then, represent points and hence determine a line.*

If now a set of numbers p_{ik} is given, not all of whose members are zero, and which satisfies (42.4), we can select a non-zero element p_{ik} and work with the two points of (42.6) in which it occurs. For example, suppose $p_{12} \neq 0$. Then we select the points $(0,p_{12},p_{13},p_{14})$ and $(-p_{12},0,p_{23},-p_{42})$ to play the roles of x and y, and from these determine a new set $\bar{p}_{ik}$ in the order of (42.3). By direct calculation, the sextuple $\bar{p}$ is

$$(p_{12}^2, \; p_{12}p_{13}, \; p_{12}p_{14}, \; -p_{13}p_{42} - p_{14}p_{23} = p_{12}p_{34}, \; p_{12}p_{42}, \; p_{12}p_{23}).$$

Because $p_{12} \neq 0$, $\bar{p}$ is proportional to p hence the line connecting the two selected points yields the given p_{ik} set of numbers.

Instead of determining a line by two of its points, x and y, it may be determined by two of its planes, ξ and η. Setting

(42.7) $$P_{ik} = \begin{vmatrix} \xi_i & \xi_k \\ \eta_i & \eta_k \end{vmatrix}, \qquad i,k = 1,2,3,4,$$

then by duality of the previous argument, a set of numbers P_{ik}, not all zero, and satisfying (42.4), also determine a line uniquely. There is thus a second, and equally justified, set of coordinates for the line. These systems, however, are not independent, since:

(42.8) $$P_{12} : P_{13} : P_{14} : P_{34} : P_{42} : P_{23} = P_{34} : P_{42} : P_{23} : P_{12} : P_{13} : P_{14}.$$

---

[6]Quotation marks are used because not all of the quadruples need be $\neq 0$.

To indicate how this can be established, we prove, for example, the end equation, namely that $P_{13}P_{23} - P_{14}P_{42} = 0$. The fact that x and y are each on both the planes ξ and η yields the equations:

(42.9) $x \cdot \xi = 0, \qquad x \cdot \eta = 0, \qquad y \cdot \xi = 0, \qquad y \cdot \eta = 0.$

Multiplying the first equation by $-\eta_1$ and the second by ξ_1, and adding the two, produces

$$x_2 P_{12} + x_3 P_{13} + x_4 P_{14} = 0.$$

The same operation with the last two equations gives

$$y_2 P_{12} + y_3 P_{13} + y_4 P_{14} = 0.$$

Multiplying the first of the new equations by $-y_2$ and the second by x_2 and adding gives the desired result, $P_{13}P_{23} - P_{14}P_{42} = 0$.

Since the sets p_{ik} and P_{ik} are only determined to within a factor, whenever it is convenient to do so we will assume that the proportionality factor in (42.8) is one, that is that $p_{12} = P_{34}$, $p_{13} = P_{42}$, etc.

We now consider some elementary problems of analytic geometry using line coordinates. That is, *a line M is now to be thought of as given by a sextuple m_{ik}, or alternately by a sextuple M_{ik}*, where it may be assumed that $m_{13} = M_{42}$, etc. We seek first a criterion that a given line lies in a given plane. If x and y are any two points of M, then for suitable representations of these points, and any plane ξ,

$$m_{ik}\xi_k = (x_i y_k - x_k y_i)\xi_k = x_i(y_k \cdot \xi_k) - y_i(x_k \cdot \xi_k),$$

and hence

$$\sum_{k=1}^{4} m_{ik}\xi_k = x_i(y \cdot \xi) - y_i(x \cdot \xi).$$

If M is in ξ, then x and y are also in ξ, and $y \cdot \xi = x \cdot \xi = 0$ implies $\sum_{k=1}^{4} m_{ik}\xi_k = 0$ for $i = 1,2,3,4$. Conversely, the equations $\sum_{k=1}^{4} m_{ik}\xi_k = 0$ for $i = 1,2,3,4$, express the fact that the points $x^i = (m_{i1}, m_{i2}, m_{i3}, m_{i4})$, $i = 1,2,3,4$, which are not the zero-quadruple, lie on the plane ξ. Since $m_{ii} = 0$ and $m_{ik} = -m_{ki}$, the argument of (42.6) shows that if the numbers m_{ik} are coordinates of a line, and hence not all zero, then at least two of the quadruples x^i really represent points of ξ. Therefore:

The line M lies on the plane ξ if and only if the relations

(42.10) $\sum_{k=1}^{4} m_{ik}\xi_k = 0$ *hold for $i = 1,2,3,4$. Two of these, for which the left side is not identically zero, insure that M is on ξ.*

The condition for M to pass through the point x, is, of course, that:

$$(42.11) \qquad \sum_{k=1}^{4} M_{ik}x_k = 0, \quad i = 1,2,3,4.[7]$$

Next we ask when the lines M and M' intersect. If the coordinates of the lines are determined by their respective point pairs x,y and x',y', then the lines intersect if and only if x,y,x' and y' are co-planar, that is, if and only if $|x,y,x',y'| = 0$. The relation (42.1) expresses this in the form:

(42.12) *Two lines M and M' intersect if and only if*
$$m_{12}m'_{34} + m_{13}m'_{42} + m_{14}m'_{23} + m_{34}m'_{12} + m_{42}m'_{13} + m_{23}m'_{14} = 0.$$

Since expressions of this form occur frequently, where a and b represent sextuples of the form (42.3), we introduce the notation:

$$(42.13) \quad \omega(a,b) = a_{34}b_{12} + a_{42}b_{13} + a_{23}b_{14} + a_{12}b_{34} + a_{13}b_{42} + a_{14}b_{23}.$$

Then $\omega(a,b) = \omega(b,a)$, and

$$(42.14) \qquad \omega(a,a) = 2(a_{12}a_{34} + a_{13}a_{42} + a_{14}a_{23}).$$

Also, as for any bilinear form:

(42.15)
$$\omega(\lambda a + \mu b, \lambda'a' + \mu'b')$$
$$= \lambda\lambda'\omega(a,a') + \lambda\mu'\omega(a,b') + \lambda'\mu\omega(a',b) + \mu\mu'\omega(b,b').$$

The condition (42.4) for the six numbers m_{ik}, not all zero, to be the coordinates of a line M becomes $\omega(m,m) = 0$, and the condition for the lines M and M' to intersect takes the form $\omega(m,m') = 0$.

The lines M satisfying a linear equation

(42.16) $\omega(a,m) = 0$, *where not all a_{ik} are zero,*

are said to form a *(linear) line complex.* Two cases are distinguished:

1) $\omega(a,a) = 0$.

This is the condition that the a_{ik} define a line A, since not all a_{ik} are zero, and $\omega(a,m) = 0$ states that A and M intersect. Hence:

(42.17) *If $\omega(a,a) = 0$ the complex $\omega(a,m) = 0$ consists of all lines M which intersect the line A, called the axis of the complex, which has a for its coordinates.*

This trivial type of complex is called degenerate.

2) $\omega(a,a) \neq 0$.

To picture the distribution of lines in this case we investigate those lines of the complex which are concurrent at a point x. The condition for the

---

[7] If the M_{ik} are considered as variable, then (42.11) are the equations of the lines in the bundle x. Similarly (42.10) yields the equations for the lines in the plane ξ.

line M, passing through x and $y \dotplus x$, to lie in the complex is that $m_{ik} = \begin{vmatrix} x_i & x_k \\ y_i & y_k \end{vmatrix}$ satisfy $\omega(a,m) = 0$, hence that:

$$a_{34}(x_1 y_2 - x_2 y_1) + a_{42}(x_1 y_3 - x_3 y_1) + a_{23}(x_1 y_4 - x_4 y_1)$$
$$+ a_{12}(x_3 y_4 - x_4 y_3) + a_{13}(x_4 y_2 - x_2 y_4) + a_{14}(x_2 y_3 - x_3 y_2) = 0.$$

If this is rewritten in the form $\sum\limits_{i,k=1}^{4} c_{ik} y_i x_k = 0$, then the coefficient matrix is

(42.18) $$(c_{ik}) = \begin{pmatrix} 0 & -a_{34} & -a_{42} & -a_{23} \\ a_{34} & 0 & -a_{14} & a_{13} \\ a_{42} & a_{14} & 0 & -a_{12} \\ a_{23} & -a_{13} & a_{12} & 0 \end{pmatrix}$$

Therefore the lines of the complex, which pass through x, lie in the plane ξ, where

(42.19) $$\xi_i = \sum_{k=1}^{4} c_{ik} x_k. \qquad k = 1,2,3,4.$$

From (42.1), the determinant $|c_{ik}|$ has the value

$$|c_{ik}| = (a_{12}a_{34} + a_{13}a_{42} + a_{14}a_{23})^2 = (1/4)\omega^2(a,a) \neq 0.$$

The relation (42.19) suggests some sort of spatial analogue to a plane polarity (see Section 10). Here, however, $c_{ik} = -c_{ki}$, which was impossible in the plane case since there it implied that the corresponding three-rowed determinant $|c_{ik}|$ was zero. To consider the matter more fully, a *correlation of the projective* space is defined as a one-to-one mapping of the points on the planes in the form:

(42.20) $$\xi_i' = \sum_{k=1}^{4} b_{ik} x_k, \qquad i = 1,2,3,4, \qquad |b_{ik}| \neq 0.$$

When B_{ik} is the cofactor of b_{ik} divided by $|b_{ik}|$, the induced mapping of the planes on the points is given by :

(42.21) $$x_i' = \sum_{k=1}^{4} B_{ik} \xi_k, \qquad i = 1,2,3,4.$$

The inverse transformations are:

(42.22) $$x_i = \sum_{k} B_{ki} \xi_k' \qquad and \qquad \xi_i = \sum_{k} b_{ki} x_k'.$$

The square of (42.20) is a collineation of P^3 on itself. This collineation is the identity if and only if (42.20) and the second transformation in (42.22) coincide, that is, just as in the plane case, when and only when $b_{ik} = \lambda b_{ki}$. As before, $b_{ik} = \lambda b_{ki} = \lambda^2 b_{ik}$ implies $\lambda = \pm 1$, but, as observed above, the case $\lambda = -1$ cannot now be excluded. We call a correlation of P^3 on itself of the form

$$(42.23) \qquad \xi_i = \sum_k b_{ik} x_k, \qquad |\, b_{ik}\,| \neq 0, \qquad b_{ik} = -\, b_{ki},$$

a *null system*, and a correlation of the form

$$(42.24) \qquad \xi_i = \sum_k b_{ik} x_k, \qquad |\, b_{ik}\,| \neq 0, \qquad b_{ik} = b_{ki}$$

a *polarity* of P^3.[8]

In both cases, the corresponding point and plane are called respectively *the pole (of the plane)* and the *polar plane (to the point)*. In both cases a pencil of points corresponds to a pencil of planes and conversely. Hence to a line M, as the carrier of two incident point and plane pencils, there corresponds the line M' carrying the incident image plane and point pencils. The line M' is called the polar to M. We can formulate this more intuitively:

> (42.25) *If the line M is determined by the points x and y (the planes ξ and η) then its polar line M' is determined by the planes polar to x and y (the poles of ξ and η).*

The point x is said to be *conjugate to the point y* if x is on the plane conjugate to y. Similarly, *the plane ξ is conjugate to η* if it contains the polar point to η. The conditions for conjugacy of x and y, and for ξ and η, are respectively:

$$(42.26) \qquad \sum_{i,k=1}^{4} b_{ik} y_i x_k = 0 \qquad \text{and} \qquad \sum_{i,k=1}^{4} B_{ik} \eta_i \xi_k = 0.$$

This, of course, implies that the respective conditions for x and η to be self-conjugate are $\sum_{i,k} b_{ik} x_i x_k = 0$ and $\sum_{i,k} B_{ik} \eta_i \eta_k = 0$.

In a polarity, points which are not self-conjugate always exist. For not all the numbers b_{ik} are zero. Hence for some i, say $i = 1$, there is a number $b_{ik} \neq 0$. If $b_{11} \neq 0$, then d_1 is not self-conjugate. If $b_{12} \neq 0$, then $(1,\lambda,0,0)$ is not self-conjugate for any value of λ other than a root of $b_{11} + 2\lambda b_{12} + \lambda^2 b_{22} = 0$. For a null system, however, every point is

---

[8]Some authors call (42.23) a null polarity and (42.24) an ordinary polarity.

self-conjugate since $\sum b_{ik}x_ix_k = \sum b_{ki}x_kx_i = (1/2)\sum(b_{ik} + b_{ki})x_ix_k = 0$.
Therefore:

(42.27)   *A correlation of a projective space on itself, whose square is the identity, is a null system when every point is self-conjugate, otherwise it is a polarity.*

Polarities are considered in more detail in the next section. In a null system every point is self-conjugate, hence the line joining two conjugate points x and y lies in both of their polar planes. Therefore every line determined by two conjugate points is self-polar.

If now the b_{ik} of a null system are identified with the elements c_{ik} in (42.18), then the condition for conjugacy of x and y becomes $\sum c_{ik}y_ix_k = 0$. Thus we have found:

(42.28)   *A non-degenerate (linear) line complex consists of the self-polar lines in a null system. The lines of the complex through a given point form a pencil in the plane polar to the point, and the lines of the complex in a given plane form a pencil through the point polar to the plane.*

The proofs of the following two facts are useful exercises to increase familiarity with linear complexes.

Let $a^1, a^2, \cdots, a^5$ be five distinct points such that the five planes through a^i, a^{i+1}, a^{i+2}, $i = 1,2,3,4,5$, are all different, where $a^6 = a^1$, $a^7 = a^2$, etc. The correlation which associates the plane through a^i, a^{i+1}, a^{i+2} with a^{i+1} is a null system.

It is obvious that all degenerate complexes are projectively equivalent. The same holds for non-degenerate complexes because, in an appropriate coordinate system, a given non-degenerate complex takes the form $p_{12} + p_{34} = 0$.

Before continuing the investigation of line loci, it will prove helpful, first, to develop the theory of quadrics.

43. Polarities and Quadrics

A polarity Φ had the forms:

(43.1) $\xi_i = \sum_k b_{ik}x_k$, $i = 1,2,3,4$, $b_{ik} = b_{ki}$, $|b_{ik}| \neq 0$.

(43.2) $x_i = \sum_k B_{ik}\xi_k$, $i = 1,2,3,4$, $B_{ik} = B_{ki}$, $|B_{ki}| = |b_{ik}|^{-1}$, where B_{ik} is the co-factor of b_{ik} in (b_{ik}) divided by $|b_{ik}|$. In each non-self-conjugate plane ξ, the mapping Φ induces a polarity $\Phi(\xi)$, and induces the polarity $\Phi(z)$ in the bundle z if z is not self-conjugate. For if x is any point of ξ, then ξ_x, the polar to x, is not ξ and hence cuts it in a line M_x. Then $x \to M_x$ is the plane polarity $\Phi(\xi)$. If the line M_x in ξ is given, the point x can be determined as follows. The pencil of planes on M_x maps under Φ into a pencil of points carried by the line M'_x, polar to M_x. Because ξ is not self-conjugate, M'_x does not lie in ξ and hence intersects it in the desired point x. That $x \to M_x$ is actually a polarity in ξ follows from Φ^2 being the identity. If z is not self-conjugate then $\xi_z = z\Phi$ is also not self-conjugate. In this case both polarities $\Phi(z)$ and $\Phi(\xi_z)$ exist. As in the plane case they are both elliptic, or else both hyperbolic.[§]

If x^1 is any non-self-conjugate point and ξ_{x^1} is its polar plane, consider any triple x^2, x^3, x^4 in ξ_{x^1} forming a self-polar triangle with respect to the induced polarity $\Phi(\xi_{x^1})$. In the plane polarity, x^3 and x^4 lie on the line M_{x^2}, polar to x^2, and this line is determined as the intersection of ξ_{x^1} and ξ_{x^2}. Since x^2 is on ξ_{x^1}, ξ_{x^2} contains x^1, hence x^1, x^3, x^4 determine the plane polar to x^2. Because of symmetry, a plane through any three of the vertices of the tetrahedron x^1, x^2, x^3, x^4 is the polar plane of the fourth vertex. Such *a tetrahedron is called self-polar*, and from (42.25) it is clear that in such a tetrahedron a line through two of the vertices is polar to the line through the remaining two.

As in the plane case, when the polarity has the points d_1, d_2, d_3, d_4 as the vertices of a self-polar tetrahedron, it takes the simple form:

$$(43.3) \qquad \xi_i = b_i x_i, \qquad b_i \neq 0, \qquad i = 1,2,3,4. \text{[§]}$$

The conditions for point and plane conjugacy are respectively:

$$(43.4) \qquad \sum_{i=1}^{4} b_i x_i y_i = 0, \qquad \sum_{i=1}^{4} b_i^{-1} \eta_i \xi_i = 0$$

and those for self-conjugacy are:

$$(43.5) \qquad \sum_{i=1}^{4} b_i x_i^2 = 0, \qquad \sum_{i=1}^{4} b_i^{-1} \eta_i^2 = 0.$$

Consider now a line M with coordinates m_{ik} determined by the points x and y. If $x \to \xi$ and $y \to \eta$ under the above polarity, then the line M', polar to M, has coordinates

$$M'_{ik} = \begin{vmatrix} \xi_i & \xi_k \\ \eta_i & \eta_k \end{vmatrix} = \begin{vmatrix} b_i x_i & b_k x_k \\ b_i y_i & b_k y_k \end{vmatrix} = b_i b_k m_{ik}.$$

From (42.8) then,

$$m'_{12} : m'_{13} : m'_{14} : m'_{34} : m'_{42} : m'_{23}$$
$$= M'_{34} : M'_{42} : M'_{23} : M'_{12} : M'_{13} : M'_{14}$$
$$= b_3 b_4 m_{34} : b_4 b_2 m_{42} : b_2 b_3 m_{23} : b_1 b_2 m_{12} : b_1 b_3 m_{13} : b_1 b_4 m_{14}.$$

For M to be self-polar, m'_{ik} must be proportional to m_{ik}, hence the condition is:

$$(43.6) \qquad \frac{b_3 b_4 m_{34}}{m_{12}} = \frac{b_4 b_2 m_{42}}{m_{13}} = \frac{b_2 b_3 m_{23}}{m_{14}} = \frac{b_1 b_2 m_{12}}{m_{34}} = \frac{b_1 b_3 m_{13}}{m_{42}} = \frac{b_1 b_4 m_{14}}{m_{23}}.$$

Not all the numbers m_{ik} are zero, hence if λ is the common ratio in (43.6), and say $m_{12} \neq 0$, the equality of the first and fourth fractions implies that $\lambda^2 m_{12}^2 = b_3^2 b_4^2 m_{34}^2 = b_1 b_2 b_3 b_4 m_{12}^2$. Hence:

$$(43.7) \qquad b_1 b_2 b_3 b_4 > 0, \text{ if (real) self-polar lines exist.}$$

It follows from (42.25) that all the points on a self-polar line and all the planes through it are self-conjugate. Conversely, a line consisting of self-conjugate points is evidently self-polar.

The polarity in (43.3) can be simplified further (by a coordinate transformation of the form $x'_i = \lambda_i x_i$) to one of the following forms.

$$(43.8) \qquad \xi_1 = x_1, \quad \xi_2 = x_2, \quad \xi_3 = x_3, \quad \xi_4 = x_4.$$

$$(43.9) \qquad \xi_1 = x_1, \quad \xi_2 = x_2, \quad \xi_3 = x_3, \quad \xi_4 = -x_4.$$

$$(43.10) \qquad \xi_1 = x_1, \quad \xi_2 = x_2, \quad \xi_3 = -x_3, \quad \xi_4 = -x_4.$$

In the first case the locus of self-conjugate points, which is given by $\sum_{i=1}^{4} x_i^2 = 0$, contains no real points and is called a non-degenerate, *imaginary quadric.*[9] In the case of (43.9) the self-conjugate locus,

$$x_1^2 + x_2^2 + x_3^2 - x_4^2 = 0,$$

has real points. Because (43.7) is not satisfied, there are no self-polar lines. Hence no lines are contained in the locus and it is spoken of as a non-degenerate, *non-ruled quadric.* For the third polarity, (43.10), the corresponding locus, $x_1^2 + x_2^2 - x_3^2 - x_4^2 = 0$, which has real points and also satisfies (43.7) is called a non-degenerate, *ruled quadric.* In this case (43.6), with $\lambda = \pm 1$, shows that the self-polar lines form two families F and F'. Respective lines of the families, M and M', have coordinates of the form:

$$M : (m_{12}, m_{13}, m_{14}, m_{12}, -m_{13}, -m_{14})$$
$$M' : (m'_{12}, m'_{13}, m'_{14}, -m'_{12}, m'_{13}, m'_{14}).$$

---

[9] A quadratic surface $\sum_{i,k=1}^{4} a_{ik} x_i x_k = 0$, $a_{ik} = a_{ki}$, is called non-degenerate when $|a_{ik}| \neq 0$.

The condition (42.4) then becomes:

$$(43.11) \qquad m_{12}^2 - m_{13}^2 - m_{14}^2 = 0, \qquad m_{12}'^2 - m_{13}'^2 - m_{14}'^2 = 0.$$

Since not all the numbers m_{ik} are zero, (43.11) implies that $m_{12} \neq 0$ and $m_{12}' \neq 0$. Hence each line of F intersects every line of F', and conversely, since clearly $\omega(m,m') = 0$. However, two different lines, M and $\overline{M}$, of the same family do not intersect. For we have:

$$(43.12) \qquad \omega(m,\overline{m}) = 2(m_{12}\overline{m}_{12} - m_{13}\overline{m}_{13} - m_{14}\overline{m}_{14}).$$

Because $m_{12} \neq 0$ and $\overline{m}_{12} \neq 0$ we may suppose $m_{12} = \overline{m}_{12}$. Then $\omega(m,\overline{m}) = 0$, in conjunction with (43.11), would imply

$$- (m_{12} - \overline{m}_{12})^2 + (m_{13} - \overline{m}_{13})^2 + (m_{14} - \overline{m}_{14})^2 = 0,$$

and this, with $m_{12} = \overline{m}_{12}$, would make $M = \overline{M}$.

We will continue with this topic in the next section, but before we leave it, it may be instructive to compare the present procedure with that of elementary analytic geometry. In homogeneous coordinates, the self-conjugate locus of (43.10), for instance, would be written as

$$x_1^2 - x_3^2 = x_4^2 - x_2^2,$$

and treated in the following way. With λ an arbitrary parameter, x is a point of the locus if it satisfies the simultaneous equations:

$$\left\{ \begin{array}{l} x_1 - x_3 = \lambda(x_4 - x_2) \\ \lambda(x_1 + x_3) = x_4 + x_2 \end{array} \right\} \quad \text{or} \quad \left\{ \begin{array}{l} x_1 - x_3 = \lambda(x_4 + x_2) \\ \lambda(x_1 + x_3) = x_4 - x_2 \end{array} \right\}$$

The first and second of these equations are planes with coordinates $(1,\lambda,- 1,- \lambda)$ and $(\lambda,- 1,\lambda,- 1)$. We can now calculate directly the coordinates M_{ik}' of the line M' in which they intersect, and then use (42.8) to obtain m_{ik}'. The results are:

$$(43.13) \quad \begin{array}{l} M_{12}' = m_{34}' = -1-\lambda^2, \; M_{13}' = m_{42}' = 2\lambda, \; M_{14}' = m_{23}' = - 1 + \lambda^2, \\ M_{34}' = m_{12}' = 1 + \lambda^2, \; M_{42}' = m_{13}' = 2\lambda, \; M_{23}' = m_{14}' = \lambda^2 - 1. \end{array}$$

Comparing this with (43.11), it follows that M' is an element of F'. Also, because of (43.11) and $m_{12}' \neq 0$, every element of F' can be written in the form (43.13). A similar calculation with the second set of simultaneous equations shows that these planes intersect in the lines of F.

The dual to a (non-degenerate) point quadric is a (non-degenerate) *plane quadric*. It consists, by definition, of all self-conjugate planes in a polarity. The *tangent plane* at a point x of a point quadric is the plane ξ_x, polar to x. Similarly the *point of contact* on the plane ξ of a plane quadric is the point x_ξ, polar to ξ. Clearly, and as one can calculate in (43.9) or (43.10), the tangent planes of a point quadric form a plane quadric. In turn its contact points form the point quadric, hence, as formerly, one may drop the distinction between the two and simply speak of the quadric.

A *tangent line* of a quadric is any line lying in a tangent plane and passing through the contact point of the plane. Thus a quadric may also be considered as the locus of a (three parameter) family of lines. The equations, in line coordinates, defining the family could easily be derived from (43.9) or (43.10).

A tangent plane of a non-ruled quadric C has its contact point as its only intersection with C. For if C is defined by (43.9), the tangent plane ξ at y is given by $x_1y_1 + x_2y_2 + x_3y_3 - x_4y_4 = 0$. Since y lies on C, $y_1^2 + y_2^2 + y_3^2 - y_4^2 = 0$, *hence* $y_4 \neq 0$. If z is a point of C, assumed to be on ξ, then $z_4 \neq 0$ and we may take $z_4 = y_4$. From the equation for y to be on C, together with the equations

$$z_1^2 + z_2^2 + z_3^2 - z_4^2 = 0 \quad \text{and} \quad z_1y_1 + z_2y_2 + z_3y_3 - z_4y_4 = 0,$$

it follows that $(z_1 - y_1)^2 + (z_2 - y_2)^2 + (z_3 - y_3)^2 = 0$, and hence that $y \sim z$.

If M is a self-polar line in a ruled quadric C, then the points of M are the poles of the planes through M. Thus if x is the intersection point of the lines M and M', from the families F and F' respectively, the pole of the plane ξ, which spans M and M', must lie on both M and M', hence is x. This and the previous result can be summed up in the form:

(43.14) *If C is not ruled, then a tangent plane ξ intersects it in its contact point only. If C is ruled, then ξ cuts it in the two lines of F and F' respectively through the contact point of ξ.*

Next we consider the relation of C to a plane ξ which is not self-conjugate. The pole x of ξ is also not self-conjugate, and the induced polarities $\Phi(x)$ and $\Phi(\xi)$ are either both elliptic or both hyperbolic. In the latter case, the self-conjugate elements of the two polarities form a cone K_x and a conic C_x respectively. The construction of $\Phi(\xi)$ shows that the points of C_x are also self-conjugate with respect to Φ, and therefore lie on C. Hence C_x is the intersection of ξ with C. By duality, K_x is the intersection of the plane bundle on x with C as a plane quadric. Hence K_x is formed by the tangent planes of C passing through x, and the contact points of the planes in K_x form C_x. We have the theorem, then:

(43.15) *If ξ, with pole x, is not a tangent plane of the quadric C, defined by the polarity Φ, and if the induced polarity $\Phi(\xi)$ (and hence $\Phi(x)$) is hyperbolic, then ξ intersects C in a conic C_x and the tangent planes to C at the points of C_x coincide with the tangent planes of C which passes through x.*

In the usual way, a non-tangent plane which contains no point of C is called *non-intersecting*, and is called *intersecting* if it cuts C. A point x, not on C, through which there are tangent planes is called a *tangent-cone point*, otherwise a *no-tangent point*. Thus ξ is intersecting if and only if its pole is a tangent-cone point, that is, if and only if the polarity $\Phi(\xi)$ is hyperbolic.

Since a line intersects every plane (or lies in it), when C is a ruled quadric every plane is intersecting. As a consequence:

(43.16)   *If C is ruled, then for every point x not on C, the polarities $\Phi(x)$ and $\Phi(\xi)$, induced at x and its polar ξ, are hyperbolic.*

The form (43.9) for a non-ruled quadric shows that the plane $x_4 = 0$ is not intersecting. Hence:

(43.17)   *If C is real and not ruled, then both intersecting and non-intersecting planes, and cone-tangent and no-tangent points, exist. Also both types of induced polarities, hyperbolic and elliptic, exist.*

The fact that if two elements are conjugates with respect to a space polarity they are also conjugates with respect to an induced polarity, implies the following generalizations of (11.7) and (11.8). If x and x' are conjugate with respect to the quadric C, and if $[x,x']$ cuts C in u and v, then x, x', u and v form a harmonic set. If x is a no-tangent point of a (non-ruled) quadric C, and a variable line through x cuts C in u and v, then the locus of the 4th harmonic point to x, u and v is a plane.

The question arises, whether there is a *three-dimensional analogue to Steiner's method* of generating conics by intersecting corresponding lines in two projectively, but not perspectively, related pencils. It will be seen in the next section (see (44.10) and (44.13)) that a strict analogue exists for ruled quadrics. If imaginary elements are introduced, this may be generalized to arbitrary quadrics. For a non-ruled quadric carries two families of imaginary lines. For instance $x_1^2 + x_2^2 + x_3^2 - x_4^2 = 0$ carries the lines

$$\left\{ \begin{array}{l} x_1 + ix_3 = \lambda(x_4 + x_2) \\ \lambda(x_1 - ix_3) = x_4 - x_2 \end{array} \right\} \quad \text{and} \quad \left\{ \begin{array}{l} x_1 - ix_3 = \lambda(x_4 + x_2) \\ \lambda(x_1 + ix_3) = x_4 - x_2 \end{array} \right\}$$

However, this is very unsatisfactory from the geometric point of view. A method of obtaining all quadrics without using imaginary numbers is the following. Let Φ be a correlation between two bundles with different centers, z and z', so a line L in z corresponds to a plane $\xi = L\Phi$ in z'. Unless all the intersections $L \cap L\Phi$ lie on a plane, they form a point quadric. A plane quadric is generated by relating projectively the lines L in a bundle z with the points $x = L\Phi$ in a plane ζ. The planes $(L \wedge L\Phi)$ traverse a plane quadric unless they all pass through one point. On these questions see: Th. Reye, Die Geometrie der Lage.

For the sake of completeness, we consider the locus C given by a *degenerate quadratic equation*:

$$(43.18) \qquad \sum_{i,k=1}^{4} a_{ik}x_ix_k = 0, \quad a_{ik} = a_{ki}, \quad |a_{ik}| = 0, \quad \text{not all } a_{ik} = 0.$$

As in the plane case, the equations

$$(43.19) \qquad \sum_{k=1}^{4} a_{ik}u_k = 0, \quad i = 1,2,3,4,$$

have a non-trivial solution $u = (u_1,u_2,u_3,u_4)$. Let u be taken as d_4. Substitution in (43.19) shows, then, that $a_{14} = a_{24} = a_{34} = a_{44} = 0$, and hence that $a_{41} = a_{42} = a_{43} = 0$. Thus the quadric reduces to

$$(43.20) \qquad \sum_{i,k=1}^{3} a_{ik}x_ix_k = 0.$$

As in the discussion of cones, we may interpret x_1,x_2,x_3 as projective coordinates in the plane $x_4 = 0$. Then (43.20) represents either a degenerate or non-degenerate conic C_1 in this plane. As an equation in four variables it then represents all points which lie on the lines joining d_4 to the points of C_1. If the determinant $|a_{ik}|$, $i,k = 1,2,3$, is not zero, C_1 is non-degenerate, but may be real or imaginary. When it is imaginary, d_4 is the only (real) point of C and we call it an *imaginary cone*. When C_1 is real (and non-degenerate), C is an *ordinary cone*. On the other hand, when the determinant of the conic C_1 vanishes, C_1 degenerates to a point or to two lines which may or may not be distinct. Correspondingly C is a *line, two planes, or a single plane*.

A quadratic equation in more variables, say $x_1,x_2,\cdots,x_{n+1}$, that is,

$$(43.21) \qquad \sum_{i,k=1}^{n+1} a_{ik}x_ix_k = 0, \quad a_{ik} = a_{ki}, \quad |a_{ik}| \neq 0,$$

may be thought of as representing a *non-degenerate quadric Q in an n-dimensional projective space P^n*. Without developing the theory of these quadrics, it is often helpful, in interpreting such an equation, to use geometric terminology (compare next section). Thus a point $y = (y_1,y_2,\cdots,y_{n+1})$ of the quadric Q is one which satisfies (43.21). The tangent (hyper) plane of Q at y is $\sum_{i,k} a_{ik}x_iy_k = 0$. For any point y, this equation represents the plane polar to y. The quadric may contain straight lines or, more generally, may contain P^m, where $m < n - 1$. It cannot contain a hyperplane since $| a_{ik} | \neq 0$.[10]

44. Linear Congruences and Reguli

In discussing line loci further we will adopt the point of view just mentioned and will develop it in some detail, not so much because the results are significant for our present purpose, but because the method is an important one.

In the projective space P^3, a line M is given by the sextuple

$$m = (m_{12}, m_{13}, m_{14}, m_{34}, m_{42}, m_{23}).$$

This sextuple may also be interpreted as a point in a five-dimensional projective space P^5. Conversely, given a point p in P^5, with coordinate p_i, $i = 1, 2, \cdots, 6$, we can define $p_1 = p_{12}$, $p_2 = p_{13}$, $p_3 = p_{14}$, $p_4 = p_{34}$, $p_5 = p_{42}$, $p_6 = p_{23}$ and take $p_{ik} = - p_{ki}$. However, since the coordinates p_i are arbitrary (save that they are not all zero) it is clear that the numbers p_{ik} associated with p will not, in general, serve as the coordinates of a line in P^3. They will do so, in fact, if and only if they satisfy the equation

$$\omega(p,p) = 2(p_{12}p_{34} + p_{13}p_{42} + p_{14}p_{23}) = 0.$$

With p variable and with $p_{12} = p_1$, $p_{34} = p_4$, etc., this is a quadratic equation of the form (43.21). The term $2p_{12}p_{34}$, for instance, can be written as $p_1 p_4 + p_4 p_1$ so that, corresponding to (43.21), $a_{14} = 1$ and $a_{41} = 1$. *The equation* $\omega(p,p) = 0$, then, *defines a fixed quadric* Q *in* P^5, *which is not degenerate* since its coefficient determinant, $| a_{ik} |$, is

$$\begin{vmatrix} 0 & 0 & 0 & 1 & 0 & 0 \\ 0 & 0 & 0 & 0 & 1 & 0 \\ 0 & 0 & 0 & 0 & 0 & 1 \\ 1 & 0 & 0 & 0 & 0 & 0 \\ 0 & 1 & 0 & 0 & 0 & 0 \\ 0 & 0 & 1 & 0 & 0 & 0 \end{vmatrix} = (-1)^3 = -1 \neq 0.$$

Points of P^5 on Q, and only these, define corresponding lines in P^3. If p lies on Q, then $\omega(p,p) = \sum_{i,k} a_{ik} p_i p_k = 0$. For a fixed p, then, whether or not it lies on Q, the equation $\omega(p,q) = \sum_{i,k} a_{ik} p_i q_k = 0$ (in variable q) repre-

---

[10]A non-degenerate quadric in P^n cannot contain a P^m with $m > \dfrac{n-1}{2}$, see Van der Waerden, "Einführung in die algebraische Geometrie," pp. 31-32.

sents the hyperplane polar to p with respect to Q. Thus, to the lines of the complex $\omega(p,m) = 0$, in P^3, correspond points in P^5 on the polar plane to p. Since m is a line in P^3, its corresponding point is also on Q. More exactly, then, to the lines of the complex correspond the intersection points of Q with the hyperplane polar to p. When the complex is degenerate, that is when $\omega(p,p) = 0$, its lines correspond to the intersection points of Q with its tangent plane at p.

Lines in P^3 (and points in P^5) are given by sextuples, and *lines* in P^3 (points in P^5) are called *independent or dependent* according as the corresponding sextuples are independent or dependent. By (40.7) the maximal number of independent points in P^5 is six. On the other hand, the six points defined by the respective rows in the determinant $|\,a_{ik}\,|$ are clearly independent. By inspection, these points lie on Q, hence six independent lines exist in P^3, which is the maximal number.

Since there are not more than five independent points in a hyperplane of P^5 (see (40.5)) there cannot be more than five independent lines of P^3 in a complex. On the other hand there always are five. For consider $\omega(p,q) = 0$, where p is fixed. Since not all p_{ik} are zero, suppose for instance $p_{14} = p_3 \neq 0$. Then for q of the form $(q_1,0,0,0,0,q_6)$, $\sum_{i,k} a_{ik}p_iq_k = 0$ reduces

to $p_4q_1 + p_3q_6 = 0$, which certainly has solutions in q_1,q_6 for which $q_1 \neq 0$. Whether q_6 is zero or not, $\omega(q,q) = \sum_{i,k} a_{ik}q_iq_k = q_4q_1 + q_3q_6 = 0$, hence q

is on Q and defines a line in the complex $\omega(p,m) = 0$. Call this point $q^1 = (q_1^1,0,0,0,0,q_6^1)$, where $q_1^1 \neq 0$. In a similar way we find that $q^2 = (0,q_2^2,0,0,0,q_6^2)$, $q^4 = (0,0,0,q_4^4,0,q_6^4)$, $q^5 = 0,0,0,0,q_5^5,q_6^5)$, where $q_i^i \neq 0$, are points of Q satisfying $\omega(p,q) = 0$. For a final point, q^3 may be taken to be any point of Q satisfying $q_3 \neq 0$ and $\omega(p,q^3) = 0$. It is easily seen that such a point exists and that the five points of Q, q^i, are independent.§ Hence the arbitrary complex $\omega(p,m) = 0$ has five independent lines. Conversely, if five independent lines, M^i, $i = 1,\cdots,5$, are given, then $|\,m,m^1,m^2,m^3,m^4,m^5\,| = 0$ represents a linear complex containing the five given lines.[11] Combined with the previous remarks, this implies:

(44.1) *A linear complex consists of the lines dependent on five independent lines.*

As a consequence of (44.1), every hyperplane in P^5 intersects Q. The corresponding case for ruled quadrics in P^3 suggests that Q contains lines

---

[11] $|\,m,m^1,m^2,m^3,m^4,m^5\,|$ is the determinant whose ith row is the sextuple, m^i, namely the coordinates of M^i.

in P^5. The following theorem confirms this and shows, at the same time, to which set of lines in P^3 the points of a line in Q correspond.

If p' and p'' are distinct points of P^5, and

$$\Delta(p',p'') = \omega(p',p')\omega(p'',p'') - \omega^2(p',p''),$$

then the line $p = \lambda'p' + \lambda''p''$, $(\lambda',\lambda'') \neq (0,0)$, bears the following relationship to Q: when $\Delta(p',p'') < 0$, the line does not intersect Q;

(44.2) when $\Delta(p',p'') > 0$, the line cuts Q in two distinct points; when $\Delta(p',p'') = 0$, but at least one of its terms, $\omega(p',p')$, $\omega(p'',p'')$, $\omega(p',p'')$ is not zero, the intersection is a single point; when all the terms in $\Delta(p',p'')$ are zero, the whole line lies on Q (such a line corresponds to a plane pencil of lines in P^3).

PROOF: By direct substitution, the variable point $\lambda'p' + \lambda''p''$ lies on Q if and only if

(44.3) $$\lambda'^2\omega(p',p') + 2\lambda'\lambda''\omega(p',p'') + \lambda''^2\omega(p'',p'') = 0.$$

For $\Delta(p',p'') < 0$ this equation has no real, non-trivial solution in λ',λ'', while for $\Delta(p',p'') > 0$ it has two, which are distinct. When the coefficients in the equation are not all zero, but $\Delta(p',p'')$ is zero, there is one real solution. Finally, if all the coefficients vanish, then all pairs, λ',λ'', are solutions, hence the entire line lies in Q. In the last case, $\omega(p',p') = \omega(p'',p'') = 0$, hence the sextuples p' and p'' represent lines M' and M'' in P^3. Since $\omega(p',p'')$ is also zero, M' and M'' intersect in some point, say x. If we choose y' and y'' respectively on M' and M'', both distinct from x, we may suppose that

$$m'_{ik} = p'_{ik} = \begin{vmatrix} x_i & x_k \\ y'_i & y'_k \end{vmatrix} \quad \text{and} \quad m''_{ik} = p''_{ik} = \begin{vmatrix} x_i & x_k \\ y''_i & y''_k \end{vmatrix}.$$

Hence, to the point $\lambda'p' + \lambda''p''$ on the line $[p',p'']$ in Q, there corresponds in P^3 the line

$$\lambda'p'_{ik} + \lambda''p''_{ik} = \begin{vmatrix} x_i & x_k \\ \lambda'y'_i + \lambda''y''_i & \lambda'y'_k + \lambda''y''_k \end{vmatrix}$$

which joins x and the point $\lambda'y' + \lambda''y''$ on $[y',y'']$.

The facts of (44.2) may be obtained in terms of P^3 alone as follows. The polar plane to a point p^1 in P^5 has the equation $\omega(p^1,q) = 0$, or

$$p_1^1 q_4 + p_2^1 q_5 + p_3^1 q_6 + p_4^1 q_1 + p_5^1 q_2 + p_6^1 q_3 = 0.$$

It is natural, then, to call the hyperplanes of P^5, $\omega(p^j,q) = 0$, or the complexes of P^3, $\omega(p^j,m) = 0$, $j = 1,2,\cdots,k$, independent if the matrix (p_i^j), $j = 1,2,\cdots,k$, $i = 1,2,\cdots,6$, has rank k. From (40.3), four such independent hyperplanes, $\omega(p^j,q) = 0$, $j = 1,2,3,4$, as a simultaneous system have two (non-trivial) independent solutions, $q = p'$ and $q = p''$, and every other solution is expressible in the form $\lambda'p' + \lambda''p''$. Thus the line $p = \lambda'p' + \lambda''p''$ in P^5 is the intersection of the independent hyperplanes $\omega(p^j, q) = 0$, and must cut Q in points corresponding to the lines of intersection of the four complexes, $\omega(p^j, m) = 0$, in P^3. Hence (44.2) can be reformulated as:

(44.4)   *The intersection of four independent complexes in P^3 is empty, consists of two lines, of a single line, or is a pencil of lines.*

The intersection in P^5 of three independent hyperplanes, $\omega(p^j, q) = 0$, $j = 1, 2, 3$, consists of all points linearly dependent on three independent points q^1, q^2, q^3, that is, of all points p of the form

$$p = \lambda_1 q^1 + \lambda_2 q^2 + \lambda_3 q^3, \quad (\lambda_1, \lambda_2, \lambda_3) \neq (0, 0, 0).$$

The triple $\lambda = (\lambda_1, \lambda_2, \lambda_3)$, associated with p, may be regarded as a point in a projective plane $\overline{P^2}$. Since Q is a quadric, it is to be expected that $\overline{P^2}$ either lies in Q or cuts it in a degenerate or non-degenerate conic. Indeed, the points p on the three hyperplanes which are also on Q, correspond to the λ-triples determined by

$$(44.5) \qquad \omega(p, p) = \omega\left(\sum_{i=1}^{3} \lambda_i q^i, \sum_{k=1}^{3} \lambda_k q^k\right) = \sum_{i,k=1}^{3} \lambda_i \lambda_k \omega(q^i, q^k) = 0.$$

The plane $\overline{P^2}$ will lie in Q if (44.5) holds for all λ-triples, which is possible if and only if $\omega(q^i, q^k) = 0$ for $i, k = 1, 2, 3$. In particular, $\omega(q^i, q^i) = 0$ means that q^i represents a line M^i in P^3, $i = 1, 2, 3$. The equations $\omega(q^i, q^k) = 0$, $i \neq k$, show that the lines M^i intersect each other and so either lie in a plane ξ or pass through a point x. A pencil of lines (in P^3) corresponds to a line in P^5 (on Q). Since in P^5 each of the hyperplanes $\omega(p^i, q) = 0$ contains the lines determined in pairs by the points q^i, the pencils in P^3 corresponding to these lines belong to each of the complexes $\omega(q^i, m) = 0$. The pencils are determined by any pair of their lines, hence each $\omega(q^i, m)$ contains all the lines of ξ, or alternately, all the lines through x. Stated in terms of P^3:

(44.6)   *The intersection of three independent complexes consists of all the lines in a plane ξ or of all the lines through a point x if and only if it equals the intersection of three degenerate complexes $\omega(q^i, m) = 0$, $i = 1, 2, 3$, with $\omega(q^i, q^k) = 0$, $i, k = 1, 2, 3$.*

In terms of P^5, a plane on Q corresponds to the lines in a plane or to the lines in a bundle of P^3.

When not all the coefficients in (44.5) are zero, the equation represents a conic in P^2. When $|\omega(q^i, q^k)| = 0$, the conic is degenerate and so represents a point, a line, or two lines in $\overline{P^2}$. Correspondingly, the complexes,

$\omega(q^i,m) = 0$, $i = 1,2,3$, intersect in a line, in a pencil of lines, or in two pencils of lines with one common line.

If $|\omega(q^i,q^k)| \neq 0$, the conic is imaginary (see the discussion of (10.23)) when

$$\Delta_1 = \omega(q^1,q^1)\omega(q^2,q^2) .- \omega^2(q^1,q^2) > 0 \quad \text{and} \quad \omega(q^1,q^1) \cdot |\omega(q^i,q^k)| > 0.$$

The conic is real if

(44.7) $\Delta_1 \leqslant 0$, *or if* $\Delta_1 > 0$ *and* $\omega(q^1,q^1) \cdot |\omega(q^i,q^k)| < 0.$

The latter case is, of course, the most interesting. A family of lines, each one of which is linearly dependent on three pairwise skew (and therefore independent) lines, M^1,M^2,M^3 is said to form a *regulus* $\overline{F}$. In the case of

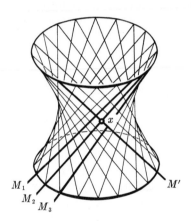

M_1 M'
M_2
M_3

Fig. 92

(44.7), the lines of P^3 corresponding to the conic form a regulus. For if r^1,r^2,r^3 are any three points of P^5 on the conic, they are not collinear and hence are independent. In P^3 they therefore represent three lines R^1,R^2,R^3, which are pairwise skew. Because the points r^i are dependent on the independent points q^i, $i = 1,2,3$, any point dependent on the triple r^i is also dependent on the triple q^i. Or, geometrically, the plane $\overline{P}^2$ is also determined by the points r^1, r^2 and r^3. The points of the conic are thus dependent on the points r^i, hence the lines of P^3 corresponding to the conic are dependent on R^1, R^2 and R^3, and form a regulus.

If M^1, M^2 and M^3 are the generators of a regulus F (Figure 92), then any line M which intersects all three of them intersects every line in the regulus. For if N is a line of the regulus, it has a representation $n = \lambda_1 m^1 + \lambda_2 m^2 + \lambda_3 m^3$ (where m^i is the Pluecker sextuple for M^i).

Then $\omega(m,n) = \omega(m, \sum_{i=1}^{3} \lambda_i m^i) = \sum_{i=1}^{3} \lambda_i \omega(m,m^i) = 0$, since $\omega(m,m^i) = 0$ is the condition that M and M^i intersect. The lines M', which intersect all lines of the regulus F, form a second *regulus* F', said to be *conjugate to F*. Clearly F is also conjugate to F'.

To see that F' is a regulus, we observe that its general line M' cuts M^i, $i = 1,2,3$, and hence belongs to the three degenerate complexes $\omega(m^i,m) = 0$, $i = 1,2,3$. Conversely, as we just saw, any line in this intersection cuts M^1, M^2 and M^3 and so belongs to F'. Since the simultaneous system $\omega(m^i,m) = 0$ has rank three, then by (40.3), it has three independent solutions, $m^{1'}, m^{2'}, m^{3'}$. These are generators for F', which consists of all lines dependent on them.

A line which cuts each of two intersecting lines must belong to the plane they determine or else pass through their intersection point. Therefore three lines, each of which cuts two intersecting lines, cannot be pairwise skew. Because each of the lines M^i, $i = 1,2,3$, cuts both lines in every pair in F', it follows that no two lines of F' intersect. By the same argument, no two lines of F intersect. Therefore:

(44.8) *Any two lines of a regulus F are skew. The lines intersecting all lines in F form a regulus F' conjugate to F. Either regulus is formed by all lines which intersect three independent lines in the other.*

To summarize, three independent hyperplanes in P^5, $\omega(p^i,q) = 0$, intersect in a projective plane $\overline{P}^2$ which contains three independent points q^i, $i = 1,2,3$. The plane $\overline{P}^2$ cuts Q in a real conic, with a corresponding regulus F in P^3, if the q^i determining $\overline{P}^2$ satisfy (44.7). In addition to (44.7), it is natural to ask *what conditions the initial coefficients p^i must satisfy for this construction to yield the regulus F in P^3*. First, suppose the construction does yield F. Let M^1,M^2,M^3 be three independent lines of F with corresponding points m^1,m^2,m^3 on the intersection of $\overline{P}^2$ with Q. A general line M' of the conjugate regulus F' satisfies $\omega(m',m^i) = 0$, $i = 1,2,3$. Because q^i is dependent on m^1, m^2, and m^3, $i = 1,2,3$, it follows that $\omega(m',q^i) = 0$, $i = 1,2,3$. Therefore m' lies on Q and the plane $\overline{P}^{2*}$ in which the three hyperplanes $\omega(p,q^i) = 0$ intersect. Since the plane $\overline{P}^{2*}$ is also determined as the plane through p^1, p^2 and p^3, this shows that the plane so determined cuts Q in the conic corresponding to the regulus F'. Therefore (44.7) holds with p^i substituted for q^i. The argument is now seen to be symmetric. That is, if (44.7) holds for p^i, $i = 1,2,3$, then the plane $\overline{P}^{2*}$ cuts Q in a conic which defines the regulus F' in P^3. The conjugate regulus F, by the previous reasoning, must then be obtainable from the original construction starting with p^i. This result may be stated in the form:

(44.9) *The intersection of three independent complexes $\omega(p^i,m) = 0$, $i = 1,2,3$ is a regulus if and only if $\Delta^* \leqslant 0$ or if $\Delta^* > 0$ and $\omega(p^1,p^1) \cdot | \omega(p^i,p^k) | < 0$, where $\Delta^* = \omega(p^1,p^1)\omega(p^2,p^2) - \omega^2(p^1,p^2)$.*

Now consider three lines M^1, M^2 and M^3 in a regulus F (Figure 92). Each point x of M^2 determines two planes, $(x \wedge M^1)$ and $(x \wedge M^3)$, and as x varies on M^2, the association $(x \wedge M^1) \to (x \wedge M^3)$ establishes a projectivity between the plane pencils on M^1 and M^3. The line in which two corresponding planes intersect must, by construction, cut all three lines, M^1, M^2 and M^3, hence is a line of the conjugate regulus F'. Therefore:

(44.10) *A regulus is the locus of the lines of intersection of corresponding planes in two projectively related plane pencils, where the axes of the pencils are skew. An arbitrary line of the regulus cuts an arbitrary line of the conjugate regulus in a point lying on corresponding planes of the pencils.*

(44.11) *The correspondence established between two lines of a regulus by associating points which lie on the same line of the conjugate regulus is a projectivity.*

Actually, (44.10) and (44.11) are duals of each other. The converse of one will therefore imply the converse of the other. We state them simultaneously.

(44.12) *The intersections (connections) of the corresponding planes (points) in two projective plane (point) pencils, on skew lines, form a regulus.*

The non-parenthetical statement may be established by showing that the point set, carried by the intersection locus, is a ruled quadric. By the results of the previous section, it is then a regulus. Together with (44.10) this will show:

(44.13) *The carrier of a regulus is a ruled quadric and conversely.*

For the proof, let the coordinate system be chosen so that the axes of the two projective pencils are given by the planes

$$x_3 = 0,\ x_4 = 0 \quad \text{and} \quad x_1 = 0,\ x_2 = 0$$

respectively. Take d_1 and d_2 on the planes $x_3 = x_4 = 0$, and d_3,d_4 on the planes $x_1 = x_2 = 0$, so that, in the projectivity, $x_3 = 0$ corresponds to $x_1 = 0$, and $x_4 = 0$ corresponds to $x_2 = 0$. Finally, choose $e = (1,1,1,1)$ on two corresponding elements so that the plane $x_3 - x_4 = 0$, through d_1, d_2 and e, corresponds to the plane $x_1 - x_2 = 0$, through d_3, d_4 and e. The three specific planes $x_3 = 0$, $x_4 = 0$, $x_3 - x_4 = 0$ form with the general plane of the pencil $\lambda x_3 - \mu x_4 = 0$ a set with cross ratio λ/μ. The corresponding planes of the second pencil are $x_1 = 0$, $x_2 = 0$, $x_1 - x_2 = 0$ and (from the preservation of cross ratio) $\lambda x_1 - \mu x_2 = 0$. Since $\lambda x_3 - \mu x_4 = 0$ corresponds to $\lambda x_1 - \mu x_2 = 0$, for all $(\lambda, \mu) \neq (0,0)$, points on the intersection locus satisfy the equation:

$$(44.14) \qquad\qquad x_1 x_4 - x_2 x_3 = 0.$$

This is a non-degenerate quadric, since its determinant has the value $(- 1/2)^4$, which completes the proof.

In the polarity (43.10) the self-conjugate locus was $x_1^2 + x_2^2 - x_3^2 - x_4^2 = 0$. Setting $x_4 = 1$, in this equation and also in (44.14), yields two normal forms for ruled quadrics, namely:

$$x_1^2 + x_2^2 - x_3^2 = 1, \text{ and } x_1 = x_2 x_3.$$

If x_1, x_2, x_3 are interpreted as rectangular coordinates we recognize these as a hyperboloid of one sheet and a hyperbolic paraboloid. As was the case with conics, however, these quadrics cannot be distinguished projectively.

We complete this section in considering the intersection of two independent complexes $\omega(a^1, m) = 0$ and $\omega(a^2, m) = 0$. The corresponding hyperplanes in P^5 intersect in a three-dimensional projective space $\bar{P}^3$, hence we expect $\bar{P}^3$ to cut Q in a degenerate or non-degenerate quadric. The extreme cases, where $\bar{P}^3$ does not intersect Q or else lies entirely in Q, cannot occur. Because Q contains infinitely many projective planes (corresponding to the planes and bundles of lines in P^3), and since each of these must intersect $\bar{P}^3$ in at least one point, Q and $\bar{P}^3$ must intersect. The following discussion will show that $\bar{P}^3$ cannot lie in Q (compare with the footnote on page 266).

The intersection of two distinct (linear) line complexes, called a (linear) *line congruence*, may also be defined as the locus of lines dependent on four independent lines. A line common to both $\omega(a^1, m) = 0$ and $\omega(a^2, m) = 0$ also lies in all the complexes:

$$(44.15) \qquad \omega(\lambda_1 a^1 + \lambda_2 a^2, m) = \lambda_1 \omega(a^1, m) + \lambda_2 \omega(a^2, m) = 0.$$

In this set, the degenerate complexes are those for which

$$(44.16) \qquad \begin{aligned} &\omega(\lambda_1 a^1 + \lambda_2 a^2, \lambda_1 a^1 + \lambda_2 a^2) \\ &= \lambda_1^2 \omega(a^1, a^1) + 2\lambda_1 \lambda_2 \omega(a^1, a^2) + \lambda_2^2 \omega(a^2, a^2) = 0. \end{aligned}$$

Assume, first, that there are two solutions to this equation, λ_1', λ_2' and λ_1'', λ_2'' such that $\lambda_1' : \lambda_2' \neq \lambda_1'' : \lambda_2''$. Let $m' = \lambda_1' a^1 + \lambda_2' a^2$ and $m'' = \lambda_1'' a^1 + \lambda_2'' a^2$ indicate the axes of the corresponding complexes. There are, then, two cases.

1)   *The lines m' and m'' intersect.*

In this case $\omega(m',m'') = 0$, which together with $\omega(m',m') = 0$ and $\omega(m'',m'') = 0$ implies that (44.16) holds for all pairs λ_1,λ_2 of the form $\lambda_1 = \lambda_1' + \sigma\lambda_1''$, $\lambda_2 = \lambda_2' + \sigma\lambda_2''$. This, in turn, implies $\omega(a^i,a^k) = 0$, $i,k = 1,2$, and hence that all the complexes in (44.16) are degenerate. In particular, a^1 and a^2 represent lines, which we can take to be m' and m'', and the line congruence consists of all lines in the plane $(a^1 \wedge a^2)$, and all lines through the intersection of a^1 and a^2.

2)   *The lines m' and m'' do not intersect.*

Then exactly one line of the congruence passes through any point x not on m' or m''.

3)   *If (44.16) has no real roots, then the equations $\omega(a^i,m) = 0$, $i = 1,2$, together with the equations $\sum\limits_{k=1}^{4} m_{ik}x_k = 0$, $i = 1,2,3,4$, determine exactly one line m of the congruence through the point x.*

4)   *Suppose, finally, that (44.16) has a double root $\lambda_1' : \lambda_2'$, and take $m' = \lambda_1'a^1 + \lambda_2'a^2$ as the axis of the corresponding degenerate complex.*

Observing that for any pair λ_1'',λ_2'' not proportional to λ_1',λ_2', the pencil

$$\lambda_1\omega(\lambda_1'a^1 + \lambda_2'a^2,m) + \lambda_2\omega(\lambda_1''a^1 + \lambda_2''a^2,m) = 0$$

represents exactly the same complexes as (44.15), it follows that $\lambda_1 = 1$, $\lambda_2 = 0$ yields the only degenerate complex.
Putting $m'' = \lambda_1''a^1 + \lambda_2''a^2$, the equation,

$$\lambda_1^2\omega(m',m') + 2\lambda_1\lambda_2\omega(m',m'') + \lambda_2^2\omega(m'',m'') = 0,$$

does not vanish identically and has $\lambda_1 = 1$, $\lambda_2 = 0$ as a double root. Hence $\omega(m',m') = 0$ and $\omega(m',m'') = 0$ but $\omega(m'',m'') \neq 0$. This means that the complex $\omega(m'',m) = 0$ contains m', but is not degenerate. The congruence, therefore, *consists of all lines in the non-degenerate complex* $\omega(m'',m) = 0$ *which intersect the line m' of the complex.*

Further information on the geometry of lines may be found in K. Zindler, "Liniengeometrie mit Anwendungen." A concise treatment of the representation of an n-dimensional projective space in terms of coordinates of r-dimensional linear subspaces is found in B. L. van der Waerden, "Einführung in die Algebraische Geometrie."

45. Affine Geometry in Space

An n-dimensional, affine space originates from an n-dimensional projective space by singling out or deleting one hyperplane, called the hyperplane at infinity. Restricting ourselves to three dimensions, we will always choose $x_4 = 0$ as the plane ζ at infinity. For points in the affine space, then, $x_4 \neq 0$, and when the coordinates of $x = (x_1, x_2, x_3 x_4)$ are so normalized that $x_4 = 1$, the numbers x_1, x_2, x_3 are called the *affine coordinates* of x.

A collineation of the projective space which carries $x_4 = 0$ into itself is called an *affinity*. By the same argument as in the plane case, it has the form

$$x_1' = \sum_{k=1}^{3} a_{ik}x_k, \quad i = 1,2,3, \quad x_4' = x_4, \quad |a_{ik}| \neq 0, \quad i,k = 1,2,3.\,§$$

Therefore, the general form of an affinity, in affine coordinates, is:

$$(45.1) \qquad x_i' = \sum_{k=1}^{3} a_{ik}x_k + a_i, \quad i = 1,2,3, \quad |a_{ik}| \neq 0.$$

The special affinities which leave the plane $x_4 = 0$ pointwise invariant have the form:

$$(45.2) \qquad x_i' = bx_i + a_i, \quad i = 1,2,3, \quad b \neq 0,$$

and are called *similitudes*. A similitude, which is not the identity, has a fixed point in the affine space if and only if the equations,

$$x_i = bx_i + a_i, \quad i = 1,2,3,$$

have a solution, that is, if and only if $b \neq 1$. The fixed point, when $b \neq 1$, has the affine coordinates $a_i/(1 - b)$, $i = 1,2,3$, and in this case the similitude is a homology with $a/(1 - b)$ as its center and the plane at infinity as its axial plane. It is a harmonic homology when its square,

$$x_i'' = b(bx_i + a_i) + a_i = b^2 x_i + a_i(1 + b), \quad i = 1,2,3,$$

is the identity, that is, when $b = -1$. In that case it is also called the reflection in the point $a/(1 - b) = a/2$.

When $b = 1$, the similitude

$$(45.3) \qquad x_i' = x_i + a_i, \quad i = 1,2,3,$$

if it is not the identity, has no finite fixed points. It is then (compare (15.5)) an elation with $x_4 = 0$ as its axial plane, and $(a_1, a_2, a_3, 0)$ as its center. The mappings (45.3), which are called *translations*, form an Abelian group.

Lines which intersect at a point of $x_4 = 0$ are called *parallel lines* and planes which intersect in a line of $x_4 = 0$ are called *parallel planes*. Thus, affinities preserve parallelism, while under a similitude a line and its image line are parallel, as is the case with every plane and its transform. As before, affine geometry deals with those theorems which remain true under affinities.

The *affine ratio* $A(a,b,c)$ of three collinear points a,b,c is defined as in the two-dimensional case. If the line M carrying a, b and c intersects the plane ζ in the point p_∞, then by definition (see (15.12)),

$$A(a,b,c) = R(p_\infty,a,b,c) = \frac{c_i - a_i}{b_i - a_i}, \qquad i = 1,2,3.$$

The point c is the affine center of a and b when $A(a,b,c) = 1/2$. In that case a, b, p_∞, and c form a harmonic quadruple and the reflection in c interchanges a and b.

The *quadrics are classified*, from the affine point of view, by the nature of their intersection with the plane at infinity. Let Q be a non-degenerate quadric. Then the following cases are possible.

1) ζ *is not a tangent plane of Q.*
        a) *Q is imaginary and called an imaginary ellipsoid.*
        b) *Q is non-ruled and does not intersect ζ. Then Q is a real ellipsoid.*
        c) *Q is non-ruled and intersects ζ. Then Q is a hyperboloid of two sheets.*
        d) *Q is ruled (and therefore intersects ζ). Then Q is a hyperboloid of one sheet.*

2) ζ *is a tangent plane of Q (hence Q is not imaginary).*
        a) *Q is non-ruled and called an elliptic paraboloid.*
        b) *Q is ruled and called a hyperbolic paraboloid.*

Normal forms for these types are easily obtained, either by geometric or analytic arguments. For the cases under 1), ζ is not self-conjugate, hence projective coordinates can be so chosen that the coordinate tetrahedron is self-polar and with ζ still $x_4 = 0$. Then Q has the form $\sum_{i=1}^{4} b_i x_i^2 = 0$, $b_i \neq 0$, $i = 1,2,3,4$. With $b_4 = -1$ and $x_4 = 1$, this takes the affine form,

$$\pm c_1 x_1^2 \pm c_2 x_2^2 \pm c_3 x_3^2 = 1, \qquad c_i > 0, \qquad i = 1,2,3.$$

The affinity $x_i' = \sqrt{c_i} x_i$, $i = 1,2,3$, then takes Q into one of the following four types (with the primes dropped):

(45.4)
$$
\begin{aligned}
x_1^2 + x_2^2 + x_3^2 &= -1 \quad \text{\textit{imaginary ellipsoid,}} \\
x_1^2 + x_2^2 + x_3^2 &= 1 \quad \text{\textit{real ellipsoid,}} \\
x_1^2 + x_2^2 - x_3^2 &= 1 \quad \text{\textit{hyperboloid of one sheet,}} \\
x_1^2 + x_2^2 - x_3^2 &= -1 \quad \text{\textit{hyperboloid of two sheets.}}
\end{aligned}
$$

Replacing 1 by x_4^2 on the right gives a check on this statement. The last equation, for example, becomes $x_1 + x_2 - x_3^2 + x_4^2 = 0$ which is a non-ruled quadric intersecting $x_4 = 0$ in the real conic $x_1^2 + x_2^2 - x_3^2 = 0$.

For the reduction of the quadrics under 2) to the types indicated, we employ the following generalization of (16.4).

(45.5) *If p is the pole of ζ with respect to a real (non-degenerate) quadric Q, the intersectors of Q in a pencil of parallel planes, whose axis M does not contain p, intersect Q in central conics whose centers lie on N, the polar line to M.*

For, let M be any line in ζ, not through p. Then M is not self-polar and its polar line N passes through p. Any plane ξ through M which intersects Q intersects it in a conic C_ξ, and the point p_ξ in which N intersects ξ is the pole of M with respect to the conic C_ξ, and is thus the center of C.

The forms in (45.4) may be obtained from (45.5) if Q is not imaginary. For the assumption that M, in ζ, does not pass through p is always satisfied if ζ is not a tangent plane of Q. Hence, p is the affine center of Q (which is then called a central quadric). If p is selected as the origin of the affine coordinate system, then since the reflection in p, namely the affinity $x_i' = -x_i$, carries Q into itself, the equation representing Q cannot possess any linear terms. Next, a plane ξ through p which intersects Q may be chosen for the plane $x_1 = 0$ and coordinates x_2, x_3 selected in ξ so that C_ξ has the form $x_2^2 \pm x_3^2 = 1$. With the x_1 – axis chosen as the line polar to the intersection of ξ with ζ, (45.5) shows that $x_1' = x_1$, $x_2' = -x_2$, $x_3' = -x_3$ carries Q into itself, hence the equation of Q has no mixed terms involving x_1. With a proper choice of a unit on the x_1-axis the last three types in (45.4) are then obtained.

If p lies on ζ, let ξ be any plane which intersects Q and whose intersection, M, with ζ does not pass through p. In ξ select coordinates x_2, x_3 such that the intersection of ξ with Q has the form $x_2^2 \pm x_3^2 = 1$ and take N, the polar to M, as the x_1-axis. Since $x_1' = x_1$, $x_2' = -x_2$, $x_3' = -x_3$ again carries Q into itself, there are no mixed terms containing x_1 in the equation of Q. Also, because the point with projective coordinates (1,0,0,0) lies on Q, the coefficient of x_1^2 must vanish. Hence Q has the form

$$2a_{14}x_1 + x_2^2 \pm x_3^2 = 1, \qquad a_{14} \neq 0.$$

The affinity $x_1' = a_{14}x_1 - 1/2$, $x_2' = x_2$, $x_3' = x_3$ leads to the normal forms:

(45.6) $$2x_1 + x_2^2 + x_3^2 = 0, \quad \textit{elliptic paraboloid,}$$
$$2x_1 + x_2^2 - x_3^2 = 0, \quad \textit{hyperbolic paraboloid.}$$

The first of these is not ruled since the plane $x_1 = c$, $c > 0$, does not intersect it. The second is ruled because it carries the two families of lines,

$$x_2 - x_3 = \lambda, \quad -2x_1 = \lambda(x_2 + x_3), \quad \text{and} \quad x_2 + x_3 = \lambda, \quad -2x_1 = \lambda(x_2 - x_3).$$

The affine classification of degenerate quadrics is left to the reader.§ We only observe that *cones* and *cylinders* are distinguished by the apex being at a finite or infinite distance.

Besides affine geometry, *equiaffine geometry of space* deals with the properties invariant under affinities of the form

(45.7) $$x'_i = \sum_{k=1}^{3} a_{ik}x_k + a_i, \qquad i = 1,2,3, \qquad |a_{ik}| = \pm 1.$$

In equiaffine geometry, volume can be defined in a manner analogous to that used for area in the plane case, but the area of plane figures can no longer be defined.

46. Convex Sets in Space

Corresponding to the distinct points a and b in space, the *open and closed segments* $S^*(a,b)$ and $S(a,b)$ are defined, as in the plane, by the sets $\theta a + (1 - \theta)b$, where $0 < \theta < 1$ and $0 \leqslant \theta \leqslant 1$ respectively. A set K in the affine space is called *convex* if for any pair of points a,b in K the segment $S(a,b)$ is also in K. Many statements about spatial convex sets can be reduced to facts about plane convex sets since every plane which intersects a spatial convex set cuts it in a plane convex set. The converse holds even in the stricter form:

(46.1)   *If the intersection of a (non-empty) set K with every plane through a fixed point p is either empty or convex, then K is convex.*

For any two points, a and b, of K lie in at least one plane through p. By assumption $S(a,b)$ lies in the intersection of this plane with K and hence lies in K.

As an application of (46.1) we show that the no-tangent points of either an ellipsoid E or an elliptic paraboloid E' form a convex set. Let p be a no-tangent point of E (or E'). Then the polarity which defines the quadric induces a hyperbolic polarity on any plane ξ which passes through p. The intersection of E (or E') with ξ is the conic defined by the induced polarity which also has p as a no-tangent point. When E is used, the conic has no points at infinity and is therefore an ellipse. For E', the conic can have at most one ideal point and so is either an ellipse or a parabola. In either case the intersection with ξ consists of the no-tangent points of an ellipse or a parabola and these, in (17.1) where shown to be convex sets. Using (46.1) and (17.2) we obtain:

(46.2)   *The no-tangent points of an ellipsoid or of an elliptic paraboloid, alone or together with the quadric, form a convex set.*

Consider now $L(x) = a_1x_1 + a_2x_2 + a_3x_3 + a_4$, where a_1,a_2,a_3 is not the null triple. Then the *two sides of the plane* $L(x) = 0$ are given by $L(x) > 0$ and $L(x) < 0$ while the *two half-spaces* correspond to $L(x) \geqslant 0$ and $L(x) \leqslant 0$. The same argument as in the plane case (see (17.3)), or (46.1), shows that each of *the sides* of $L(x) = 0$, *and each of the half-spaces* is a convex set.[§] Since the intersection of any aggregate of convex sets is either empty or else convex (compare (17.5)), convex sets may be constructed by intersecting sets which are known to be convex. For instance, let a^1,a^2,a^3,a^4 be four non-co-planar points. The plane through a^2,a^3,a^4 has two sides and two half-spaces, and we choose L_1^* and L_1, to denote that side and that half-space respectively which contains a^1. Similarly L_i^* and L_i contain a_i and are defined with respect to the plane through the remaining triple. The four convex sets L_i^* intersect in a convex set called an open tetrahedron, denoted by $T^*(a^1,a^2,a^3,a^4)$, while the convex intersection of the four L_i sets, denoted by $T(a^1,a^2,a^3,a^4)$, is called a closed tetrahedron. If the points a_i, $i = 1,2,3,4$, belong to a convex set K, then $T(a^1,a^2,a^3,a^4) \subset K$.

A point is defined as an *interior point* of a convex set K if it belongs to an open tetrahedron, T^*, whose vertices lie in K.

(46.3)   *The point p is an interior point of the convex set K, if three non-coplanar open segments*, $S^*(\overline{a^i,b^i})$, $i = 1,2,3$, *exist which lie in K and contain p.*

PROOF: Let $a^i \varepsilon S^*(p,\overline{a^i})$, $b^i \varepsilon S^*(p,\overline{b^i})$, $i = 1,2,3$ (Figure 93). The points a^1,a^2,a^3 are not co-planar, nor are the points b^1,b^2,b^3, because the segments $S^*(\overline{a^i,b^i})$ are given non-co-planar. Because K is convex and contains the points a^i and b^i, $i = 1,2,3$, the triangular sets $T(a^1,a^2,a^3)$ and $T(b^1,b^2,b^3)$ belong to K. Let a be an interior point of $T(a^1,a^2,a^3)$. Then the line $[a,p]$ intersects the set $T(b^1,b^2,b^3)$ in one of its interior points, b. The tetrahedron $T(b,a^1,a^2,a^3)$ has its vertices in K, and since p belongs to $T^*(b,a^1,a^2,a^3)$ it is interior to K.

Obviously, if p is an interior point of a convex set K, it is also an interior point of the plane convex set in which any plane through p intersects K. As a corollary of (46.3), two planes suffice for the converse, that is:

(46.4)   *The point p is an interior point of a convex set K if it is an interior point of two distinct plane sections of K.*

The points of a convex set K which are not interior points are called *boundary points*. If a convex set K has no interior points, then it lies in a plane. For if K contained four non-co-planar points, the interior of the tetrahedron spanned by these points would consist of points interior to K.

A *supporting plane* ξ of a convex set K is a plane which contains points of K but is such that K lies entirely in one of the half-spaces bounded by ξ. Clearly a supporting plane of K cannot contain an interior point of K. When K has no interior points and so lies in a plane ξ, then ξ is a supporting plane of K and, in a trivial way, every boundary point of K lies in a supporting plane (namely ξ). However, this important property is true in general.

(46.5)   *Every boundary point of a convex set K lies on a supporting plane.*[12]

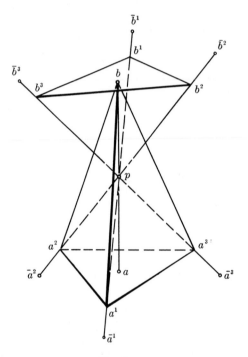

Fig. 93

PROOF: From the previous remarks, it may be supposed that K has an interior point p. Let q be any boundary point (Figure 94). Every plane through p, and in particular every plane ξ through p and q, intersects K in a plane convex set K_ξ which has p as an interior point and q (see (17.6)) as a boundary point. Let M_ξ' and M_ξ'' denote the extreme supporting lines of K_ξ at q. Clearly K_ξ lies in one of the half-planes of M_ξ' and in one of the

---

[12]This theorem will not be used later.

half-planes of M''_ξ. We denote by V_ξ the intersection of those two half-planes. For each plane ξ in the pencil $L(p,q)$ there is then a set V_ξ and the union of all the plane convex sets V_ξ is a convex set V.

For, suppose V is not convex. Then points x^1 and x^2 exist in V such that $S(x^1,x^2)$ contains a point x^3 not in V. By hypothesis, x^1 is in a set V_{ξ^1} and x^2 is in a set V_{ξ^2}, and since those sets are convex, ξ^1 and ξ^2 must be distinct planes. This in turn implies that x^3 is not on $[p,q]$, hence p,q and x^3 determine a plane ξ^3, and x^3 is not in V_{ξ^3}. Now if x^1 and x^2 are not interior points of the sets V_{ξ^1} and V_{ξ^2}, then points y^1 and y^2 exist in V_{ξ^1} and V_{ξ^2} respectively which are interior points, and such that y^i is arbitrarily close to x^i, $i = 1,2$. (We could take y^i on $S^*(p,x^i)$ for example.) If y^3 denotes the intersection of $S(y^1,y^2)$ with ξ^3, then for y^i sufficiently near x^i, $i = 1,2$, y^3 can be kept so near x^3 that y^3 also is not in V_{ξ^3}. From the construction of

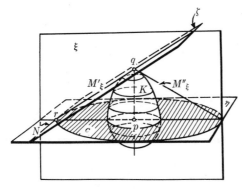

Fig. 94

the sets V_ξ, the fact that y^3 is not in V_{ξ^3} implies that for $0 \leqslant \theta < 1$ none of the points $y^3_\theta = (1 - \theta)y^3 + \theta q$ is in V_{ξ^3}. On the other hand all the points $y^i_\theta = (1 - \theta)y^i + \theta q$, $0 \leqslant \theta < 1$, lie in V_{ξ^i}, $i = 1,2$. But since y^1 and y^2 are interior points of V_{ξ^1} and V_{ξ^2}, and these sets are bounded by the extreme supporting lines of K_{ξ^1} and K_{ξ^2} it follows that for $\theta = \bar\theta$ sufficiently near 1, $y^1_{\bar\theta}$ and $y^2_{\bar\theta}$ are points of K_{ξ^1} and K_{ξ^2} respectively and so belong to K. Hence $S\left(y^1_{\bar\theta}, y^2_{\bar\theta}\right)$ is in K. Therefore the intersection point of $S\left(y^1_{\bar\theta}, y^2_{\bar\theta}\right)$ and $S^*(y^3,q)$ is in K and so is in V_{ξ^3}, which contradicts the previous conclusion that no point of $S^*(y^3,q)$ is in V_{ξ^3}. We conclude therefore that V is convex.

Now let r be any point of V, distinct from q, and lying on one of the supporting lines M'_ξ or M''_ξ, say M'_ξ. A plane η through r and p, but not through q, intersects V in a plane convex set C. Because $S(p,r)$ lies in V, but no point following p on the ray $R(p,r)$ lies in V, r is a boundary point of V (compare (17.6)). In η let N be a supporting line of C through r. Then the plane ζ through N and q is clearly a supporting plane of V. Because V contains K, and q is in K, this plane is also a supporting plane of K at q, which establishes (46.5).

As in the plane case, this proof yields more than the stated theorem. The set W of boundary points of V consists of rays, originating at q, and lying on lines M'_ξ and M''_ξ. The set W is called the supporting cone of K at q. This cone becomes a plane if and only if for two distinct planes ξ^1, ξ^2 the lines $M'_{\xi i}$ and $M''_{\xi i}$ coincide, $i = 1,2$. In that case it is called the tangent plane of K at q and is the only supporting plane through q.

As before, to exclude uninteresting convex sets, we define a *convex domain* in affine space as a convex set with interior points which is not the whole space and which contains $S(a,b)$ if it contains $S^*(a,b)$. When K is a convex domain, other than the region between two parallel planes, its boundary S is called a *convex surface* and the supporting planes of K are also called the supporting planes of S. As just shown, a convex surface has at least one supporting plane at each of its points. If S contains no (proper) segments, or if each of its supporting planes contacts it in only one point, it is called *strictly convex*. It is *differentiable* if it has a unique supporting plane (i.e., a tangent plane) at each of its points.

47. **Three–Dimensional Projective Metrics**

The definition of a projective metric for an n-dimensional space was given in Section 20. *Here we are interested in the case $n = 3$.* The space R is then a subset of P^3 which does not lie in a plane and is so metrized that the strict triangle inequality $xy + yz > xz$ holds for any distinct triple x,y,z not collinear in the sense of P^3. Moreover, for any two different points of R, the intersection, $\lambda(x,y)$, of $[x,y]$ with R is a metric straight line or a great circle.

According to Hamel's theorem, the sets $\lambda(x,y)$ are either all straight lines or they are all great circles of the same length.[13] In the second case R spans the entire space P^3, while in the first instance there is at least one plane of P^3 which does not intersect R. With this plane as the distinguished plane of an affine space, A^3, the space R becomes a convex subset of A^3 and has only interior points.[§] For any two points, x and y, of R there is exactly one metric segment connecting them and this segment coincides with $S(x,y)$.[§]

The space R is either the whole A^3, is the region between (but not

---

[13]Hamel's theorem for the plane implies Hamel's theorem for the space.[§]

including) two parallel planes, or its boundary is a convex surface S_R. As in the plane case:

(47.1)   *An open projective metric, xy, is equivalent to $\pi(x,y)$, (or $e(x,y)$).*

PROOF: Let $\pi(a^i,a) \to 0$ (Figure 95). To show that $a^i a < \varepsilon$ for all i greater than some $N(\varepsilon)$, choose a line L, in R, through a, and on L take b^1 and b^2 on opposite sides of a so that $ab^i < \varepsilon/5$, $i = 1,2$. In a plane through b^2, but not through L, a triangular set $T(c^1,c^2,c^3)$ can be taken which has b^2 as an interior point. The proof for the plane case showed that we can also take the points c^i so that x in $T(c^1,c^2,c^3)$ implies $xb^2 < \varepsilon/5$. The point a is an interior point of the tetrahedron $T(b^1,c^1,c^2,c^3)$ hence a^i lies in this tetra-

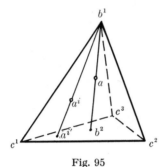

Fig. 95

hedron for $i > N$ (which depends indirectly on ε). Let the line through b^1 and a^i, $i > N$, intersect $T(c^1,c^2,c^3)$ in $a^{i'}$. Then,

$$a^i a \leqslant a^i a^{i'} + a^{i'} b^2 + b^2 a \leqslant b^1 a^{i'} + a^{i'} b^2 + b^2 a$$
$$\leqslant 2a^{i'} b^2 + b^1 b^2 + b^2 a < 2\varepsilon/5 + 2\varepsilon/5 + \varepsilon/5 = \varepsilon.$$

A similar modification of the corresponding proof for the plane shows that $a^i a \to 0$ implies $\pi(a^i,a) \to 0.§$

The following are strict analogues of the corresponding facts for the plane established in Section 22.

(47.2)   *A motion of a closed, projective-metric three-space is a collineation. A motion of an open, projective-metric three-space is induced by one and only one collineation.§*

(47.3)   *A motion of an open, projective-metric three-space which leaves two points fixed leaves every point of the line through these points fixed. A motion which leaves three non-collinear points fixed leaves every point in the plane through these points fixed. If a motion leaves four, non-co-planar points fixed it is the identity.§*

This theorem has the following two corollaries:

(47.4) *Four non-co-planar points and their images completely determine a motion of an open, projective-metric three-space.*

(47.5) *In a motion (not the identity) of an open, projective-metric three-space, there are no fixed points, or else the fixed points consist of a single point, all the points of a line, or all the points of a plane.*

An involutory motion, Φ, of an open space always has fixed points, for if $a\Phi \not\backsim a$, then the center c of a and $a\Phi$ is fixed. According as the fixed point set is a point, line, or plane, Φ is called the *reflection in a point, a line, or a plane*. The definite article is again justified by the uniqueness of such a reflection. For a reflection in a determines the image x' of $x \not\backsim a$ by the fact that a is the center of x and x'. If Φ is a reflection in the line M (the plane ξ) and x is not on M (not on ξ) then the center a of x and $x' = x\Phi$ must lie on M (on ξ) and be the unique foot of x on M (on ξ). For if y is any point of M (of ξ) distinct from a, then

$$xx' = 2ax < xy + x\Phi y\Phi = 2xy$$

implies $ax < xy$. In either case, then, x' is unique.

The line M is called perpendicular to the set μ *at* f *if* f *is in* $\mu \cap M$ *and every point of M has f as foot on* μ. Therefore the line which connects a point a to its image $a' \not\backsim a$ under a reflection in a line M or a plane ξ is perpendicular to M, or to ξ, at the center of a and a'. Combining these statements with earlier results in this chapter yields:

(47.6)

The reflection in x of the open three-space, R, is induced by a harmonic homology with center x, whose axial plane does not intersect R.

The reflection of R in a line L is induced by a biaxial involution, one of whose axes is L and whose second axis M does not intersect R. The lines intersecting both L and M are perpendicular to L.

The reflection of R in a plane ξ is induced by a harmonic homology whose axial plane is ξ but whose center x is not in R. The perpendiculars to ξ are the lines through x.

If the reflection, Φ, in the plane ξ exists, and a is not in ξ, then, as in the plane case, ξ is the locus $B(a,a')$ of points equidistant from a and $a' = a\Phi$. Theorem (23.7) allows the construction of the locus $B(a,a')$ for any distinct pair of points, a,a'. In particular we can construct c_1 and c_2 distinct from c, the center of a and a', and such that c, c_1 and c_2 are not collinear. If reflections in all planes are known to exist, the plane through c, c_1 and c_2 must then be the locus $B(a,a')$ and the reflection in this plane must interchange a and a'. The following is the analogue to (23.11):

(47.7) *Corresponding to every pair of congruent quadruples, there will exist a motion carrying one into the other if and only if reflections in all planes exist. If reflections in all planes exist, then every motion is the product of four (or fewer) such reflections.*

The second part of the theorem and the sufficiency statement in the first part may be proved by the same type of reasoning as in the plane case. If a_1, a_2, a_3, a_4 and b_1, b_2, b_3, b_4 are given congruent quadruples, and if reflections in all planes exist, let Φ_1 be the reflection in $B(a_1, b_1)$, if $a_1 \nsim b_1$, or the identity if $a_1 \sim b_1$. Choose Φ_2 to be the reflection in $B(a_2\Phi_1, b_2)$, if $a_2\Phi_1 \nsim b_2$, and the identity if $a_2\Phi_1 \sim b_2$. The first motion takes a_1 into b_1 and the product of the first and second takes a_2 into b_2. Because $a_1 a_2 = (a_1\Phi_1)(a_2\Phi_1) = b_1(a_2\Phi_1) = b_1 b_2$, b_1 is on $B(a_2\Phi_1, b_2)$ and hence the product $\Phi_1\Phi_2$ leaves a_1 at b_1. Continuing in this way, it is clear that after four steps the first quadruple is mapped on the second by a product of motions $\Phi_1\Phi_2\Phi_3\Phi_4$, where Φ_i is a reflection or the identity. On the other hand, if a motion Φ is given, then any non-coplanar set a_i is mapped by Φ into a non-coplanar set b_i, $i = 1,2,3,4$. By (47.4) the mapping $a_i \to b_i$ determines Φ, which can then be expressed as the product of four (or fewer) reflections.

Assume, now, that for any two congruent quadruples a motion exists which maps one on the other. Let ξ be any plane and M any one of its lines. In ξ choose a and b on M, and c and c' on opposite sides of M, so that a,b,c and a,b,c' are congruent triples. Applying the above assumption to the quadruples a,b,c,c and a,b,c',c' yields a motion which carries ξ into itself and induces in ξ a reflection about M. Thus every plane admits (plane) reflections in all of its lines. Now suppose x is any point which is not in the arbitrary plane ξ. The minimum properties of a continuous function of two variables imply that a foot f of x on ξ exists.[§] If a_1 and a_2 are distinct points of ξ, not collinear with f, then f is the foot of x on the line M_i joining f and a_i, $i = 1,2$. Choose x' so that f is the center of x and x'. The plane determined by x, f and a_i has a reflection Φ_i in M_i, $i = 1,2$, hence $x' = x\Phi_i$, $i = 1,2$, and the quadruples x,f,a_1,a_2 and x',f,a_1,a_2 are congruent in that order. Therefore a motion Ψ exists mapping the first quadruple on the second. Since Ψ induces Φ_i in the plane of x,f,a_i, $i = 1,2$, it follows that Ψ^2 leaves x,f,a_1,a_2 fixed and is therefore the identity. Because Ψ leaves f, a_1 and a_2 fixed, it leaves all points of ξ fixed. It is thus the reflection in ξ. This concludes (47.7).

As in the plane case this theorem implies:

(47.8) *If reflections in every plane exist and if Ψ is an isometry of the set μ in R on the set μ' in R, then a motion Φ of R exists which coincides with Ψ on μ.[§]*

As in the plane case, (47.7) is not true if "reflections in all planes" is replaced by "reflections in all points." Again, Minkowskian geometry will furnish an example. It is not difficult to prove, however, that it suffices to postulate the existence of reflections in all lines.

48. Minkowskian Geometry in Space

Three-dimensional Minkowskian geometry is defined in the same way as in the two-dimensional case. *Its projective metric is defined in the whole affine space, and has the property that affine ratio equals distance ratio.*

Obviously in a Minkowskian three-space the induced geometry in any plane is also Minkowskian. Conversely, if a three-dimensional projective metric induces a Minkowskian geometry in every plane of the affine space, then distance ratio equals affine ratio, hence the space geometry is Minkowskian. It follows that opposite sides in any parallelogram have equal lengths and this implies, as in the plane case, that *a Minkowskian metric is invariant under the translations of the affine space.* Also an affine reflection in a point is the reflection in the point for the Minkowski metric, in the sense of the preceding section. More generally:

(48.1)   *The similitude,* $x_i' = \lambda x_i + a_i$, $i = 1,2,3$, $\lambda \neq 0$, *multiplies all Minkowskian distances by* $|\lambda|$.

This follows from (24.8) and from the fact that $x_i'' = \lambda x_i$ induces a similitude in every plane through the origin, $z = (0,0,0)$. Expressing a general similitude as the product of this special similitude with a translation gives the stated fact.

If $K(p,\delta)$ (with $\delta > 0$ understood) denotes the sphere with center p and radius δ, that is, the locus of points x satisfying $m(x,p) = \delta$, then the intersection of $K(p,\delta)$ with any plane through p is a strictly convex curve with p as affine center. From (46.1) it follows that $K(p,\delta)$ is a strictly convex surface with p as affine center, and (48.1) shows then that all spheres are homothetic. In particular, $x_i' = \delta x_i$ transforms the unit sphere $K(z,1)$ into $K(z,\delta)$. Hence if $K(z,1)$ is known, then $m(x,y)$ is determined for any pair, x,y. Since a projective metric in A^3 which induces a Minkowski metric in every plane is itself Minkowskian, (24.12) yields:

(48.2)   *In the affine space A^3 let K be a closed, strictly convex surface with affine center c. For any two points a and b in A^3 define $m(a,b) = |A(a,y,b)|$, where y is the intersection of the ray from a through b with the image of K under the mapping*

$$x_i' = x_i + (a_i - c_i), \quad i = 1,2,3.$$

Then $m(a,b)$ is the Minkowski metric for which $K = K(c,1)$.

If $K(z,1)$ is given in a definite affine coordinate system, x_1, x_2, x_3, then

$$F(x_1, x_2, x_3) = F(x) = m(z,x)$$

has the following properties:

(48.3)   For $x \neq z$, $F(x) > 0$ and $F(\lambda x) = |\lambda| F(x)$ for all λ. Also $F(x) = 1$ is a strictly convex surface.[§]

Conversely,

If a function $F(x)$ has the properties of (48.3), then

(48.4) $$m(x,y) = F(x - y) = F(x_1 - y_1, x_2 - y_2, x_3 - y_3)$$

is a Minkowski metric.

The proof is the same as for (25.8).[§]

While the reflection in a given point x exists, in general reflections in planes do not exist. In fact, the following analogue of (25.2) holds.

(48.5)   The reflection in the plane ξ exists if and only if a sphere with center on ξ has one of its family of parallel chords bisected by ξ.

PROOF: Let the reflection Φ in the plane ξ exist. Then Φ maps on itself any sphere $K(p,\delta)$, where p is on ξ. It also maps on itself every line perpendicular to ξ, and hence maps on itself any chord of $K(p,\delta)$ which is perpendicular to ξ. All such chords are parallel since the lines carrying them pass through the center of the harmonic homology induced by Φ, and this center, not being in R, is on the plane at infinity (compare (47.6)).

Conversely, let ξ bisect F, a family of parallel chords in a sphere $K(p,\delta)$ with center on ξ. Define Φ as the mapping of A^3 on itself which leaves every point of ξ fixed and which maps x not on ξ into x' determined by the conditions that $[x,x']$ is parallel to the lines carrying F and that the center of x and x' is on ξ. Since Φ is obviously an involution, it will be the reflection in ξ if it preserves distance. But any pair of points x,y and their images $x\Phi$ and $y\Phi$ lie in a plane η parallel to the direction of F. If g is any point of $N = [\xi,\eta]$, the chords of $K(g,\delta)$ with the direction of F are also bisected by ξ since $K(g,\delta)$ can be obtained from $K(p,\delta)$ by a translation. By the proof of (25.2), Φ induces in η a reflection about N, hence $m(x,y) = m(x\Phi, y\Phi)$, q.e.d.

If $K(p,\delta)$, and therefore every sphere, is an ellipsoid, a plane ξ through p bisects all chords whose lines pass through the pole of ξ (see (16.4)), hence the reflection in ξ exists, and by the same token reflections in all planes exist. As in the two-dimensional case, the converse holds, hence:

(48.6)   A Minkowski geometry admits a reflection in every plane if and only if its spheres are ellipsoids.

To establish the necessity we could follow the procedure of the plane case and use the three-dimensional analogue of Loewner's theorem (18.7). But there is a simpler way. We make the definition:

> A Euclidean geometry is a Minkowskian geometry possessing reflections in all planes.

Then, (48.6) is equivalent to:

(48.7)   A Minkowskian geometry is Euclidean if and only if its spheres are ellipsoids.

We need only prove the necessity, so assume that reflections in all planes exist. From (47.7), any quadruple can be moved into a given congruent one, and the second part of the proof in (47.7) shows that every plane possesses reflections in all its lines (this can also be deduced from (47.6)). The metric is therefore Euclidean in every plane, hence the intersection of $K(z,1)$ with any plane through z is an ellipse. What we wish to prove is that this implies that $K(z,1)$ is an ellipsoid. Since we need this fact later, we interrupt the present proof to state it explicitly.

(48.8)   A convex surface K with center z and the property that every plane through z intersects it in an ellipse is an ellipsoid.

For let an affine coordinate system be chosen so that z is the origin, with the x_3-axis perpendicular to the x_1,x_2 plane, and with $m[z,(0,0,1)] = 1$. Further, take the x_1,x_2-axes so that their plane cuts $K(z,1)$ in the ellipse $x_1^2 + x_2^2 = 1$. If, now, $x = (x_1,x_2,x_3)$ is an arbitrary point of $K(z,1)$, the line through x parallel to the x_3-axis is perpendicular to the x_1,x_2 plane and intersects this plane in $\bar{x} = (x_1,x_2,0)$, the foot of x. Since the metric is Euclidean in the plane determined by x and the x_3-axis,

$$1 = m^2(z,x) = m^2(z,\bar{x}) + m^2(\bar{x},x).$$

The choice of $m[z,(0,0,1)]$ implies $m(\bar{x},x) = |x_3|$, while the choice of the x_1,x_2 coordinates implies $m^2(z,\bar{x}) = x_1^2 + x_2^2$. Since the coordinates of x satisfy $x_1^2 + x_2^2 + x_3^2 = 1$, $K(z,1)$ is an ellipsoid, and (48.6), (48.7) and (48.8) are established.

The general expression for a Euclidean metric is now easily obtained. In projective coordinates $K(z,1)$ is a non-degenerate quadric,

$$\sum_{i,k=1}^{4} a_{ik}x_i x_k = 0, \qquad a_{ik} = a_{ki}, \qquad |a_{ik}| \neq 0,$$

with center at the origin $z = (0,0,0,1)$. Hence $x_4 = 0$ is the polar plane to $(0,0,0,1)$. Therefore the plane $\sum_{i=1}^{4} a_{i4}x_i = 0$ coincides with $x_4 = 0$, which implies $a_{14} = a_{24} = a_{34} = 0$ and $a_{44} \neq 0$. From $a_{ik} = a_{ki}$ we also have $a_{41} = a_{42} = a_{43} = 0$. Dividing by $-a_{44}$, an affine representation of $K(z,1)$ is obtained in the form

$$\sum_{i,k=1}^{3} g_{ik}x_ix_k = 1, \qquad g_{ik} = g_{ki}, \qquad |g_{ik}| \neq 0.$$

Because every line $x_i = ty_i$, $y \neq (0,0,0)$, through the origin intersects $K(z,1)$ for some non-zero value of t, it follows that $\sum_{i,k=1}^{3} g_{ik}y_iy_k = 1/t^2 > 0$, which means that the form $\sum_{i,k} g_{ik}x_ix_k$ is positive definite. Moreover, when this condition is satisfied, then $F(x) = \left[\sum_{i,k=1}^{3} g_{ik}x_ix_k\right]^{1/2}$ satisfies (48.4). Thus we have found:

The general expression for a Euclidean metric $e(x,y)$ is

(48.9)
$$e(x,\, y) = \left[\sum_{i,k=1}^{3} g_{ik}(x_i - y_i)\,(x_k - y_k)\right]^{1/2},$$

where $\sum_{i,k=1}^{3} g_{ik}x_ix_k$ is a positive definite quadratic form. [14]

The proof of the three-dimensional analogue to (25.5) is exactly the same as for two dimensions, hence:

(48.10)   *Two Minkowskian geometries in A^3, with respective unit spheres $K(z,1)$ and $K'(z',1)$, are congruent, if and only if an affinity of A^3 exists which carries $K(z,1)$ into $K'(z',1)$.*§

Because any ellipsoid can be put in the form $x_1^2 + x_2^2 + x_3^2 = 1$ by an affinity, any ellipsoid can be carried into any other by an affinity. This, with (48.10), implies that *all Euclidean, three-dimensional geometries are congruent* and we may again speak of *the* Euclidean geometry instead of *a* Euclidean geometry.

Volume is defined in a Minkowskian geometry much as area was for the plane. The solid unit sphere $m(z,x) \leqslant 1$ is denoted by U, and affine

---

[14] The condition $|g_{ik}| \neq 0$ follows from the definiteness of the form. The symbols g_{ik} have been chosen for the coefficients in conformity with the usage of Riemannian geometry.

coordinates are chosen in such a way that $\iiint\limits_{U} dx_1\, dx_2\, dx_3 = 4\pi/3$. For any region D, the volume $V(D)$ is then defined by

$$V(D) = \iiint\limits_{D} dx_1\, dx_2\, dx_3.$$

By the same argument as for the plane, Minkowskian motions belong to the equiaffine group and so leave this volume invariant.[§]

The expression for volume is extended to arbitrary affine coordinates x_1, x_2, x_3 as follows. If

$$\sigma = \iiint\limits_{U} dx_1\, dx_2\, dx_3,$$

then $x_1' = x_1$, $x_2' = x_2$, $x_3' = 4\pi x_3/3\sigma$ are coordinates for which the integral over U equals $4\pi/3$, hence

(48.11)
$$V(D) = \iiint\limits_{D} dx_1'\, dx_2'\, dx_3' = \frac{4\pi}{3\sigma} \cdot \iiint\limits_{D} dx_1\, dx_2\, dx_3$$

is the expression for volume in general affine coordinates.

The explicit evaluation of σ, if at all possible, is usually difficult. Even when the unit sphere is an ellipsoid in the general form $\sum g_{ik}x_i x_k = 1$, the value of σ is most easily found by means of algebra instead of by direct integration. The affinity

$$x_i = \sum_{i=1}^{3} b_{ij}y_j, \qquad i = 1,2,3, \qquad |b_{ii}| \neq 0,$$

transforms the above quadratic form to

$$\sum_{i,k=1}^{3} g_{ik}\left(\sum_{j=1}^{3} b_{ij}y_j\right)\left(\sum_{n=1}^{3} b_{kn}y_n\right) = \sum_{i,k,j,n=1}^{3} g_{ik}b_{ij}b_{kn}y_j y_n$$
$$= \sum_{j,n=1}^{3} \bar{g}_{jn}y_j y_n, \text{ where } \bar{g}_{jn} = \sum_{i,k=1}^{3} g_{ik}b_{ij}b_{kn}.$$

Setting $b_{ji}' = b_{ij}$, and $d_{jk} = \sum_i g_{ik}b_{ij} = \sum_i b_{ji}'g_{ik}$, the general rule for matrix multiplication, namely,

$$(a_{ij})(b_{jk}) = \left(\sum_j a_{ij}b_{jk}\right),$$

shows that in terms of matrices we have

$$(d_{jk}) = (b'_{ji})(g_{ik}) \quad \text{and} \quad (\bar{g}_{jn}) = (d_{jk})(b_{kn}).$$

From

$$(\bar{g}_{jn}) = (b'_{ji})(g_{ik})(b_{kn})$$

it follows then that

(48.12) $$\qquad\qquad |\bar{g}_{jn}| = |g_{ik}| \cdot |b_{kn}|^2.$$

By a proper choice of the equiaffinity $x_i = \sum_j b_{ij} y_j$ (compare (18.4)),

the ellipsoid $\sum g_{ik} x_i x_k = 1$ can be transformed into $\sum_i y_i^2 = a^2$. The

integral over $\sum_i y_i^2 \leqslant a^2$ has the value $4\pi a^3/3$. If this ellipsoid is expressed

in the form $\sum_{j,n} \bar{g}_{jn} y_j y_n = 1$, then

$$\frac{4\pi a^3}{3} = \frac{4\pi}{3} \cdot |\bar{g}_{jn}|^{-\frac{1}{2}}.$$

Thus it follows from (48.12) that for $\sum g_{ik} x_i x_k \leqslant 1$

$$\sigma = \frac{4\pi}{3} |g_{ik}|^{-\frac{1}{2}},$$

and hence by (48.11) that:

(48.13) $$V(D) = |g_{ik}|^{\frac{1}{2}} \iiint_D dx_1 \, dx_2 \, dx_3 \text{ expresses the Euclidean volume}$$

when distance is given in the general form (48.9).

Although the *area* of the plane figures is not defined in the equiaffine geometry of space, there is *a natural way (and only one)* to introduce it in *Minkowski space*. The reason for this is that the induced geometry in any plane is Minkowskian and we know what area means in a Minkowski plane.

Since area is defined for every polyhedron it can be extended to surfaces by a limit process. However, the explicit expression for area, even in the Euclidean case, becomes very complicated when distance has the general form (48.9). To evaluate the area element, the methods of Section 26 may be used by setting

$$\Phi(x,y) = \sum_{i,k} g_{ik} x_i y_k.$$

The same argument as there shows that the cosine of the angle between rays from the origin to the points x and y, distinct from the origin, is

(48.14) $$\cos (x,y) = \Phi(x,y)\Phi^{-\frac{1}{2}}(x,x)\Phi^{-\frac{1}{2}}(y,y),$$

and so,

(48.15) $$\sin^2 (x,y) = [\Phi(x,x)\Phi(y,y) - \Phi^2(x,y)] \cdot \Phi^{-1}(x,x)\Phi^{-1}(y,y).$$

If now a surface is given in the parametric form $x_i = x_i(u,v)$, $i = 1,2,3$, and if $x_u = (\partial x_1/\partial u,\ \partial x_2/\partial u,\ \partial x_3/\partial u)$, $x_v = (\partial x_1/\partial v,\ \partial x_2/\partial v,\ \partial x_3/\partial v)$, then the area of the "infinitesimal" parallelogram spanned by $x_u\,du$ and $x_v\,dv$ is

$$\Phi(x_u\,du,\ x_u\,du)\Phi(x_v\,dv,\ x_v\,dv) \sin (x_u,x_v),$$

and so, by (48.15), the area element has the form

$$[\Phi(x_u,x_u)\Phi(x_v,x_v) - \Phi^2(x_u,x_v)]^{\frac{1}{2}}\,du\,dv.$$

When $\Phi(x,x) = \sum x_i^2$ and x_{iu} denotes $\partial x_i/\partial u$, this can be expressed as

$$\left[\left(\sum_i x_{iu}^2\right)\left(\sum_i x_{iv}^2\right)^2 - \left(\sum_i x_{iu}x_{iv}\right)^2\right]^{\frac{1}{2}}du\,dv.$$

Another form, sometimes used in calculus, can be obtained from this by using the Lagrange identity of vector calculus which states that the bracketed expression above equals

$$(x_u \times x_v)^2 = \begin{vmatrix} x_{2u} & x_{3u} \\ x_{2v} & x_{3v} \end{vmatrix}^2 + \begin{vmatrix} x_{3u} & x_{1u} \\ x_{3v} & x_{1v} \end{vmatrix}^2 + \begin{vmatrix} x_{1u} & x_{2u} \\ x_{1v} & x_{2v} \end{vmatrix}^2.$$

49. Euclidean Geometry in Space

We now assume that Euclidean distance has the standard form:

$$e(x,y) = [(x_1 - y_1)^2 + (x_2 - y_2)^2 + (x_3 - y_3)^2]^{\frac{1}{2}},$$

corresponding to the simplest form of the unit sphere, $\sum x_i^2 = 1$.

First we discuss the *theory of quadrics*. As in the case of conics, the axes of a quadric present a new phenomenon, where an *axis* is defined as a straight line such that a reflection in this line carries the quadric into itself. If M is an axis of the non-degenerate quadric Q, and the plane ξ perpendicular to M at y intersects Q in a conic C_y, then y is the center of C_y. Hence ξ intersects ζ, the plane at infinity, in the line N_y which is the polar of y with respect to C_y. Therefore N_y is not tangent to Q and so does not pass through p, the pole of ζ, even if p lies on ζ. From (45.5), then:

(49.1)   *The line M is an axis of the non-degenerate quadric Q if and only if the planes through the polar line of M are perpendicular to M.*

To obtain Q in normal form, we first take the case where Q has a center p which is then the pole of ζ. We wish to show that at least one axis of Q exists, that is, that at least one line M is perpendicular to the planes through its polar line M'. With the point g_M at infinity on M we associate the point at infinity, g_ξ, in which the (parallel) perpendiculars to the polar plane to g_M intersect ζ. Since g_ξ is the pole of M' in the plane polarity induced by $\sum x_i^2 = 1$ on ζ, the mapping $g_M \to g_\xi$ of ζ on itself is a projectivity and so (compare (9.2)) has a fixed point g_{M_1}. Then M_1 is perpendicular to all the corresponding (parallel) planes ξ_1. Among these planes, we choose one which intersects Q in a conic C_ξ, and in this plane introduce rectangular coordinates, x_2, x_3 so that C_ξ takes the standard form

$$c_2 x_2^2 \pm c_3 x_3^2 = 1, \qquad c_i > 0, \qquad i = 1,2.$$

Now let M_1 be the x_1-axis. Because Q goes into itself under the reflection, $x_1' = x_1$, $x_2' = -x_2$, $x_3' = -x_3$, the equation of Q contains no mixed terms involving x_1 and so has the form:

$$c_1 x_1^2 + c_2 x_2^2 \pm c_3 x_3^2 + 2c_{14} x_1 = 1.$$

That $c_1 \neq 0$ follows from the fact that the point $(1,0,0,0)$, which is not on Q, would otherwise satisfy the projective equation of Q. If $c_{14} \neq 0$ then the translation $x_1' = x_1 + c_{14}/c_1$, $x_2' = x_2$, $x_3' = x_3$ reduces the equation to $\sum_i c_i x_i'^2 = 1 + c_{14}^2/c_1$, where the right side is not zero because Q is non-degenerate. Dividing through by $1 + c_{14}^2/c_1$, and relabeling the coefficients, yields the standard Euclidean form for a central quadric,

$$(49.2) \qquad \sum_{i=1}^{3} a_i x_i^2 = 1,$$

where not all the a_i are negative since Q is real. Clearly all of the coordinate axes are axes of Q (in this form) hence every central quadric has at least one triple of mutually perpendicular axes.

If p lies in ζ, then M_p, the line polar to p in the elliptic polarity induced by $\sum x_i^2 = 1$, does not pass through p. The line M_p', polar to M_p with respect to Q, passes through p because M_p lies in the polar plane to p. In any plane ξ perpendicular to M_p' and intersecting Q in a conic C_ξ let rectangular coordinates x_2, x_3 be introduced so that C_ξ again has the form $c_2 x_2^2 \pm c_3 x_3^2 = 1$, where $c_1, c_2 > 0$. With M_p' taken as the x_1-axis, the point $(1,0,0,0)$ lies on Q hence the resulting equation of Q reduces to the normal form:

$$(49.3) \qquad 2x_1 + a_2 x_2^2 \pm a_3 x_3^2 = 1.$$

In addition to axes, a quadric has planes of symmetry such that a reflection in one of these planes takes Q into itself. For a central quadric, (49.2) shows the existence of at least three such planes, while the paraboloid (49.3) has at least two.

We next consider Γ_e, the group of motions of Euclidean three-space. The study of this group is more involved than its plane counterpart, but fortunately much of the work is already contained in Section 34, because of the following argument. A motion Ψ in Γ_e is an affinity of the form

$$\Psi : x_i' = \sum_{k=1}^{3} a_{ik}x_k + a_i, \qquad i = 1,2,3, \qquad |a_{ik}| \neq 0.$$

Clearly Ψ is the product of

$$\Phi : x_i'' = \sum_k a_{ik}x_k, \qquad i = 1,2,3, \qquad |a_{ik}| \neq 0$$

and the translation

$$T : x_i' = x_i'' + a_i, \qquad i = 1,2,3.$$

Since Ψ and T are motions, $\Phi = \Psi T^{-1}$ is also a motion. Under Φ the origin z is fixed, hence Φ leaves invariant the distance from z to x and also its square, $\Omega_1(x,x) = \sum x_i^2$. But in Section 34 we saw that if Φ leaves $\Omega_1(x,x)$ fixed it also has $\Omega_1(x - y, \ x - y) = e^2(x,y)$ as an invariant and is therefore a motion on Euclidean three-space. Hence:

> The group of motions of Euclidean space, where distance has the standard form
>
> $$e(x,y) = \left[\sum_{i=1}^{3} (x_i - y_i)^2 \right]^{1/2},$$
>
> is represented in a one-to-one way by the transformations

(49.4)
$$x_i' = \sum_{k=1}^{3} a_{ik}x_k + a_i, \qquad i = 1,2,3,$$

> for which

$$\sum_{k=1}^{3} a_{ki}^2 = 1, \ i = 1,2,3, \quad \text{and} \quad \sum_{k=1}^{3} a_{ki}a_{kj} = 0, \ i,j = 1,2,3, \ i \neq j.$$

We also know (see (34.3) that $|a_{ik}| = \pm 1$ and that

$$\sum_{k=1}^{3} a_{ki}^2 = 1, \quad i = 1,2,3, \qquad \sum_{k=1}^{3} a_{ik}a_{jk} = 0, \qquad i,j = 1,2,3, \quad i \neq j.$$

To obtain the inverse of Ψ, defined above, we observe that $\Psi = \Phi T$ yields $\Psi^{-1} = T^{-1} \cdot \Phi^{-1}$. Since T^{-1} is given by $x_i'' = x_i' - a_i$, and Φ^{-1}, from (34.2), is the mapping $x_i = \sum_k a_{ki} x_k'$, it follows that Ψ^{-1} is:

$$(49.5) \quad \Psi^{-1} : x_i = \sum_k a_{ki}(x_i' - a_i) = \sum_k a_{ki} x_i' - \sum_k a_{ki} a_i, \qquad i = 1,2,3.$$

A motion Φ which leaves the point p fixed induces a projectivity in the bundle p and so, by the space dual to (9.2), carries at least one plane ξ through p into itself. Therefore it also leaves invariant the line M which is perpendicular to ξ at p. The mapping induced by Φ on M is the identity or the reflection in p, while, by (26.17), the induced mapping on ξ is the identity, a rotation about p, or the reflection in a line N through p. Combined, we have the following six possibilities, 1abc and 2abc.

1. *Identity on M*: $\left\{ \begin{array}{l} (a)\ \textit{identity on } \xi, \\ (b)\ \textit{rotation of } \xi \textit{ about } p, \\ (c)\ \textit{reflection of } \xi \textit{ in } N. \end{array} \right.$
2. *Reflection of M in p*:

For (1a), Φ is clearly the space identity, and for (2a) is the reflection in ξ. Case (1b) is called a *rotation of the space about the axis M*. It includes the reflection in M as a special case when the rotation of ξ about p is the reflection of ξ in p. In (1c), every point of M and of N is fixed, hence every point of the plane η which they determine is fixed. Because Φ is not the identity it must be the reflection in η. The case (2b), after the reflection of M in p, becomes (1b) and is therefore the product of a rotation about M and a reflection in ξ. This includes the reflection in p as a special case. Finally (2c) leaves every point of N fixed and is easily seen to be the space reflection in N.

If M is taken for the x_1-axis and ξ for the x_2, x_3 plane, then, for the cases under 1, Φ has the form

$$x_1' = x_1, \qquad x_2' = a_{22} x_2 + a_{23} x_3, \qquad x_3' = a_{32} x_2 + a_{33} x_3,$$

and for the remaining cases Φ becomes

$$x_1' = - x_1, \qquad x_2' = a_{22} x_2 + a_{23} x_3, \qquad x_3' = a_{32} x_2 + a_{33} x_3.$$

Putting $\Delta_1 = \begin{vmatrix} a_{22} & a_{23} \\ a_{32} & a_{33} \end{vmatrix}$, the determinant Δ of the motion has the value Δ_1 in the first case and $- \Delta_1$ in the second. Hence $\Delta = 1$ for the cases 1ab and 2c and $\Delta = - 1$ for the other cases.

If M, and p on M, are arbitrary, a suitable motion Ψ will take M into the x_1-axis and ξ into the the x_2, x_3-plane. Then $\Psi^{-1} \Phi \Psi$ leaves z and the x_1-axis fixed, and the value of its determinant Δ is ± 1 corresponding to the possibilities described above. But if $\bar{\Delta}$ is the determinant of Ψ and Δ' that of Φ, then from $\Delta = \Delta'\bar{\Delta}^2$ we conclude:

(49.6) *A motion which has a fixed point, and whose determinant is 1, is the identity or a rotation, the latter including reflections in a line. If its determinant is − 1, it is either the reflection in a plane or a rotation about a line followed by a reflection in a plane perpendicular to the line. This last case includes reflections in a point.*

If Ψ is a motion without a fixed point it carries the origin z into some point $a = z\Psi$. Then the product of Ψ with the translation T, $x_i' = x_i - a_i$, leaves z fixed and hence is one of the motions described in (49.6). Since $\Psi = \Phi T^{-1}$, and T^{-1} is again a translation, we see:

(49.7) *Every motion of Euclidean space is the product of one of the motions (49.6) and a translation.*[15]

In particular, a motion with positive determinant is the product of a rotation Φ and a translation T, either of which may be the identity. If the rotation is not the identity and the axis of rotation is taken for the x_1-axis, the motion has the form

$$x_1' = x_1 + a_1,$$
$$x_2' = a_{22}x_2 + a_{23}x_3 + a_2, \quad \begin{pmatrix} a_{22} & a_{23} \\ a_{32} & a_{33} \end{pmatrix} \neq \begin{pmatrix} 1 & 0 \\ 0 & 1 \end{pmatrix},$$
$$x_3' = a_{32}x_2 + a_{33}x_3 + a_3,$$

where $\Delta_1 = 1$. This may be written as the product of

$$\Phi_1 : x_1'' = x_1, \qquad x_2'' = a_{22}x_2 + a_{23}x_3 + a_2, \qquad x_3'' = a_{32}x_2 + a_{33}x_3 + a_3$$

with

$$T_1 : x_1' = x_1'' + a_1, \qquad x_2' = x_2'', \qquad x_3' = x_3''.$$

Then Φ_1 maps every plane $x_1'' = c$, in particular $x_1'' = 0$, into itself, whereas, by (26.15), the induced mapping on $x_1'' = 0$ is a rotation about a point y. Hence Φ_1 is a rotation about the line perpendicular to the plane $x_1'' = 0$ at y, while T_1 is a translation (possibly the identity) parallel to the x_1''-axis. Thus we have:

(49.8) *A motion with positive determinant is the identity, a rotation, a translation, or the product of a rotation and a translation parallel to the axis of rotation.*

---

[15]Because the translations form an invariant subgroup of the affinities§, hence also of the Euclidean motions, the given motion is also the product of a (generally) *different* translation and the *same* motion (49.6) (compare (15.9)).

50. Hilbert's Geometry in Space. Hyperbolic Geometry

Let K be a closed, strictly convex surface in affine space and D be its interior. *Hilbert geometry* in D is defined exactly as in the plane case. If a and b are any two points of D and if x and y are the respective intersections of the rays $R(a,b)$ and $R(b,a)$ with K, then the Hilbert distance of a from b is defined to be

$$h(a,b) = (k/2) \log R(a,b,x,y), \qquad k > 0.$$

With this distance, the points of $S^*(x,y)$ form a metric line (compare Section 28). The induced geometry in the intersection of a plane with D is a plane Hilbert geometry. If c is any point of D, not on $S(x,y)$, then this remark applied to the plane through a, b and c shows that

$$h(a,b) + h(b,c) > h(a,c).$$

Thus the conditions for a projective metric are satisfied.

If $K(p,\delta)$, where $\delta > 0$, denotes the sphere about p with radius δ, that is the locus of points whose Hilbert distance from p is δ, then the previous remark, together with (46.1) and (28.10), implies that $K(p,\delta)$ is a (closed) strictly, convex surface in D.

A motion of the Hilbert geometry is induced by a collineation of the projective space which carries D into itself; conversely every such collineation induces a motion of the Hilbert geometry because it leaves cross ratio invariant.

If a plane ξ intersects D and if Φ, the reflection in ξ of the Hilbert space, exists, then Φ is induced by a harmonic homology, with ξ as axial plane and with center x not in D (see (47.6)). The lines through x carry the perpendiculars to ξ. In any plane η through x, Φ induces a reflection in the line $[\eta,\xi]$. It follows from the construction of perpendiculars (Section 28) that x cannot lie on K. Corresponding to (29.2) we have:

(50.1)   *When the boundary K of the domain D of a Hilbert geometry is an ellipsoid, reflections in all planes exist. When reflections in all the planes through one fixed point of D exist, then K is an ellipsoid.*

PROOF: Let K be an ellipsoid. Take any plane ξ intersecting D and let x denote its pole with respect to K. The harmonic homology, Φ, with center x and axial plane ξ carries K, and therefore D, into itself. It is therefore the reflection in ξ for the Hilbert geometry defined in D.

Conversely, let the Hilbert geometry defined in D, with boundary K, possess the reflection in any plane through a point z of D. Choose ξ to be any plane through z and let M denote the perpendicular to ξ at z. Any line N in ξ, and through z, is perpendicular to M. This follows from (23.6) because the reflection in ξ induces in the plane of M and N a reflection in N. Therefore the reflection in any plane η through M leaves fixed every point of $L = [\eta, \xi]$ and carries into itself the line of ξ perpendicular to L at z. Hence it induces in ξ the reflection in L. Since η can be taken as any plane through M, the line L can be taken as any line of ξ through z and it follows from (29.2) that ξ intersects K in an ellipse, The arbitrariness of ξ implies that every plane through z cuts K in an ellipse. We now choose the plane at infinity ζ so that it contains the center of the harmonic homology, induced by the space reflection in ξ, and also contains the polar line of z with respect to the ellipse in which ξ intersects K. Then ζ does not intersect K or D, and z becomes the affine center of K. Now (48.8) shows that K is an ellipsoid, q.e.d.

A *Hilbert geometry which possesses reflections in all planes is called hyperbolic.* We may then express (50.1) as follows:

(50.2)　　*The Hilbert geometry defined in D is hyperbolic if and only if D is the interior of an ellipsoid.*

By theorems (47.4) and (47.7), hyperbolic geometry enjoys the maximum possible degree of mobility. If two congruent, non-co-planar quadruples are given, then exactly one hyperbolic motion exists carrying one into the other. Thus, as in the plane case, hyperbolic geometry satisfies the congruence axioms of Euclidean geometry, but not the parallel axiom. Clearly, there is nothing to prevent the extension of these considerations to n-dimensions.

Though in principle the development of spatial hyperbolic geometry is now straightforward, the technical difficulties in dealing with length, area, and volume become considerable. Riemannian geometry provides the adequate tools, and we restrict ourselves here to the simplest problems.

Consider first the general case where $h(a,b)$ is any Hilbert distance defined in D. If N_0, cutting K at x_0 and y_0, is an arbitrary line through a point p, and if the variable points a and b tend to p in such a way that the line $[a,b]$ converges to N_0, then the same argument as in the plane case (Section 29) shows that

(50.3)　　$$\lim_{a,\,b \to p} \frac{h(a,b)}{e(a,b)} = \frac{k}{2}\left[\frac{1}{e(p,x_0)} + \frac{1}{e(p,y_0)} \right] = \Phi(p,N_0),$$

where $e(a,b) = \left[\sum (a_i - b_i)^2 \right]^{1/2}$, and the affine coordinate system is such that ζ does not intersect K or D. From (29.5) and (46.1) we conclude:

(50.4)　　*If on every line N, of the bundle through p, two points, z_1 and z_2, are selected so that $e(p,z_i) = \Phi^{-1}(p,N)$, $i = 1,2$, then the locus of the points z_i is a strictly convex surface, K_p, with center p.*

If $m(a,b)$ is the Minkowski metric with K_p as unit sphere, the limit (50.3) becomes

$$(50.5) \qquad \lim_{a,b \to p} \frac{h(a,b)}{m(a,b)} = 1.$$

Because of (50.3), on an arbitrary (continuously differentiable) curve in D, expressed as $x(t) = (x_1(t), x_2(t), x_3(t))$, arc length is given by

$$\int \Phi(x(t), N(t)) \, (x_1'^2 + x_2'^2 + x_3'^2)^{1/2} dt,$$

where $N(t)$ denotes the line tangent to the curve at the point corresponding to t. One may also say that in Hilbert geometry the line element ds, with point of origin x and direction N, has length:

$$(50.6) \qquad dS_h = \Phi(x,N) \, (dx_1^2 + dx_2^2 + dx_3^2)^{1/2}$$

(the primes indicating differentiation).

The expression for volume is obtained by considerations analogous to those at the end of Section 29. If U_p is the domain bounded by K_p, then, as before, let $\sigma(p)$ be defined by:

$$(50.7) \qquad \sigma(p) = \frac{4\pi}{3} \left[\iiint_{U(p)} dx_1 \, dx_2 \, dx_3 \right]^{-1}$$

According to (50.5), *in the neighborhood of p the Hilbert distance is approximated by the Minkowski distance corresponding to the unit sphere K_p.* Hence, in a small region β containing p it is reasonable to postulate that the Minkowski volume of β, namely $\sigma(p) \iiint_{\beta} dx_1 \, dx_2 \, dx_3$, is a good approximation for the Hilbert volume of β. Using this principle, the usual limit procedure for integrals yields

$$(50.8) \qquad \iiint_{B} \sigma(x) dx_1 \, dx_2 \, dx_3$$

as the Hilbert volume of a general region B in which x is a variable point. The same principle determines surface area in Hilbert geometry through the use of Minkowski area.

We now turn to hyperbolic geometry, and choose affine coordinates x_1, x_2, x_3 such that the boundary ellipsoid K takes the form $x_1^2 + x_2^2 + x_3^2 = 1$.

Again, the Euclidean metric, $e(x,y) = \left[\sum_{i=1}^{3} (x_i - y_i)^2 \right]^{1/2}$ will play an auxiliary role. Any motion, in terms of the Euclidean metric, which leaves the origin $z = (0,0,0)$ fixed, carries K into itself. Since the motion is induced by a collineation it is also a motion for the hyperbolic space. Hence a sphere about z with (Euclidean) radius $\bar{r} < 1$ is also a hyperbolic sphere whose radius r (again) satisfies

(50.9) $$\bar{r} = \tanh (r/k).$$

In the plane case, the local Minkowskian unit circle K_p was seen to be an ellipse with its minor axis on the line $p \times z$. With $\bar{r} = e(p,z)$, the Euclidean lengths of the semi-minor and semi-major axes were obtained as $(1 - \bar{r}^2)k$ and $(1 - \bar{r}^2)^{1/2}/k$ respectively. Because the Euclidean rotations about $[p,z]$ are also hyperbolic rotations, the Minkowskian sphere K_p is generated by the rotation of the circle K_p about $[p,z]$. Therefore the volume of this sphere is

$$\iiint\limits_{U(p)} dx_1\, dx_2\, dx_3 = \frac{4\pi}{3} \cdot \frac{(1 - \bar{r}^2)}{k} \cdot \left[\frac{(1 - \bar{r}^2)^{1/2}}{k} \right]^2 = \frac{4\pi}{3} \frac{(1 - \bar{r}^2)^2}{k^3}$$

so that, by (50.7) and (50.9),

$$\sigma(p) = k^3(1 - \bar{r}^2)^{-2} = k^3 \cosh^4 (r/k).$$

To find the volume of a sphere with hyperbolic radius r, its center can be placed at z so that $\bar{r} = \tanh (r/k)$. If ρ, φ, θ denote spherical coordinates, with $0 \leqslant \varphi \leqslant \pi$ and $0 \leqslant \theta < 2\pi$, the two coordinate systems are connected by the standard relations

$$x_1 = \rho \cos \theta \sin \varphi, \qquad x_2 = \rho \sin \theta \sin \varphi, \qquad x_3 = \rho \cos \varphi.$$

The Euclidean element of volume has the form:

$$dx_1\, dx_2\, dx_3 = \rho^2 \sin \varphi\, d\rho\, d\varphi\, d\theta.$$

Therefore,

$$\iiint\limits_{U(z,\, r)} \sigma(x)\, dx_1\, dx_2\, dx_3 = k^3 \int_0^{2\pi} d\theta \int_0^{\pi} \sin \varphi\, d\varphi \int_0^{\bar{r}} \rho^2(1 - \rho^2)^{-2}\, d\rho$$

$$= 2\pi k^3 \left[\frac{\bar{r}}{1 - \bar{r}^2} - \text{area } \tanh \bar{r} \right],$$

which, simplified, yields

(50.10)   *In a hyperbolic space the volume of a sphere with radius r is*

$$2\pi k^3 \left[\sinh \frac{r}{k} \cosh \frac{r}{k} - \frac{r}{k} \right].$$

The surface area of this sphere can be found without explicit integration. If x is any point of the sphere, the plane ξ through x and normal to $[z,x]$ cuts the ellipsoid K_p in a (Euclidean) circle C with center x and radius

$(1 - \bar{r}^2)^{1/2}/k$. Hence $(1 - \bar{r}^2)/k^2$ is the ratio of the area element at x of ξ, as a plane in the Euclidean x_1,x_2,x_3-space, to the area element of the Euclidean geometry with C as unit circle. Since this ratio is independent of x, the Euclidean area of the sphere, $4\pi\bar{r}^2$, multiplied by the factor $k^2/(1 - \bar{r}^2)$ gives the hyperbolic area, or:

(50.11)   *The area of a sphere, with radius r, is $4\pi k^2 \sinh^2(r/k)$.*

The general expression for the *hyperbolic line element* in the present x_1,x_2,x_3-coordinates is obtained as in the plane case. The function $\Phi(x,N)$ is given by

$$\Phi(x,N) = k(1 - \bar{r}^2 \sin^2 \omega)^{1/2}/(1 - \bar{r}^2)$$

(compare with (31.2)), where $\bar{r}$ is the Euclidean distance of x from the origin, and ω is the angle between N and the line $[x,z]$. If x_1',x_2',x_3' are direction numbers of N, then it follows from Section 48 that

$$\sin^2 \omega = \left[\begin{vmatrix} x_2 & x_3 \\ x_2' & x_3' \end{vmatrix}^2 + \begin{vmatrix} x_3 & x_1 \\ x_3' & x_1' \end{vmatrix}^2 + \begin{vmatrix} x_1 & x_2 \\ x_1' & x_2' \end{vmatrix}^2 \right] (x_1^2 + x_2^2 + x_3^2)(x_1'^2 + x_2'^2 + x_3'^2).$$

Hence, from (50.6), the line element, originating at x with direction N, determined by dx_1, dx_2 and dx_3, has the form:

$$(50.12) \quad dS_h^2 = k^2 \frac{\left(\sum dx_i^2 \right) - \begin{vmatrix} x_1 & x_2 \\ dx_1 & dx_2 \end{vmatrix}^2 - \begin{vmatrix} x_2 & x_3 \\ dx_2 & dx_3 \end{vmatrix}^2 - \begin{vmatrix} x_3 & x_1 \\ dx_3 & dx_1 \end{vmatrix}^2}{(1 - x_1^2 - x_2^2 - x_3^2)^2}.$$

Since the hyperbolic angular measure coincides with that of the local Euclidean geometry, the angle between two directions at a point can be computed from (50.12) in conjunction with (26.5) (compare with (31.12)).

51. Hyperbolic Spheres, Limit Spheres and Equidistant Surfaces

Consider two points p and g, on the sphere $K(z,r)$, $r > 0$, and let α denote the hyperbolic angle at z formed by the segments $S(z,p)$ and $S(z,g)$, where $0 \leqslant \alpha \leqslant \pi$. If z is the origin, then α is both the Euclidean and the hyperbolic measure of the angle. The plane (or a plane) through z, p and g intersects $K(z,r)$ in a circle, and, by (31.5), the smaller arc connecting p and g has the length $\alpha k \sinh(r/k)$. This is also the Euclidean length of arc intercepted by the same rays on the sphere with the Euclidean radius $k \sinh(r/k)$. Hence:

(51.1)   *The hyperbolic metric induces on a sphere of radius r the metric of the Euclidean sphere with radius $k \sinh(r/k)$.*

If $K(y,r)$ is a sphere whose center is not z, a hyperbolic motion which carries z into y carries $K(z,r)$ into $K(y,r)$. Since it is induced by a collineation, the motion takes a quadric into a quadric so that $K(y,r)$, which is a closed surface in D, appears in the Euclidean space as an ellipsoid. Moreover, since the Euclidean rotations about $[y,z]$ are also hyperbolic rotations, $K(y,r)$ is an ellipsoid of revolution about the axis $[y,z]$.

If, now, N is a line through a point x which intersects K at x_N. and if y is a variable point on N in D moving toward x_N, then as $h(y,x) \to \infty$ the spheres $K(y,h(y,x))$ tend to an ellipsoid $\Lambda(N,x)$ of revolution about N, touching K at x_N and passing through x. The surface $\Lambda(N,x)$ is called the *limit sphere through x with radius N*. The intersection of $\Lambda(N,x)$ with a plane ξ through N is the limit circle in ξ which passes through x and has N as a radius. From (32.7), and the definition of "corresponding points," it follows that:

(51.2) $\Lambda(N,x)$ *is the locus of corresponding points to x on the bundle of asymptotes to N.*

The same arguments as in Section 32 shows that $\Lambda(N,x)$ depends only on x and x_N, and the latter is called the center of the limit sphere. The analogue of (32.9) also holds.

(51.3)   *If x' is the point of $\Lambda(N,x)$ on N', an asymptote to N, then*
$$\Lambda(N,x) = \Lambda(N',x').\S$$

The geometry induced on $K(y,h(y,x))$ is, by (51.1), that of the Euclidean sphere with radius $k \sinh h(y,x)/k$. Since $\sinh t \to \infty$ as $t \to \infty$, and since the geometry on a Euclidean sphere tends to that of the Euclidean plane as the radius of the sphere becomes infinite, it follows that:

(51.4)   *The metric of hyperbolic space induces the Euclidean metric on a limit sphere.*

The spheres $K(y,r)$ are the surfaces orthogonal (in the hyperbolic sense) to all lines through an ordinary point y, and the limit sphere $\Lambda(N,x)$ is orthogonal to all lines through the point x_N on K. Now let p be a point outside of K with ξ as its polar plane (with respect to K). Then all lines of D perpendicular to ξ are carried by lines through p. Let M be such a line, perpendicular to ξ at x. In both directions from x on M lay off equal segments whose hyperbolic length is α, and let x_α^1 and x_α^2 denote the opposite endpoints. As x varies on ξ, x_α^1 and x_α^2 traverse C_α^ξ, *the two equidistant surfaces to ξ at a distance α.* If ξ passes through z, and K_ξ denotes the intersection of K with ξ, then the initial results in Section 32 show that the two surfaces C_α^ξ together with K_ξ form an ellipsoid of revolution about the perpendicular to ξ at z, and that this ellipsoid touches K at K_ξ. More generally, the two surfaces C_α^ξ, at a distance α from ξ, together with K_ξ form an ellipsoid of revolution about the line through z perpendicular to ξ (in the hyperbolic or Euclidean sense). The surfaces C_α^ξ are orthogonal to the lines through p, the polar to ξ.

Since the induced metric is spherical on $K(x,r)$, and Euclidean on $\Lambda(N,x)$, it is natural to expect that it is hyperbolic on C_α^ξ. This is immediately confirmed by (32.2). If C is one of the surfaces C_α^ξ, say the one consisting of points x_α^1 in the previous construction, and if x_α^1 is mapped on x, its foot on ξ, it follows from (32.2) that the ratio of the length of a curve in C to the length of the image curve in ξ has the constant value $\cosh(\alpha/k)$. A shortest connection on C of two points x_α^1, y_α^1 must therefore correspond to a shortest connection in ξ of the feet x, y, that is, to the segment $S(x,y)$.§ Thus:

(51.5) *The hyperbolic metric $h(x,y)$ induces on a surface C, equidistant at distance α from ξ, the hyperbolic metric $h(x,y)\cosh(\alpha/k)$. With reference to this metric, the straight lines of C are its intersections with the planes perpendicular to ξ (or the planes through the pole of ξ).*

To put these results another way, let y be a variable point on a ray issuing from z, and let S_y denote the surface through z orthogonal (in the hyperbolic sense) to all lines through y which enter D. A *visualization of the results in* (39.13) *and* (39.14) is obtained in the phenomenon that for $0 < e(y,z) < 1$ the metric induced on S_y is spherical, for $e(y,z) = 1$ it is Euclidean, and for $e(y,z) > 1$ it is hyperbolic.

If hyperbolic geometry had been developed first, the discovery of a non-hyperbolic geometry would have been immediate, since the induced geometry on a limit sphere provides a hyperbolic model for a non-hyperbolic geometry, namely Euclidean geometry.

It is natural to ask, in analogy, *if there is not in Euclidean space a surface on which the Euclidean metric induces a hyperbolic geometry,* and so provides us with a model of it. The answer, whose proof requires tools not available here, is as follows: on a surface with constant negative curvature the geometry is hyperbolic in the sense that domains exist on the surface which are congruent, in terms of the induced metric, to sub-domains of the hyperbolic plane. But *hyperbolic geometry cannot be realized in the large.* There are no surfaces in E^3 which, in terms of the induced metric, are congruent to the whole hyperbolic plane.[16]

To return to the general discussion, the result (51.4) leads to coordi-

---

[16]See W. Blaschke, "Vorlesungen über Differentialgeometrie."

nates in which the hyperbolic line element takes a very simple form. Let $\Lambda = \Lambda(N,p)$ be any limit sphere (where p is on N). Since the induced metric on Λ is Euclidean, ordinary rectangular coordinates, u_1,u_2, with p as origin, can be introduced on Λ. On N, a third coordinate u_3 can be taken indicating hyperbolic distance from p, and taken positive for the direction toward the center of $\Lambda(N,p)$. If u is any point of D, the radial line N_u of Λ which passes through u cuts Λ in a point u' with rectangular coordinates (u_1,u_2) on Λ, and the limit sphere $\Lambda(N_u,u)$ intersects N at a (signed) distance u_3 from p. Then (u_1,u_2,u_3) are the so-called *limit sphere coordinates* of u. By (32.10), the length of the segment on N_u intercepted by Λ and $\Lambda(N_u,u) = \Lambda_u$ is also $|u_3|$.

The length of the line element dS_h from u to $u + du$ may be determined in the following way. The length of the line element from $(u_1,u_2,0)$ to $(u_1 + du_1,u_2 + du_2,0)$ is $(du_1^2 + du_2^2)^{1/2}$ because u_1 and u_2 are rectangular coordinates on Λ. From (32.14), the length of the element on Λ_u from (u_1,u_2,u_3) to $(u_1 + du_1,u_2 + du_2,u_3)$ is $e^{-u_3/k}(du_1^2 + du_2^2)^{1/2}$. Because hyperbolic geometry is locally Euclidean, and because limit spheres are orthogonal to their radii, the triangle with vertices

$$(u_1,u_2,u_3), \quad (u_1 + du_1,u_2 + du_2,u_3), \quad \text{and} \quad (u_1 + du_1,u_2 + du_2,u_3 + du_3)$$

can be regarded as a Euclidean right triangle. From the theorem of Pythagoras, then:

$$(51.6) \qquad dS_h^2 = e^{-2u_3/k}(du_1^2 + du_2^2) + du_3^2.$$

If dS_h^2 is written in the form $\displaystyle\sum_{i,k=1}^{3} g_{ik}\,du_i\,du_k$, it follows that

$$|g_{ik}(u)| = \begin{vmatrix} e^{-2u_3/k} & 0 & 0 \\ 0 & e^{-2u_3/k} & 0 \\ 0 & 0 & 1 \end{vmatrix} = e^{-4u_3/k},$$

hence, from (48.13), the volume element in limit sphere coordinates is:

$$(51.7) \qquad dV = e^{-2u_3/k}\,du_1\,du_2\,du_3.$$

We use this formula to evaluate the volume of a cylindrical set Z of the following type. In Λ let B denote any set which has a finite area $A(B) = \displaystyle\iint_B du_1\,du_2$, and let Z be the region bounded by B and the radii of Λ to the boundary of B. Then (51.7) yields:

$$V(Z) = \iint_B du_1\,du_2 \int_0^\infty e^{-2u_3/k}\,du_3 = \frac{k}{2}\,A(B),$$

so $V(Z)$ is proportional to $A(B)$.

Equidistant surfaces may also be used to define a coordinate system. In a plane ξ let u_1, u_2 be any system of coordinates, with origin p, so that the hyperbolic line element in ξ has the general form

$$d\sigma^2 = E(u_1, u_2)du_1^2 + 2F(u_1, u_2)du_1\,du_2 + G(u_1, u_2)du_2^2.$$

On the line N, normal to ξ at p, let u_3 designate the hyperbolic distance of a point from p, taken positive for points on one side of ξ and negative for points on the other side. If u is any point of D, the foot of u on ξ has coordinates u_1, u_2 in ξ and the equidistant surface to ξ which passes through u cuts N at a signed distance u_3 from p. Then (u_1, u_2, u_3) are equidistant-surface coordinates of u. To obtain the line element dS_h from u to $u + du$ in these coordinates, the fact that $d\sigma$ is the line element from $(u_1, u_2, 0)$ to $(u_1 + du_1, u_2 + du_2, 0)$, together with (32.2), implies that $d\sigma \cosh u_3/k$ is the line element from (u_1, u_2, u_3) to $(u_1 + du_1, u_2 + du_2, u_3)$. As in the previous case, this element and dS_h and du_3 approximate a Euclidean right triangle, hence

$$dS_h^2 = \cosh^2(u_3/k)d\sigma^2 + du_3^2.$$

Therefore,

$$|\,g_{ik}(u)\,| = \begin{vmatrix} E\cosh^2(u_3/k) & F\cosh^2(u_3/k) & 0 \\ F\cosh^2(u_3/k) & G\cosh^2(u_3/k) & 0 \\ 0 & 0 & 1 \end{vmatrix} = (EG - F^2)\cosh^4 u_3/k,$$

and the volume element is

$$dV = (EG - F^2)^{1/2}\cosh^2(u_3/k)du_1\,du_2\,du_3.$$

If now B is any domain of ξ with a finite hyperbolic area,

$$A(B) = \iint_B (EG - F^2)^{1/2}\,du_1\,du_2,$$

and if Z' is the set of points which lie on the normals to ξ at points of B, and which lie between ξ and an equidistant surface C_α^ξ, then

$$V(Z') = \iint_B (EG - F^2)^{1/2}du_1\,du_2 \int_0^\alpha \cosh^2(u_3/k)du_3$$

$$= \frac{1}{2} A(B)[k \sinh(\alpha/k)\cosh(\alpha/k) + \alpha].$$

As $k \to \infty$, the bracketed expression tends to 2α, so that $V(Z')$ becomes, as expected, $\alpha \cdot A(B)$, the Euclidean formula for the volume of a cylinder with base $A(B)$ and altitude α.

The normals at B project the region B into a region B_α on C_α^ξ. To obtain $A(B_\alpha)$, we need only observe that the ratio of the lengths of corresponding line elements on B_α and B is $\cosh \alpha/k$. Since the area element coincides with the local, Euclidean area element, and since multiplying lengths by a factor multiplies Euclidean area by the square of the factor, it follows that

$$A(B_\alpha) = \cosh^2(\alpha/k) \cdot A(B).$$

Two planes, ξ and η, in D whose line of intersection contains no point of D or K are called *hyperparallel*. There is exactly one line which is perpendicular to both ξ and η, as may be seen with the use of equidistant surfaces. If ξ is taken as the x_1,x_2-plane and its intersection with K is denoted by K_ξ, then the equidistant surfaces C_α^ξ, together with K_ξ, form an ellipsoid of revolution, E_α, about the x_3-axis. For small α, the ellipsoid is close to ξ and therefore does not intersect η. For large α, η and E_α intersect in a curve. Hence there is exactly one value α_0, greater than zero, for which E_{α_0} is tangent to η. If p is the point of contact and f is its foot on ξ then $[p,f]$ is clearly a common perpendicular to ξ and η. As x ranges over ξ and y over η, the minimum of $h(y,x)$ is α_0. There is no other common perpendicular.§ Moreover, if the line of intersection of two planes contains points of D or K the planes have no perpendicular in common.§

52. The Hyperbolic Group of Motions

In homogeneous coordinates, the ellipsoid K has the form:

(52.1) $$x_1^2 + x_2^2 + x_3^2 - x_4^2 = 0,$$

and is defined by the polarity,

(52.2) $$x_1 = \xi_1, \qquad x_2 = \xi_2, \qquad x_3 = \xi_3, \qquad x_4 = -\xi_4.$$

The argument for the plane case, (13.3), generalizes to establish:

(52.3) *If a (real, non-degenerate) quadric Q is defined by the hyperbolic polarity γ, a collineation Φ of the space carries Q into itself if and only if $\Phi\gamma = \gamma\Phi$.*§

Therefore the collineations which leave the ellipsoid (52.1) invariant are precisely those which commute with the polarity (52.2).

To establish some results for the elliptic and hyperbolic cases simultaneously, let $\Omega_\varepsilon(x,y)$ be defined by

$$\Omega_\varepsilon(x,y) = x_1y_1 + x_2y_2 + x_3y_3 + \varepsilon x_4y_4, \qquad \varepsilon = \pm 1,$$

and, with the same meaning for ε, let γ denote the polarity:

$$x_1 = \xi_1, \qquad x_2 = \xi_2, \qquad x_3 = \xi_3, \qquad x_4 = \varepsilon\xi_4.$$

The collineations which commute with γ form a group $\Gamma\gamma$.§ The argument for this, and for the following results, exactly parallels that given in Section 34 for the plane analogues.

The collineation

(52.4) $$\Phi : x_i' = \sum_{k=1}^{4} a_{ik}x_k, \qquad i = 1,2,3,4, \qquad |a_{ik}| = \pm 1,$$

commutes with γ *if and only if the coefficients satisfy the relations:*

(52.5)
$$a_{1i}^2 + a_{2i}^2 + a_{3i}^2 + \varepsilon a_{4i}^2 = 1, \qquad i = 1,2,3,$$
$$a_{14}^2 + a_{24}^2 + a_{34}^2 + \varepsilon a_{44}^2 = \varepsilon,$$
$$a_{1i}a_{1k} + a_{2i}a_{2k} + a_{3i}a_{3k} + \varepsilon a_{4i}a_{4k} = 0, \quad i,k = 1,2,3,4, \quad i \neq k.^\S$$

(52.6)   *The transformations* (52.4) *leave* $\Omega_\varepsilon(x,y)$ *invariant if and only if the coefficients,* a_{ik}, *satisfy* (52.5).§

In the two-dimensional case, the condition $|a_{ik}| = 1$ could be imposed, since replacing a_{ik} by $- a_{ik}$ changed the sign of $|a_{ik}|$ without affecting the collineation. With this condition, the form corresponding to (52.4) gave a one-to-one representation of the elements in $\Gamma\gamma$. In the present situation, changing a_{ik} to its negative does not alter the sign of $|a_{ik}|$. For an odd number of dimensions, and the elliptic case $\varepsilon = 1$, there is no natural way of obtaining a one-to-one representation, and we will leave the form ambiguous. In the hyperbolic case, where $\varepsilon = - 1$, the image of the origin is $(a_{14},a_{24},a_{34},a_{44})$ and hence $a_{44} \neq 0$. To impose the condition $a_{44} > 0$ is therefore possible, but is desirable only if it does not destroy the group property of the collineations in (52.4). The analogue to (34.2) shows that the coefficient of x_4' in Φ^{-1} is also a_{44}, and is therefore positive. Hence, the group property will be maintained if whenever a collineation

$$\Psi : x_j'' = \sum_{i=1}^{4} b_{ji}x_i'$$

satisfies (52.5), and $b_{44} > 0$, it follows that the coefficient of x_4 in $\Phi\Psi$ is also positive. By direct calculation, this coefficient is

$$c_{44} = b_{41}a_{14} + b_{42}a_{24} + b_{43}a_{34} + b_{44}a_{44}.$$

From (52.5) and the analogue to (34.3) it follows that

$$\left(\sum_{i=1}^{3} b_{4i}^2\right) - b_{44}^2 = - 1, \qquad \text{and} \qquad \left(\sum_{i=1}^{3} a_{i4}^2\right) - a_{44}^2 = - 1.$$

Hence the desired property can be shown by establishing the lemma:

If $\Omega_{-1}(x,x) = - 1$, $\Omega_{-1}(y,y) = - 1$, $x_4 > 0$ *and* $y_4 > 0$, *then*

(52.7) $$x_4 y_4 \geqslant 1 + \left| \sum_{i=1}^{3} x_i y_i \right|.$$

By hypothesis,

$$x_4^2 = 1 + \sum_{i=1}^{3} x_i^2 \qquad \text{and} \qquad y_4^2 = 1 + \sum_{i=1}^{3} y_i^2,$$

hence the elementary relation $a^2 + b^2 \geqslant 2\,|\,ab\,|$ and the Cauchy-Schwarz inequality (19.1) yield

$$x_4^2 y_4^2 = (1 + \sum x_i^2)(1 + \sum y_i^2)$$
$$= 1 + \sum (x_i^2 + y_i^2) + \left(\sum x_i^2\right)\left(\sum y_i^2\right) \geqslant 1 + 2\sum |x_i y_i| + \left(\sum x_i y_i\right)^2$$
$$\geqslant 1 + 2\left|\sum x_i y_i\right| + \left|\sum x_i y_i\right|^2$$
$$= \left(1 + \left|\sum x_i y_i\right|\right)^2.$$

Since $x_4 > 0$ and $y_4 > 0$ the lemma follows.

The normalization $x_4 > 0$ can now be introduced, and because $x_4^2 = 1 + \sum_{1}^{3} x_i^2$ it implies automatically that $x_4 \geqslant 1$. Since

$$\Omega_{-1}(x,x) = -1 \qquad \text{and} \qquad a_{41}^2 + a_{42}^2 + a_{43}^2 - a_{44}^2 = -1,$$

the lemma implies that

$$a_{44} x_4 \geqslant 1 + |\,a_{41} x_1 + a_{42} x_2 + a_{43} x_3\,|,$$

hence $x_4' = \sum a_{4i} x_i \geqslant 1$ and the normalization is invariant. These results may be summarized:

The transformations

$$x_i' = \sum_{k=1}^{4} a_{ik} x_k, \qquad i = 1,2,3,4, \qquad |\,a_{ik}\,| = \pm 1,$$

with $a_{44} > 0$, and satisfying

(52.8)
$$\sum_{k=1}^{3} a_{ki}^2 - a_{4i}^2 = 1, \qquad i = 1,2,3, \qquad \sum_{k=1}^{3} a_{k4}^2 - a_{44}^2 = -1,$$
$$\sum_{j=1}^{3} a_{ji} a_{jk} - a_{4i} a_{4k} = 0, \qquad i,k = 1,2,3,4, \qquad i \neq k,$$

give a one-to-one representation of the group of motions of hyperbolic three-space. For $\Omega_{-1}(x,x) = -1$, all elements of the group leave the inequality $x_4 > 0$ invariant.

With the condition $x_4 > 0$, it follows from (52.7) that $\Omega_{-1}(x,y) \leqslant -1$. The analogue to (35.2) is obtained, by the same argument, but in the form:

(52.9)   *For any two points, x,y, in D,*

$$h(x,y) = k \text{ Area cosh } [-\Omega_{-1}(x,y)], \qquad \Omega_{-1}(x,y) = -\cosh h(x,y)/k.$$

The x_i normalized by $\Omega_{-1}(x,x) = -1$ and $x_4 > 0$ are called *Weierstrass coordinates*.

The equation of a plane may be written in the form

$$\Omega_{-1}(x,\xi) = \sum_{i=1}^{3} x_i \xi_i - x_4 \xi_4 = 0.$$

For the plane to belong to D it is necessary that its Euclidean distance from the origin be less than one, namely that $|\xi_4|(\xi_1^2 + \xi_2^2 + \xi_3^2)^{-1/2} < 1$, so that $\Omega_{-1}(\xi,\xi) > 0$. The numbers ξ_i, so normalized that $\Omega_{-1}(\xi,\xi) = 1$, are the *Weierstrass plane coordinates*, and are determined to within a factor of ± 1. In the present case, there is no natural way of removing this ambiguity. As was true for the plane, the transformation in plane coordinates which (52.8) induces is the same as the point transformation, that is,

$$\xi_i' = \sum_{k=1}^{4} a_{ik}\xi_k, \qquad i = 1,2,3,4.\S$$

Because of (52.6), the expressions $|\Omega_{-1}(x,\xi)|$ and $|\Omega_{-1}(\xi,\eta)|$ are invariant under motions and change of sign in the plane coordinates, and so must have geometric meanings. Since $\Omega_{-1}(x,\xi) = 0$ means that x lies on ξ, the case of interest is $\Omega_{-1}(x,\xi) \neq 0$. In that case, coordinates may be selected so that ξ is the plane $x_3 = 0$ and x is on the positive half of the x_3-axis. Then the Weierstrass coordinates of ξ are $(0,0,1,0)$, while x has a representation of the form $\left(0,0,x_3,\sqrt{1 - x_3^2}\right)$. In the plane $x_2 = 0$, x_1, x_3 and x_4 are Weierstrass coordinates so, from (35.5),

$$\Omega_{-1}(x,\xi) = x_3 = \sinh h(x,z)/k,$$

where z is the origin. Since z is also the foot of x on ξ, this implies:

(52.10) $|\Omega_{-1}(x,\xi)|$ *is the hyperbolic sine of the distance of the point x from the plane ξ.*

The angle between two planes, ξ and η, is defined as in the Euclidean case. If N is the line of intersection of ξ and η, a plane normal to N cuts ξ and η in two lines, N_ξ and N_η respectively, and the smaller angle between these lines is taken as the smaller angle between ξ and η. To find the geometric meaning of $\Omega_{-1}(\xi,\eta)$, let $N = [\xi,\eta]$ be the x_3-axis, so that the representations of ξ and η are $(\xi_1, \xi_2, 0, \xi_4)$ and $(\eta_1, \eta_2, 0, \eta_4)$ respectively. Take $x_3 = 0$ as a plane normal to N and take N_ξ and N_η as the intersections of this plane with ξ and η. Any point $(x_1, x_2, 0, x_4)$ on N satisfies the relation

$$\Omega_{-1}(x, \xi) = x_1 \xi_1 + x_2 \xi_2 - x_4 \xi_4 = 0,$$

which shows that in the plane $x_3 = 0$, the numbers ξ_1, ξ_2, ξ_4 are Weierstrass coordinates of the line N_ξ. Similarly, in $x_3 = 0$, the numbers η_1, η_2, η_4 are the Weierstrass coordinates of N_η. Hence, by (35.8),

$$|\Omega_{-1}(\xi, \eta)| = \xi_1 \eta_1 + \xi_2 \eta_2 - \xi_4 \eta_4 = \cos \measuredangle (N_\xi, N_\eta) = \cos \measuredangle (\xi, \eta).$$

When ξ and η are hyperparallels, a similar procedure shows that $|\Omega_{-1}(\xi, \eta)|$ represents the hyperbolic cosine of the length of their common perpendicular. By a limit process, this, or the preceding case, shows that $|\Omega_{-1}(\xi, \eta)| = 1$ means that $\xi \sim \eta$ or else that $[\xi, \eta]$ is tangent to K. Summarized,

(52.11) $|\Omega_{-1}(\xi, \eta)| = 1$ *if the planes* ξ, η *are identical or if their intersection is tangent to* K. *When* ξ *and* η *intersect in* D, $|\Omega_{-1}(\xi, \eta)|$ *is the cosine of the smaller angle between them, and when they are hyperparallels, it represents the hyperbolic cosine of the distance between them (measured on their common perpendicular).*

An immediate consequence of (52.9) is:

(52.12) $\Omega_{-1}(x, c) + \cosh (\delta/k) = 0$ *is the equation of the sphere with center* c *and radius* δ.

Similarly, (52.10) yields:

(52.13) $\Omega_{-1}^2(x, \xi) = \sinh^2 (\alpha/k)$ *is the equation of the two surfaces equidistant to* ξ *at distance* α.

The equation for a limit sphere may be obtained by a familiar modification of the plane case. If N is a line through the origin z, its direction is determined by its direction cosines, $\cos \alpha_i$, where α_i is the angle N makes with the x_i-axis, $i = 1, 2, 3$. Then, as in the plane case:

If N *passes through the origin* z, *and* p *lies on* N, *the equation of* $\Lambda(N, p)$ *is*

(52.14) $$(x_1 \cos \alpha_1 + x_2 \cos \alpha_2 + x_3 \cos \alpha_3 - x_4)^2 = e^{\pm 2s/k}$$

where $\cos \alpha_i$ *is a direction cosine of* N, $s = h(p, z)$, *and the plus or minus sign is chosen according as* p *precedes or follows* z *on* N.§

We conclude this section with a few observations on the case $\epsilon = 1$. The theorems corresponding to (52.4) through (52.6) are:

The transformations

(52.15)　　$\Phi : x_i' = \sum_{i=1}^{4} a_{ik} x_k, \qquad i = 1,2,3,4, \qquad |a_{ik}| = \pm 1,$

commute with the polarity $x_i = \xi_i$, $i = 1,2,3,4$, if and only if the coefficients satisfy:

(52.16)　　$\sum_{k=1}^{4} a_{ki}^2 = 1, \quad i = 1,2,3,4, \quad \sum_{k=1}^{4} a_{ki} a_{kj} = 0, \quad i,j = 1,2,3,4, \quad i \neq j.$

(52.17)　　*The transformations (52.15) leave $\Omega_1(x,y)$ invariant if and only if the coefficients a_{ik} satisfy (52.16).*

As in the plane case, we can give these results two different interpretations. If Φ satisfies (52.16), then, since $\Omega_1(x - y, x - y) = \sum_{1}^{4}(x_i - y_i)^2$ is invariant, Φ is a *motion of the 4-dimensional Euclidean space E^4* with the metric $e(x,y) = \left[\sum_{1}^{4}(x_i - y_i)^2 \right]^{1/2}$. Because of (52.17), it is the most general motion of E^4 which leaves the origin fixed. An arbitrary motion of E^4 is then the product of such a motion with a translation (see (49.7)), and so has the form

$$x_i' = \sum_{k=1}^{4} a_{ik} x_k + a_i, \qquad i = 1,2,3,4, \qquad |a_{ik}| = \pm 1,$$

where the coefficients a_{ik} satisfy (52.16).

For the second interpretation, since Φ leaves $\Omega_1(x,x)$ fixed *it carries the sphere $S_r^3 : \sum x_i^2 = r^2$ into itself.* The spherical distance between two points x and y on S_r^3 can be expressed by

(52.18)　　　　$d(x,y) = r \text{ Arc cos } [r^{-2}\Omega_1(x,y)].$

If a definite sphere is selected, the ratio of the coordinates of a point x on it determine the point uniquely, hence the coordinates may be normalized so that $\sum_{i=1}^{4} x_i^2 = 1$. The spherical distance then becomes

(52.19)　　　　$d(x,y) = r \text{ Arc cos } \Omega_1(x,y),$

and *the transformations $x_i' = \sum a_{ik} x_k$ which satisfy (52.16) represent in a one-to-one way the group of motions of S_r^3 with the metric of either (52.18) or (52.19).*

53. Elliptic Geometry in Space

Three-dimensional elliptic geometry originates from a metrization of the entire projective space P^3 in the same manner as for the plane. Point and plane coordinates are normalized so that

$$\Omega_1(x,x) = 1 \quad \text{and} \quad \Omega_1(\xi,\xi) = 1.$$

Each point and each plane has thus two sets of coordinates which differ only by sign. The elliptic distance $\varepsilon(x,y)$ is defined by

(53.1) $$\varepsilon(x,y) = k \text{ Arc cos } |\Omega_1(x,y)|.$$

Because of (52.15) and (52.17) the mappings which leave $|\Omega_1(x,y)|$ invariant, that is the motions of the space,[17] are the transformations:

$$\Phi : x_i' = \sum_{k=1}^{4} a_{ik}x_k, \quad i = 1,2,3,4, \quad |a_{ik}| = \pm 1,$$

(53.2) *where*

$$\sum_k a_{ki}^2 = 1, \quad i = 1,2,3,4, \quad \sum_k a_{ki}a_{kj} = 0, \quad i,j = 1,2,3,4, \quad i \neq j.$$

However, these transformations do not represent the motions in a one-to-one manner because the mapping

(53.3) $$\Psi : x_i' = -x_i, \quad i = 1,2,3,4,$$

regarded as a linear transformation, is not the identity, while as a collineation it is the identity. Therefore, Φ and $\Phi\Psi$ represent the same motions. It is easily seen that if Φ is a motion, the only other transformation which satisfies (53.2) and represents the same collineation as Φ is $\Phi\Psi$.[§] Thus:

(53.4) *The pairs of transformations, Φ and $\Phi\Psi$, satisfying (53.2), are in one-to-one correspondence with the motions of elliptic space.*

Or, in the language of modern algebra, the factor group of the group (53.2), modulo the group consisting of 1 and Ψ, is isomorphic to the group of motions of elliptic space, Γ_ε.

From the analogue of (34.2) it can be seen that $A_{ik} = a_{ik}$ in the collineation Φ of (53.2). The induced representation of Φ in plane coordinates has therefore the same form, namely $\xi_i' = \sum a_{ik}\xi_k$. Because of (52.17),

---

[17]Strictly speaking, since we defined motion only for metric spaces, the use of the term motion is not justified before the triangle inequality has been established. Until then, we mean mappings which leave $\varepsilon(x,y)$ invariant.

Φ leaves $|\Omega_1(x,\xi)|$ and $|\Omega_1(\xi,\eta)|$ invariant. The smaller of the angles formed by the planes ξ and η can therefore be defined by

$$(53.5) \qquad \sphericalangle(\xi,\eta) = \text{Arc cos } |\Omega_1(\xi,\eta)|.$$

It is still to be shown that $\varepsilon(x,y)$ as defined in (53.1) is a projective metric.[18] Under the polarity $\gamma_e : x_i = \xi_i$, the plane ξ_y, polar to a point y, has the equation

$$\xi_y : \Omega_1(x,y) = 0, \quad \text{or} \quad \varepsilon(x,y) = k\pi/2,$$

and is therefore the locus of points x at a maximum distance from y. The harmonic homology Φ, with center y and having ξ_y for its axial plane, commutes with γ_e. Because of (52.4), then, Φ has the form (53.2) and so leaves $\varepsilon(a,b)$ invariant. Let z be a point distinct from y, and not on ξ_y and take z^* as the intersection of ξ_y with $[z,z']$, where $z' = z\Phi$. Since $y\Phi = y$, and $z^*\Phi = z^*$,

$$(53.6) \qquad \varepsilon(z,z^*) = \varepsilon(z',z^*) \quad \text{and} \quad \varepsilon(y,z) = \varepsilon(y,z'),$$

hence Φ *may be called either the reflection in y or the reflection in ξ_y.* By the definition of Φ, the points y, z^*, z and z' form a harmonic quadruple. If now x and x' are given as distinct points, then the same argument as in Section 36 shows that points y_1 and y_2 exist on $[x,x']$, which correspond in the involution induced by γ_e on $[x,x']$, and are such that y_1,y_2,x,x' form a harmonic set. The reflection in y_i, or in ξ_{y_i}, $i = 1,2$, is a motion which takes x into x'. Thus:

(53.7)   *Given two distinct points, x and x', reflections in two planes (and in two points) exist, each of which maps x on x'.*

By duality, for two distinct planes, η and η', reflection in two points (or two planes) exist, each carrying η into η'.

As in the case of open metrics, the line N is *called a perpendicular* to ξ_y *if* it passes through y (the pole of ξ_y). If η is any plane through y, then γ_e induces in η the reflection in the line $M = [\eta,\xi_y]$. Hence N is also perpendicular to all lines in ξ_y which pass through the intersection point of N and ξ_y.

In the plane $x_4 = 0$ the induced distance between points x and y is $x_1 y_1 + x_2 y_2 + x_3 y_3$, which coincides with the elliptic distance for two dimensions. The previous remarks show that an arbitrary plane ξ can be mapped into the plane $x_4 = 0$ by a collineation Φ which leaves $\varepsilon(a,b)$ invariant, hence the metric induced in any plane is elliptic. Since any three points lie in a plane, this implies that $\varepsilon(x,y)$ is a projective metric in the space.

---

[18]The discussion that follows is nearly identical with that given in Section 36.

The motions in the elliptic group Γ_z which leave a point p fixed form a subgroup Γ_p. For any pair of points, p and g, the groups Γ_p and Γ_g are isomorphic (see (22.7)). In particular, the elements of Γ_z, where $z = (0,0,0,1)$, also leave fixed the plane $x_4 = 0$ polar to z. From affine geometry it follows that the condition for a collineation of the form (53.2) to belong to Γ_z is that $a_{14} = a_{24} = a_{34} = a_{41} = a_{42} = a_{43} = 0$. Then $\sum\limits_{i=1}^{4} a_{i4}^2 = a_{44}^2 = 1$, and if a_{44} is taken as 1, the coefficients a_{ik} are uniquely determined. Thus Γ_z is represented in a one-to-one way by the transformations

$$\Gamma_z : \left\{ \begin{array}{l} x_i' = \sum\limits_{k=1}^{3} a_{ik}x_k, \qquad i = 1,2,3, \qquad x_4' = x_4 \\[2mm] \sum\limits_{k=1}^{3} a_{ki}^2 = 1, \qquad \sum\limits_{k=1}^{3} a_{ki}a_{kj} = 0, \qquad i,j = 1,2,3, \qquad i \neq j, \end{array} \right.$$

hence Γ_z is isomorphic to the group of motions on the two-sphere.

In particular, the foregoing implies that if x and y are two points equidistant from z, then a motion in Γ_z exists which carries x into y. It is now seen easily that:

(53.8)   *If x,y and x',y' are pairs of points such that $\varepsilon(x,y) = \varepsilon(x',y')$, then a motion exists which maps x on x' and y on y'.*

For, from (53.7) there is a motion Φ taking x into x'. Let $y\,\Phi = \bar{y}$. Since $\varepsilon(x',\bar{y}) = \varepsilon(x,y) = \varepsilon(x',y')$, the points $\bar{y}$ and y' are equidistant from x', so a motion Φ' in $\Gamma_{x'}$ exists which maps $\bar{y}$ on y'. Then $\Phi\Phi'$ is a motion which takes x into x' and y into y'. *The corresponding statement for congruent triples is, of course, not correct for elliptic geometry* (compare Section 37).

The most interesting new phenomena of spatial, as compared to plane, elliptic geometry occur in the geometry of lines. As a point x traverses a line N, the plane polar to x under γ_e traverses the plane pencil through N' polar to N. Since N' lies in the polar plane to x, for every x on N, its distance from every point of N is $k\pi/2$.

(53.9)   *If the lines N and N' are polar to each other, then every point of one is at a distance $k\pi/2$ from every point of the other.*

The perpendiculars to N form the congruence of lines intersecting both N and its polar N'. Through a point y, not on N or N', there is exactly one line perpendicular to both N and N'.

If the line N is taken as $[d_3, d_4]$, N' is the intersection of the planes

$x_3 = 0$ and $x_4 = 0$, the respective polars of d_3 and d_4, hence $N' = [d_1, d_2]$. The biaxial involution with N and N' as axes (see Section 41),

$$x_1' = x_1, \qquad x_2' = x_2, \qquad x_3' = -x_3, \qquad x_4' = -x_4,$$

satisfies (53.2) and is therefore a motion Ψ. If M is an arbitrary line, with polar M', (53.8) implies the existence of a motion Φ which takes N into M. Since, by (53.9), $M' = N'\Phi$, the mapping $\Phi^{-1}\Psi\Phi$ leaves every point of M and M' fixed. It is also an involution, because

$$(\Phi^{-1}\Psi\Phi)(\Phi^{-1}\Psi\Phi) = \Phi^{-1}\Psi^2\Phi = \Phi^{-1}\Phi = 1,$$

and (41.15) shows that it is the bi-axial involution having M and M' as axes. Since every point of N is at a distance $k\pi/2$ from every point of N', the same is true of M and M'. Because it is a product of motions, $\Phi^{-1}\Psi\Phi$ is itself a motion. Therefore:

(53.10)   *A biaxial involution, whose axes are polar to each other, is an elliptic motion (called the reflection in either of the axes).*

For $0 < \alpha < k\pi/2$, let F_α^N denote the locus of points whose distance from the line N is α. Any point x of the locus lies on exactly one line which is perpendicular to both N and its polar N'. The distance from x to N' is therefore $\dfrac{k\pi}{2} - \alpha$, hence:

(53.11) $$F_\alpha^N = F_{\alpha'}^{N'}, \text{ where } \alpha' = \frac{k\pi}{2} - \alpha.$$

A rotation about a line is defined in an obvious way,[§] and it is clear that F_α^N goes into itself under all rotations about N. But (53.11) shows that F_α^N also goes into itself under rotations about N'. Thus we have the strange phenomenon of *a surface of revolution with two different axes*.

A motion which carries N into M takes F_α^N into F_α^M. In studying the properties of F_α^N we may therefore assume that $N = [d_3, d_4]$ on $x_1 = 0$ and $x_2 = 0$. The intersection of F_α^N with the plane $x_2 = 0$ consists of the two, plane, equidistant curves C_α^N. In that plane, as previously observed, x_1, x_3, x_4 serve as the coordinates of plane elliptic geometry. The coordinates of N, in the plane, are $(1,0,0)$ so that (37.9) gives

$$|\Omega_1(x, N)| = |x_1| = \sin \alpha/k$$

as the equation of C_α^N. To interpret this relation geometrically, consider its homogeneous form

$$x_1^2 = \sin^2(\alpha/k)(x_1^2 + x_3^2 + x_4^2), \quad \text{or} \quad x_1^2 - \tan^2(\alpha/k)(x_3^2 + x_4^2) = 0.$$

In Γ_z, the rotations about N form the subgroup of motions which leave every point of $[d_3, d_4]$ fixed and so have the form[§]

$$(53.12) \qquad \begin{aligned} x_1' &= x_1 \cos \beta + x_2 \sin \beta, & x_2' &= -x_1 \sin \beta + x_2 \cos \beta, \\ x_3' &= x_3, & x_4' &= x_4. \end{aligned}$$

Since $x_1'^2 + x_2'^2 = x_1^2 + x_2^2$, for all β, the revolution of C_α^N about N yields F_α^N as:

$$(53.13) \qquad x_1^2 + x_2^2 - \tan^2 (\alpha/k) (x_3^2 + x_4^2) = 0.$$

The surface (53.13) is a ruled quadric carrying the two conjugate reguli:

$$(53.14) \qquad \begin{aligned} F_1 &: \begin{aligned} x_1 + \tan (\alpha/k)x_3 &= -t(x_2 + \tan (\alpha/k)x_4), \\ t(x_1 - \tan (\alpha/k)x_3) &= x_2 - \tan (\alpha/k)x_4. \end{aligned} \\[1em] F_2 &: \begin{aligned} x_1 + \tan (\alpha/k)x_3 &= t(x_2 - \tan (\alpha/k)x_4), \\ -t(x_1 - x_3 \tan \alpha/k) &= x_2 + \tan (\alpha/k)x_4. \end{aligned} \end{aligned}$$

Through an arbitrary point y of F_α^N there is one line of each regulus. Because F_α^N contains all points at a distance α from N, it follows that:

(53.15)   *If the distance α from a point y to a line N is such that $0 < \alpha < k\pi/2$ then through y there are exactly two lines M_1 and M_2 equidistant from N (i.e., every point of M_i is at a distance α from N).*

Named after their discoverer, W. K. Clifford (1845–1879), the lines M_1 and M_2 are called the *Clifford parallels to N through y*.

If R denotes the line through y perpendicular to N, the plane ξ through y and N cuts F_α^N in an equidistant curve C_α^N whose tangent L_1 at y is perpendicular to R. The plane η through y which is normal to N cuts F_α^N in a circle whose tangent L_2 at y is also perpendicular to R. Consequently, the plane tangent to F_α^N at y contains both L_1 and L_2 and is normal to R. Because the Clifford parallels M_1 and M_2 lie in this tangent plane they are likewise perpendicular to R. This shows:

(53.16)   *If M is a Clifford parallel to N, the perpendicular to N from a point of M is also perpendicular to M, hence N is a Clifford parallel to M (at the same distance).*

The rotation, about a line N, through an angle θ induces, in any plane η normal to N, a rotation through θ. But all such planes contain the polar N', and the rotation of η is a translation along N' through $k\theta$. Thus F_α^N goes into itself under the translations along N', each point x of the surface moving on the curve equidistant to N' in the plane $(x \wedge N')$. Because $F_\alpha^N = F_{\alpha'}^{N'}$, (53.11), and $F_{\alpha'}^{N'}$ goes into itself under the translations along N, we obtain:

(53.17) F_α^N *goes into itself under the translations along N.*

Because both the rotations about N and the translations along N are motions which leave N fixed, either type of motion takes a Clifford parallel M to N, into M', a Clifford parallel to N at the same distance. Since M and M' cannot intersect they belong to the same regulus.[§] Moreover, if y and y' are any two points of F_α^N, and ξ and ξ' denote the respective planes $(y \wedge N)$ and $(y' \wedge N)$, then a suitable rotation will take ξ into ξ' and carry y into some point y'' of ξ'. A translation along N can then be found which maps y'' on y'. Under the product of these two motions the two Clifford parallels to N at y go into the Clifford parallels to N at y', hence the angle between the first pair is the same as that between the second.

(53.18)   *The angle θ between two Clifford parallels to N, through a point y at a distance α from N, depends only on α.*

The angle is therefore a function of α. The explicit form of the function will be given in the next section.

54. The Line Element of Elliptic Space

Without any essential change, the argument of Section 38 shows that in the present variables the line element of elliptic space is:[§]

$$(54.1) \qquad dS_e^2 = k^2 \Omega_1(dx, dx) = k^2 \sum_{i=1}^{4} dx_i^2,$$

where $\Omega_1(x,x) = 1$.

We apply this to find the line element on F_α^N, where the surface is given by (53.13), that is:

$$(54.2) \qquad x_1^2 + x_2^2 - \tan^2 \alpha'(x_3^2 + x_4^2) = 0, \qquad \alpha' = \alpha/k.$$

Together with $\sum_1^4 x_i^2 = 1$, this relation implies

$$1 - x_3^2 - x_4^2 = \tan^2 \alpha'(x_3^2 + x_4^2), \quad \text{or} \quad x_3^2 + x_4^2 = \cos^2 \alpha'.$$

It follows now, from (54.2), that

$$x_1^2 + x_2^2 = \cos^2 \alpha' \tan^2 \alpha' = \sin^2 \alpha'.$$

Introducing u and v as parameters, the last two equations are satisfied if:

$$(54.3) \qquad \begin{array}{ll} x_1 = \sin \alpha' \cos v, & x_2 = \sin \alpha' \sin v, \\ x_3 = \cos \alpha' \cos u, & x_4 = \cos \alpha' \sin u. \end{array}$$

Conversely, every quadruple x_1, x_2, x_3, x_4, satisfying (54.2) and $\Omega_1(x,x) = 1$, can be represented in this form, hence (54.3) is a parametric representation of F_α^N. The line element of F_α^N is now obtained from (54.1) as:

(54.4) $$dS^2 = k^2[\cos^2 (\alpha/k) \, du^2 + \sin^2 (\alpha/k) \, dv^2].$$

Because $E = k^2 \cos^2 (\alpha/k)$, $F = 0$, and $G = k^2 \sin^2 (\alpha/k)$ are independent of u and v, dS^2 represents a Euclidean metric in the general form discussed in Section 25. Thus we have the surprising result:

(54.5)   *In the small, the geometry induced on F_α^N is Euclidean.*

The phrase "in the small" indicates the obvious fact that F_α^N cannot be congruent to the whole Euclidean plane, if for no other reason, because it is a closed surface. The situation is similar to that of the sphere and the elliptic plane: sufficiently small parts are congruent, whereas the whole surfaces are not congruent.

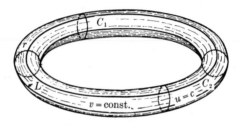

Fig. 96

The shape of F_α^N is easily seen. For a fixed value of u, or of v, the equations (54.3) represent a simple closed curve, actually a circle if x_1 and x_2, or x_3 and x_4 are interpreted as plane rectangular coordinates. The ordinary u,v plane may be generated by taking one line, say $v = 0$, and letting the lines $u = u_0$ sweep out the plane as the point $(u_0, 0)$ moves on the line $v = 0$. In the same way, if the circle $v = 0$ is selected on F_α^N, then the circles $u = u_0$ generate F_α^N as the point $(u_0, 0)$ traverses $v = 0$. Hence F_α^N has the shape of a torus, that is, of the surface T in E^3 obtained by revolving a circle about a line which lies in the same plane but does not intersect the circle (Figure 96). However, the metric induced on T by the Euclidean metric of E^3 is not the same as that on F_α^N. This may be seen from the fact that on T the circles $v =$ const (see figure) do not all have the same length.[19]

---

[19]Those familiar with differential geometry will see that T has positive curvature on the outer region bounded by C_1 and C_2 and negative curvature on the inner region, hence the metric is not Euclidean.

Though there is no torus shaped surface in E^3 on which the induced metric is Euclidean, it is nonetheless quite simple to understand how *a torus, with a metric which is Euclidean in the small, can be constructed*. First, as indicated in the diagrams (Figure 97), if a finite cylinder Z is cut along a generator L, Z can be unrolled into a rectangular region Z' in the plane. Both of the parallel boundaries of Z', L_1 and L_2, correspond to L. Conversely, starting with the region Z' the cylinder Z can be obtained by identifying L with both L_1 and L_2. In the small, distances on the cylinder are the same as the Euclidean distances in the plane. Now a torus T can be obtained from the cylinder through the same principle, the two circles r_1 and r_2 on Z corresponding to the single circle r of T. Points of T are associated, through Z, with points of the rectangular strip Z'. If the metric of Z' is transferred to T, distance on the torus is locally Euclidean.

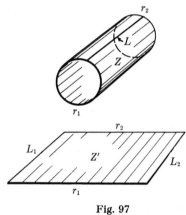

Fig. 97

We now *evaluate the function* $\theta(\alpha)$ defined in (53.18). Since θ is independent of the position of y on F_α^N, y may be chosen as the point of intersection of the lines M_1 and M_2 from the conjugate reguli in (53.14) corresponding to $t = 0$. Substituting in the resulting equations (53.14) from (54.3) yields:

(54.6)
$$M_1 : \begin{array}{l} \sin \alpha' \cos v + \tan \alpha' \cos \alpha' \cos u = 0, \\ \sin \alpha' \sin v - \tan \alpha' \cos \alpha' \sin u = 0. \end{array}$$
$$M_2 : \begin{array}{l} \sin \alpha' \cos v + \tan \alpha' \cos \alpha' \cos u = 0, \\ \sin \alpha' \sin v + \tan \alpha' \cos \alpha' \sin u = 0. \end{array}$$

From the first pair, M_1 corresponds to $u + v = \pi$, on which $du : dv = 1 : -1$. The second set of equations gives M_2 as $u - v = \pi$ on which $du : dv = 1 : 1$. (The choice of y was therefore $u = \pi$, $v = 0$.) Now (26.5) yields:

$$\cos \theta = \cos \sphericalangle (M_1, M_2) = \frac{k^2(\cos^2 \alpha' - \sin^2 \alpha')}{[k^2(\cos^2 \alpha' + \sin^2 \alpha')]^{1/2} [k^2(\cos^2 \alpha' + \sin^2 \alpha')]^{1/2}}$$
$$= \cos (2\alpha/k).$$

Observing that the lines M_1 and M_2 tend to N as $\alpha \to 0$ and tend to N' as $\alpha \to k\pi/2$, the measure of the smaller angle between M_1 and M_2 is taken for θ when $\alpha \leqslant k\pi/4$. When α passes $k\pi/4$, the angular domain measured by θ goes continuously into the domain with the larger angular measure, and this measure is taken for θ. Thus:

(54.7)　　*The angle between two Clifford parallels to N, at a distance α from N, is $2\alpha/k$.*

The Clifford parallels have attracted the interest of many mathematicians, since the very existence of these parallels is surprising. For a concise account of their properties see Coxeter's "Non-Euclidean Geometry."

In view of the aims of this book, a few additional remarks must suffice; these are easily verified geometrically or by using the formulas (54.3) and (54.4).

The translations along, and the rotations about, a line N generate an Abelian group Γ_N, which consists of all those motions of the elliptic space which induce an orientation preserving mapping of N on itself. On F_α^N these motions induce the translations, described in terms of (54.3) by $u' = u + a$, $v' = v + b$. Given two points x and y of F_α^N there is exactly one element of Γ_N which carries x into y.

If R_α^1 and R_α^2 denote the two reguli carried by F_α^N, every element of Γ_N carries R_α^i into itself, $i = 1,2$. The elements of Γ_N which carry a particular line L_1 of R_α^1 into itself form a subgroup Γ_N' of Γ_N. An element Φ of Γ_N' carries every element of R_α^1 into itself, and carries the element L_2 of R_α^2 through the point x on L_1 into the element $L_2\Phi$ of R_α^2 through $x\Phi$. If L_1' is any other element of R_α^1, and y denotes the intersection of L_2 with L_1', then Φ carries y into the intersection of $L_1 = L_1\Phi$ and $L_2\Phi$. Since Φ is a motion, $xy = x\Phi y\Phi$, therefore:

(54.8)　　*The rulings of one of the reguli R_α^i intercept segments of equal lengths on any two rulings of the other.*

Moreover, if x' is the foot of x on L_1', then $y' = y\Phi$ must be the foot of $x\Phi$ on L_1', therefore $xx' = yy'$. Since for suitable Φ in Γ_N' the point y' takes any position on L_1', it follows that L_1 is a Clifford parallel to L_1':

(54.9)　　*Any two rulings in the regulus R_α^i are Clifford parallels to each other, $i = 1,2$.*

Thus the two reguli R_α^1 and R_α^2 form on F_α^N a net which is in many respects similar to the net formed in the Euclidean plane by two families of parallel lines. Because of (54.7) the net is rectangular for $\alpha = \dfrac{k\pi}{4}$.

It should be observed that the above Φ will not carry a line L of $R^i_{\alpha'}$ into itself when $\alpha' \neq \alpha$, that is, neither regulus $R^i_{\alpha'}$ will stay linewise fixed. Indeed, the line M_1 in (54.6) will go into itself under $u' = u + a, v' = v + b$, only if the numbers a and b satisfy an easily established relation which depends on α.

Let ρ be the distance of a point x from the point d_4. Analogous to spherical coordinates in Euclidean space, spherical coordinates ρ, θ, φ can be introduced in elliptic space, with $0 \leqslant \theta < 2\pi, \ 0 \leqslant \varphi \leqslant \pi$. They are related to the x_i-coordinates by

$$x_1 = \sin \rho' \sin \theta \sin \varphi, \qquad x_2 = \sin \rho' \sin \varphi \cos \theta,$$
$$x_3 = \sin \rho' \cos \varphi, \qquad x_4 = \cos \rho', \quad \rho' = \rho/k.$$

Then,

$$dx_1 = \cos \rho' \sin \varphi \ \sin \theta \ d\rho' + \sin \rho' \cos \varphi \sin \theta \ d\varphi + \sin \rho' \sin \varphi \cos \theta \ d\theta$$
$$dx_2 = \cos \rho' \sin \varphi \cos \theta \ d\rho' + \sin \rho' \cos \varphi \cos \theta \ d\varphi - \sin \rho' \sin \varphi \sin \theta \ d\theta$$
$$dx_3 = \cos \rho' \cos \varphi \ d\rho' - \sin \rho' \sin \varphi \ d\varphi$$
$$dx_4 = - \sin \rho' d\rho'.$$

The line element, $dS^2 = k^2 \sum dx_i^2$ becomes

$$dS^2 = d\rho^2 + k^2 \sin^2 \rho' \ (d\varphi^2 + \sin^2 \varphi \ d\theta^2),$$

hence,

$$| \ g_{ik}(\rho, \varphi, \theta) \ | = k^4 \sin^4 (\rho/k) \sin^2 \varphi.$$

The volume element is therefore

$$dV = k^2 \sin^2 (\rho/k) \sin \varphi \ d\rho \ d\varphi \ d\theta.$$

For a sphere with radius $\rho \leqslant k\pi/2$, the volume is then

$$k^2 \int_0^{2\pi} d\theta \int_0^{\pi} \sin \varphi \ d\varphi \int_0^{\rho} \sin^2 t/k \ dt = 2\pi k^3 [(\rho/k) - \sin (\rho/k) \cos (\rho/k)].$$

(54.10)   *The volume of a sphere with radius ρ is*

$$2\pi k^3 [(\rho/k) - \sin (\rho/k) \cos (\rho/k)].$$

In the same way as in the hyperbolic case it is seen that:

(54.11)   *The area of a sphere with radius ρ is $4\pi k^2 \sin^2 (\rho/k)$.*

As $\rho \to k\pi/2$ the volume of the sphere, in (54.10), approaches $\pi^2 k^3$, the volume of the whole elliptic space. Under the same limit process, the sphere with radius ρ tends to the plane polar to d_4, so that (54.11) seems to yield $4\pi k^2$ for the area of the elliptic plane instead of $2\pi k^2$ as previously obtained. As in the plane case, however, it is seen that as $\rho \to k\pi/2$ a pair of points antipodal on the sphere tend to a single point of the plane. Hence, the sphere tends to the plane traversed twice which accounts for the apparent contradiction.

It is, of course, even more difficult to "picture" three-dimensional elliptic space than that of two dimensions. As before, *pairs of diametrically opposite points of the three-dimensional spherical space are in one-to-one correspondence with elliptic space.* The latter may therefore be thought of as originating from spherical space by means of identifying diametrically opposite points.

The interior and exterior regions of a torus F_α^N consist respectively of the points whose distance from N is less than α and those whose distance is greater than α. The torus F_α^N together with its interior and exterior fills the space. By (53.11), the exterior of F_α^N is at the same time the interior of $F_{\alpha'}^{N'}$. Hence *elliptic space may also be obtained from two solid tori by identifying their boundaries.* A thorough understanding of all these considerations requires a knowledge of topology. We merely mention therefore, without further explanation, that *three-dimensional projective space does not have the property of one-sidedness.* It is, in topological language, orientable. Generally, all odd-dimensional projective spaces have this property while the even-dimensional spaces are non-orientable.

Bibliography

The following list contains the books quoted in the text. In addition, a few books are listed which either treat some of the present subjects from a different point of view, or indicate their relation to geometry as a whole.

W. Blaschke, Vorlesungen über Differentialgeometrie. I. 3rd edition, Berlin, 1930 and New York, 1945.

H. S. M. Coxeter, Non-Euclidean Geometry. 2nd edition, Toronto, 1947.

H. S. M. Coxeter, The Real Projective Plane. New York, 1949.

G. Darboux, Principes de Géométrie Analytique. Paris, 1917.

H. G. Forder, Geometry. London, 1950.

W. C. Graustein, Introduction to Higher Geometry. New York, 1930.

F. Klein, Vorlesungen über höhere Geometrie. 3rd edition, Berlin, 1926.

H. Liebmann, Nichteuclidische Geometrie. 1st edition, Leipzig, 1905 ; 3rd edition, Leipzig, 1923.

Th. Reye, Die Geometrie der Lage. 6th edition, Leipzig, 1923.

G. B. Robinson, The Foundations of Geometry. Toronto, 1940.

G. Salmon, A Treatise on Conics. 6th edition, London, 1879.

B. L. Van der Waerden, Einführung in die Algebraische Geometrie. Berlin, 1939 and New York, 1945.

O. Veblen and J. W. Young, Projective Geometry. Vol. 1, Boston, 1910 ; Vol. 2, Boston, 1918.

H. E. Wolfe, Introduction to Non-Euclidean Geometry. New York, 1948.

K. Zindler, Liniengeometrie mit Anwendungen. Vol. 1, Leipzig, 1902; Vol. 2, Leipzig.

INDEX

The following abbreviations for n-dimensional spaces are used : E^n for Euclidean space, P^n for projective space, H^n for hyperbolic space, ε^n for elliptic space, and S_r^n for spherical space. Numbers, by themselves, refer to pages.

A

Abelian (or commutative) group, 22.
Affine center, 87.
Affine coordinates, in A^2, sec. 15.
  in A^3, 275.
  in A^n, 275.
Affine plane, A^2, 83.
Affine ratio, in A^2, 86.
  in A^3, 276.
Affine space, A^n, 275.
Affinity, 83, 275.
Altitudes of triangle in H^2, 175.
Angle, as cross ratio, 233.
  between Clifford parallels, 317, 319.
  in elliptic geometry, 212.
  in Euclidean geometry, sec. 26.
  in hyperbolic geometry, 175.
  of parallelism, 175, 198.
  between planes in ε^3, 313.
  between planes in H^3, 310.
Apollonious, 61
Area, in equiaffine geometry, sec. 18.
  of a domain partially bounded,
    by limit circles, 235.
    by an equidistant curve, 189.
  of equidistant surface, 306.
  in Euclidean geometry, 144.
  in Hilbert geometry, 167.
  in hyperbolic geometry, sec. 31.
  in Minkowskian geometry, 138,
    144, 291.
  of circles, in ε^2, 225.
    in E^2, 144.
    in H^2, 182.
  of sphere, in ε^3, 321.
    in H^3, 301.
  of triangle, in ε^2, 226.
    in H^2, 183.

Associated regulus, 270.
Associative law, 21.
Asymptote, to a hyperbola, 92.
  to a line in H^2, 161, 209.
Axial plane of homology, 247.
Axis of, biaxial involution, 249.
  conic in E^2, 151.
  elation in P^2, 47.
  homology in P^2, 45.
  perspectivity in P^2, 35.
  projectivity of conic, 71.
  quadric, 292.

B

Base point of pencil of conics, 77.
Biaxial involution, 249, 315.
Bolyai, 164.
Boundary point of convex set,
  in A^2, 94.
  in A^3, 279.
Bounded convex set in A^2, 98.
Brianchon, 67.
Brianchon point, 82.
Bundle, 244.

C

Cauchy-Schwartz inequality, 106.
Cayley, 227.
Center of, conic in A^2, 90.
  elation, in P^2, 47.
    in P^3, 248, 249.
  homology, in P^2, 45.
    in P^3, 247.
  perspectivity in P^2, sec. 7.
  projectivity of conic, 71.
  quadric in A^3, 277.
Central reflection, see Reflection.